Springer Series on Environmental Management

For further volumes:
http://www.springer.com/series/412

M. Luisa Martínez · Juan B. Gallego-Fernández
Patrick A. Hesp
Editors

Restoration of Coastal Dunes

Springer

Editors
Dr. M. Luisa Martínez
Red de Ecología Functional
Instituto de Ecología, A.C.
Xalapa, Veracruz
Mexico

Dr. Juan B. Gallego-Fernández
Departamento de Biología Vegetal y Ecología
Universidad de Sevilla
Sevilla
Spain

Dr. Patrick A. Hesp
School of the Environment
Faculty of Science and Engineering
Flinders University
Adelaide, SA
Australia

ISSN 0172-6161
ISBN 978-3-642-44675-7 ISBN 978-3-642-33445-0 (eBook)
DOI 10.1007/978-3-642-33445-0
Springer Heidelberg New York Dordrecht London

Cover photograph: J. C. Hidalgo/Fotolia.com

Printed on acid-free paper

Springer is part of Springer Science+Business Media (www.springer.com)

Preface

The continuously growing human population along the coasts of our world will exacerbate the impact of human activities on all coastal environments. Therefore, restoration activities will become increasingly important. In particular, sandy shores and coastal dunes will require significant restoration efforts because they are preferred sites for human settlements and tourism. A major problem is that research into coastal dunes is scarce, despite the relative economic, social, and ecological importance of these ecosystems. The literature that deals with coastal dune restoration has increased significantly over the last decade, but the few books published on the subject mostly emphasize mid-latitude dune systems. As of January 2011, the ISI Web of Science database contained more than 60,000 articles on restoration, but less than 100 of them (<0.1 %) focus on coastal dunes, beaches, and slacks (or deflation basins and plains). In general, the information available on the ecology of coastal sand dunes is very uneven, and broad geographical syntheses are rare. Examples of local coverage generally come from Europe and North America. There is a general lack of attention given to low-latitude coastal environments where much of the current exploitation and coastal development of tourism is occurring. To our knowledge, there are no books that deal with coastal dune restoration from a global perspective. The need for a worldwide compilation of experiences of restoration efforts is therefore evident.

The International Conference on Management and Restoration of Coastal Dunes (Santander, Spain, 3–5 October 2007) provided an impetus to compare and contrast different restoration projects around the world and initiate interdisciplinary and comparative studies. This book is the first step toward the development of international cooperation among those concerned about coastal sand dunes and their restoration.

This book is directed mainly at graduate students and colleagues who are interested in biological, ecological, geographical, and environmental sciences. This book will also be useful to those with an interest in conservation biology and coastal management who seek information on the different strategies that have been used to restore coastal dunes in different regions of the world. This goal can only be achieved after a comprehensive review and comparison of ongoing studies and restoration activities, where "successful" and "failed" studies or approaches (however they are determined) are compared and contrasted. Finally, this book will be a resource for coastal planners, as well as for local and state officials, residents of coastal communities, environmental advocates, developers, and others concerned with coastal issues.

A major product of this book is a compendium of empirical experiences showing that coastal dune restoration has many meanings, and thus, leads to many different actions. Coastal dune restoration may have the goal of increasing vegetation cover and reducing substrate mobility, but it may also aim to remobilize sandy terrains in order to counteract the negative impact of overstabilization: reduced diversity. Here, the relevance of the different goals in restoration is shown very clearly.

The meetings held to put together this book and organize the chapters received partial funding from grant no. 23669 (SEMARNAT-CONACYT), coordinated by M. Luisa Martinez. We are very grateful to Dieter Czseschlik, from Springer-Verlag, who originally invited Juan B. Gallego-Fernández to write a proposal for a book on coastal dune restoration. It was he who ignited this idea. Lawrence R. Walker read this book several times and provided very useful comments and recommendations. We are also grateful to Andrea Schlitzberger for her constant interest and support throughout the different stages of this book.

Finally, we would like to thank our families and children for bearing with us while we were writing, assembling, and reviewing all the chapters. Thank you to Graziela Miot Da Silva, Nicholas Hesp (again!), Jonathan, Phoebe and Sebastian Hesp, Chary García , María Gallego García, Octavio Pérez-Maqueo, and Valeria Pérez Martínez.

This book is dedicated to the memory of M. Anwar Maun (1935–2007), cherished friend and colleague who was a leader in coastal dune ecological studies. He is dearly missed by us and by all the coastal dune scientific community.

M. Luisa Martínez
Juan B. Gallego-Fernández
Patrick A. Hesp

Contents

Part II Restoring Inland Costal Dunes: Dunefields and Wetslacks

Contributors

Alicia Teresa Rosario Acosta Department of Environmental Biology, Università di Roma Tre, V.le Marconi 446, 00146 Rome, Italy, e-mail: acosta@uniroma3.it

Camiel Aggenbach Department of Energy and Environmental Studies, Faculty of Mathematics and Natural Sciences, Energy and Sustainability Research Institute Gron, University of Groningen, Nijenborgh 4, 9747 AG Groningen, The Netherlands

Sebastiaan M. Arens Bureau for Beach and Dune Research, Amsterdam, The Netherlands, e-mail: arens@duinonderzoek.nl

Fabiola López Barrera Red de Ecología Funcional, Instituto de Ecología, A.C., Antigua carretera a Coatepec No. 351, El Haya, 91070 El Haya, Xalapa, Veracruz, Mexico

Nir Becker Department of Economics and Management, Tel Hai College, 12210 Upper Galilee, Israel, e-mail: nbecker@telhai.ac.il

Renee Bekker Department of Energy and Environmental Studies, Faculty of Mathematics and Natural Sciences, Energy and Sustainability Research Institute Gron, University of Groningen, Nijenborgh 4, 9747 AG Groningen, The Netherlands

Adolfo Campos Red de Ecología Funcional, Instituto de Ecología, A.C., Antigua carretera a Coatepec No. 351, El Haya, 91070 El Haya, Xalapa, Veracruz, Mexico

Gonzalo Castillo-Campos Instituto de Ecología A.C, Red de Biodiversidad, Xalapa, Veracruz, México

Bikila W. Dullo Department of Energy and Environmental Studies, Faculty of Mathematics and Natural Sciences, Energy and Sustainability Research Institute Gron, University of Groningen, Nijenborgh 4, 9747 AG Groningen, The Netherlands

Verónica E. Espejel González Red de Ecología Funcional, Instituto de Ecología, A.C., Antigua carretera a Coatepec No. 351, El Haya, 91070 El Haya, Xalapa, Veracruz, Mexico

Rusty Feagin Spatial Sciences Laboratory, Department of Ecosystem Science and Management, Texas A&M University, College Station, TX 77845, USA, e-mail: feaginr@tamu.edu

Juan B. Gallego-Fernández Departamento de Biología Vegetal y Ecología, Universidad de Sevilla, Ap.1095, 41080 Sevilla, Spain, e-mail: galfer@us.es

Luc H. W. T. Geelen Research and Development, Waternet, Vogelenzang, The Netherlands

Ab Grootjans Department of Energy and Environmental Studies, Faculty of Mathematics and Natural Sciences, Energy and Sustainability Research Institute Gron, University of Groningen, Nijenborgh 4, 9747 AG Groningen, The Netherlands, e-mail: a.p.grootjans@rug.nl

Harrie G. J. M. Van der Hagen Dunea, Voorburg, The Netherlands

Patrick A. Hesp School of the Environment, Faculty of Science and Engineering, Flinders University, GPO Box 2100, Adelaide, SA 5001, Australia, e-mail: Patrick.hesp@flinders.edu.au

Michael J. Hilton Department of Geography (Te Ihowhenua), University of Otago (Te Whare Wananga o Otago), PO Box 56, Dunedin, New Zealand

Nancy L. Jackson Department of Chemistry and Environmental Science, New Jersey Institute of Technology, Newark, NJ 07102, USA, e-mail: jacksonn@njit.edu

Laurence Jones Centre for Ecology and Hydrology, Environment Centre Wales, Deiniol Road, Bangor LL57 2UW, Wales, UK, e-mail: LJ@ceh.ac.uk

Rod Jones Countryside Council for Wales Bangor, Gwynedd LL57 2DN, Wales, UK, e-mail: rd.jones@ccw.gov.uk

Tommaso Jucker Department of Environmental Biology, Università di Roma Tre, V.le Marconi 446, 00146 Rome, Italy

Annemieke Kooijman Department of Energy and Environmental Studies, Faculty of Mathematics and Natural Sciences, Energy and Sustainability Research Institute Gron, University of Groningen, Nijenborgh 4, 9747 AG Groningen, The Netherlands

Pua Bar Kutiel Department of Geography and Environmental Development, Ben-Gurion University, P.O.B. 653, 84105 Beer Sheva, Israel, e-mail: kutiel@bgu.ac.il

David Lehrer The Arava Institute for Environmental Studies, 88840 Kibbutz Ketura, D.N. Hevel Eilot, Israel, e-mail: david.lehrer@arava.org

D. Lithgow Red de Ecología Funcional, Instituto de Ecología, A.C., Antigua carretera a Coatepec No. 351, El Haya, 91070 El Haya, Xalapa, Veracruz, Mexico

Roy A. Lubke Department of Botany, Rhodes University, and Coastal and Environmental Services, Grahamstown, Eastern Cape, South Africa, e-mail: r.lubke@ru.ac.za

M. Luisa Martínez Red de Ecología Funcional, Instituto de Ecología, A.C., Antigua carretera a Coatepec No. 351, El Haya, 91070 El Haya, Xalapa, Veracruz, Mexico, e-mail: marisa.martinez@inecol.edu.mx

G. Mendoza-González Red de Ecología Funcional, Instituto de Ecología, A.C., Antigua carretera a Coatepec No. 351, El Haya, 91070 El Haya, Xalapa, Veracruz, Mexico

Patricia Moreno-Casasola Red de Ecología Funcional, Instituto de Ecología, A.C., Antigua carretera a Coatepec No. 351, El Haya, 91070 El Haya, Xalapa, Veracruz, Mexico, e-mail: patricia.moreno@inecol.edu.mx

J. C. Muñoz-Reinoso Departamento de Biología Vegetal y Ecología, Universidad de Sevilla, Ap.1095, 41080 Sevilla, Spain, e-mail: reinoso@us.es

Karl F. Nordstrom Institute of Marine and Coastal Sciences, Rutgers, The State University of New Jersey, New Brunswick, NJ 08901, USA, e-mail: nordstro@marine.rutgers.edu

O. Pérez-Maqueo Red de Ambiente y Sustentabilidad, Instituto de Ecología, A.C., Antigua carretera a Coatepec No. 351, El Haya, 91070 Xalapa, Veracruz, Mexico, e-mail: octavio.maqueo@inecol.edu.mx

Andrea J. Pickart US Fish and Wildlife Service, Humboldt Bay, 6800 Lanphere Rd, Arcata, CA 95521, USA, e-mail: Andrea_Pickart@fws.gov

Irene Prisco Department of Environmental Biology, Università di Roma Tre, V.le Marconi 446, 00146 Rome, Italy

Norbert P. Psuty Rutgers, The State University of New Jersey, Sandy Hook, NJ 07732, USA, e-mail: psuty@marine.rutgers.edu

I. Redondo Morales Consejera Técnica de la Dirección General de la Red de Espacios Naturales, Avda. Manuel Siurot 50, 41071 Sevilla, Spain, e-mail: isabelm.redondo@juntadeandalucia.es

Peter Rhind Countryside Council for Wales Bangor, Gwynedd LL57 2DN, Wales, UK, e-mail: p.rhind@ccw.gov.uk

Hugo López Rosas Red de Ecología Funcional, Instituto de Ecología, A.C., Antigua carretera a Coatepec No. 351, El Haya, 91070 El Haya, Xalapa, Veracruz, Mexico

C. Saavedra Azqueta Jardín Botánico Dunas del Odiel (RAJBEN, Consejería de Medio Ambiente), Sevilla, Spain, e-mail: jbotanico.dunasodiel.cma@juntaodiel.es

Lorena E. Sánchez Higueredo Red de Ecología Funcional, Instituto de Ecología, A.C., Antigua carretera a Coatepec No. 351, El Haya, 91070 El Haya, Xalapa, Veracruz, Mexico

Riccardo Santoro Department of Environmental Biology, Università di Roma Tre, V.le Marconi 446, 00146 Rome, Italy

Tanya M. Silveira Rutgers, The State University of New Jersey, Sandy Hook, NJ 07732, USA, e-mail: mendes@marine.rutgers.edu

Quirinus L. Slings nv PWN Drinking Water Company, Velserbroek, The Netherlands

Judith Vázquez Red de Ecología Funcional, Instituto de Ecología, A.C., Antigua carretera a Coatepec No. 351, El Haya, 91070 El Haya, Xalapa, Veracruz, Mexico

Peter Vestergaard Department of Biology, University of Copenhagen, Universitetsparken 15, 2100 Copenhagen Ø, Denmark, e-mail: peterv@bio.ku.dk

Chapter 1
Coastal Dunes: Human Impact and Need for Restoration

M. Luisa Martínez, Patrick A. Hesp and Juan B. Gallego-Fernández

1.1 The Heterogeneity of Coastal Dunes

1.1.1 Dune Types

Sandy coasts of the world are diverse in terms of morphology, vegetation, and dynamics. The surfzone beach types can range from dissipative to intermediate to reflective, tidal range from nontidal, micro- to macro-tidal, and they vary from accreting or prograding, stable, aggradational, to eroding or retrogradational. They may have no dunes at all, large dune systems extending over several kilometers inland and alongshore, with massive coastal dunes in excess of 100m high, or the dunes can be low (less than 1m high) and cover a small area. Coastal dunes can vary considerably and may be various types of foredunes, parabolics, blowouts, transgressive sheets, and dunefields (perhaps containing nebkha fields, deflation basins, and plains [slacks], barchans, transverse dunes, barchanoids, star dunes, and various erosional dune types) (Hesp 1991, 1999, 2012, 2011) (Fig. 1.1).

M. L. Martínez (✉)
Red de Ecología Funcional, Instituto de Ecología, A.C., Antigua Carretera a Coatepec no. 351 91070 El Haya, Xalapa, Ver., Mexico
e-mail: marisa.martinez@inecol.edu.mx

P. A. Hesp
School of the Environment, Faculty of Science and Engineering, Flinders University, GPO Box 2100 Adelaide, SA 5001, Australia
e-mail: Patrick.hesp@flinders.edu.au

J. B. Gallego-Fernández
Departamento de Biología Vegetal y Ecología, Universidad de Sevilla, Ap.1095 41080 Sevilla, Spain
e-mail: galfer@us.es

M. L. Martínez et al. (eds.), *Restoration of Coastal Dunes*, Springer Series on Environmental Management, DOI: 10.1007/978-3-642-33445-0_1,

Fig. 1.1 Pictures of coastal dune types and landforms. (**a**) Foredune-blowout complex on the St. Joseph Peninsula, Florida. (**b**) a barrier island in Florida composed of relict foredunes, comprising ridge and swale sets. The ridges are more heavily forested. (**c**) nebkha field on a transgressive dunefield, Mexico. These are often critical habitats for fauna in open, mobile dune dominated areas. (**d**) Long-walled, nested, closely spaced, multiple parabolic dunes on the California coast. (**e**) climbing and headland bypass transgressive dunefield just to the north of La Mancha, Veracruz State, MX. (all photos by P. Hesp except **d** by K. and G. Adelman [www.californiacoastline.org])

Foredunes are typically the foremost dune situated on the backshore. They are formed by aeolian sand deposition in plants above the spring high tide line. They may be quite small, and only a meter or less high; they may also be quite large and up to $\sim$20 + m high. They may have very simple symmetrical morphology and be fully vegetated, or they may be a highly erosional, scattered nebkha and remnant knobs with minimal vegetation cover (Hesp 2002).

On prograding beaches, pioneer plants grow seaward onto newly built backshore and trap sand forming incipient foredunes. Over time these grow and become the newly established foredunes, thereby making the former foredune relict and inactive. Over hundreds or thousands of years, suites of foredunes may occur over time, forming small to extensive foredune plains. Blowouts may form within the foredunes and create a foredune–blowout complex. Blowouts may be quite small to relatively large, and range in form from saucers through bowls to troughs. These too may eventually stabilize and become part of the inactive dune system (Hesp 2011).

Parabolic dunes typically form from blowouts. They are characterized by U- and V-shaped depositional lobes, trailing ridges and deflation basins or plains. They may be long-walled types (linearly extensive downwind), oblate, short types, or intermediate among these forms. They may be nested or overlapping, displaying multiple episodes of formation. The depositional lobes may range from quite simple forms to multiple-pronged digitate forms (Hesp 2004).

Transgressive dune sheets are flat to undulating sand plains. Transgressive dunefields are small to extensive sand terrains featuring a variety of dune forms on their surface. Transgressive dune sheets and dunefields are aeolian sand deposits formed by the downwind movement (across or alongshore) or transgression of sand over vegetated to semi-vegetated terrain. Such dunefields may range from quite small sheets and nebkha fields (hundreds of meters in alongshore and landward extent) to very large, unvegetated dunefields that can be similar in size to some small to moderate-sized desert dunefields (or draa) (Hesp and Thom 1990; Hesp 2011; Hesp et al. 2011).

Transgressive dunefields may be completely active (predominantly unvegetated), partially vegetated, or completely vegetated post-formation. Transgressive dunefields may be dominated by active barchans, transverses, or barchanoid dunes. The form may vary considerably depending on the age of the dunefield, dominant migration direction relative to the coast, degree of vegetation cover, wind regime, and sediment supply magnitude and variation over time. Star dunes, blowouts, parabolic dunes, nebkha and nebkha fields, deflation basins and plains, slacks, remnant knobs, bush pockets, shadow dunes, precipitation ridges, and trailing ridges, etc., may be found within the dunefields (Hesp and Walker 2012).

1.1.2 Dune Vegetation

Because they are found in all latitudes and in all climate regimes, coastal dune vegetation is very diverse and includes temperate to tropical forests, grasslands, and shrublands (Hesp 2004; Hesp and Martínez 2008). Dune species differ between dunes and among locations and have different biogeographic and phylogenetic histories (van der Maarel 1993a, 1993b; Hesp et al. 2011; Gallego-Fernández and Martínez 2011). However, in spite of these multiple differences, many plants typical of coastal dunes share their physiological and morphological responses to the environmental restrictions characteristic of coastal dunes (extreme temperatures, drought, low availability of nutrients, disturbance (e.g., wave-driven beach and dune erosion, aeolian erosion and deposition) and, most of all, salt spray and substrate mobility) (Hesp 1991; Maun 2009; Hesp and Martínez 2008; Gallego-Fernández and Martínez 2011) and are thus called psammophytes. The compendium of the vegetation of dry coastal ecosystems of the world (van der Maarel 1993a, 1993b) reveals that at least 5,000 plant species have been found growing on coastal dunes. Endemics are quite abundant. Finally, depending on vegetation cover, coastal dunes can be mobile (with zero to sparse vegetation), semi-mobile, or stabilized (fully vegetated and with limited to no substrate mobility). Coastal dunefields may also vary spatially, containing phases that are completely mobile and phases that are entirely vegetated, and vary temporally, displaying phases of dune initiation and development and phases of relative stability and little dune development.

1.2 Human Impact on Coastal Dunes and Sandy Coasts

Planet Earth is a coastal planet: its shorelines extend for more than 1.5 million km (World Resources Institute 2005), which, if added together around the equator, would circle it more than 400 times. Eighty-four percent of the world's countries have a shoreline, but only 16 % of them contain sandy beaches and coastal dunes within their territories. In addition, nearly 41 % of the world's human population lives within 100 km of the coast. Twenty-one of the 33 megacities (>10 million inhabitants) of the world are located at the coast, and it is expected that many will more than double their size within the next few decades (Martínez et al. 2007). Given this scenario, the actual and potential human impact on coastal ecosystems is more than obvious. In consequence, these ecosystems have been severely degraded as a result of excessive natural resource exploitation, chaotic demographic expansion, industrial growth, and worldwide tourism. The continuously growing human population along the coasts of our world will exacerbate the impact of human activities on these environments. Because of this, conservation activities and restoration efforts will become more and more relevant and important, to sustain coastal ecosystems. In particular, sandy shores and coastal dunes will be in great need of these efforts because they are preferred sites for human settlements and tourism. The problem is that with some notable exceptions (Nordstrom 2004, 2008; Ley-Vega de Seoane et al. 2007; Houston et al. 2001; Grootjans et al. 1997), research on coastal dune restoration is limited.

The degradation and loss of coastal dunes owing to human intervention is a consequence of different activities performed at the coast by humans (Ketchum 1972). These actions can be aggregated into **six** groups and to a major or minor extent, all of them affect coastal dunes, namely:

1. Housing and recreation
2. Industrial and commercial use
3. Waste disposal
4. Agriculture, aquiculture, and fisheries
5. Military activities

The human impact on coastal dunes has been studied and described by different authors (e.g., Ranwell 1972; Ranwell and Boar 1986; Carter 1988; Nordstrom

Table 1.1 Alterations of coastal dunes by human actions

Processes		Mechanisms leading to alterations
ABIOTIC	Landforms	Significant alteration or elimination by construction of urbanizations and infrastructures; mining Reshaping, replacing or creation: altering topography (landscaping) Alteration: trampling, off road vehicle traffic
	Sediment transport	Marine—alteration of sediment resources and transport Wind flow—alteration of directions and intensity Terrestrial transport of sediment—mining, removal of beach and dune sediments; dune reactivation; dune fixation
	Quantity and quality of water	Alterations on geomorphological and biological processes Marine intrusion Water extraction Water pollution/eutrophication
BIOTIC	Composition and abundance of plant communities	Elimination of species or alteration of community structure by: Grazing Harvesting Introduction of exotic species Planting species to stabilize sands Eutrophication
	Community dynamics	Modification of community dynamics by: Affecting seed dispersal by ocean currents and colonization after cleaning the beach of flotsam and wrack Inhibiting colonization and growth after trampling Species loss or community structure altered as a result of beach nourishment and dune reconstruction Eutrophication (direct—adding nutrients to enhance plant growth; indirect—pollution). Species interactions and dominance are modified Reduction of local and regional species pool after the introduction of exotic/alien species Loss of entire communities as a result of fragmentation of regional dune landscape
	Ecosystem dynamics	Changes in ecosystem dynamics by: Eutrophication from Nitrogen deposition, leading to dominance of species Reduced litter decomposition with the introduction of exotics such as *Casuarina*

2008). The typical end result of human activities is that natural processes are altered, and then, dune and ecosystem dynamics are also modified (Table 1.1). These processes occur both in terrestrial and aquatic ecosystems and include different abiotic and biotic aspects, all on a local and regional scale. Human activities modify (both directly and indirectly) all these processes, resulting in alterations that range from minor to extensive loss and destruction (Table 1.1).

1.3 Coastal Dune Restoration: What Do We Know?

What is coastal dune restoration? The Society for Ecological Restoration (SER) defines restoration as:

The process of assisting the recovery of an ecosystem that has been degraded, damaged or destroyed. It is an intentional activity that initiates or accelerates ecosystem recovery with respect to its health (functional processes), integrity (species composition and community structure), and sustainability (resistance to disturbance and resilience (http//www.ser.org/content/guidelines_ecological_restoration.asp).

Certainly, this is a broad concept that applies to many situations and ecosystems. The problem with it is the difficulty applying this definition to coastal dunes. In general, independently of the definition of coastal dune restoration, the literature reports the following actions when dealing with coastal dune restoration:

1. *Reshaping dunes and recovering sediment dynamics*: many coastal dunes around the world, and foredunes in particular, have been built on to various degrees, modified, re-shaped, removed or gradually destroyed because of human actions. In addition, many dunes have been invaded by exotic species, or exotic species have been deliberately introduced to the dune systems (e.g., Hesp 2000; Hilton 2006; Hilton et al. 2005, 2009; Nordstrom et al. 2009). In order to restore some degree of natural dune functioning, dunes have been constructed by:
    (a) Planting vegetation and/or constructing sand fences along the backshore, which then trap sand and cause foredunes to form.
    (b) Transferring sediment from the beach or from elsewhere to the backshore and constructing a dune, including utilizing various artificial structures as part of the construction (e.g., buried sea walls, geotubes etc.).
    (c) Removing manmade landforms (e.g., clay bunds or berms, concrete revetments or promenades) and either allowing dune-forming processes to re-start, or re-creating a more natural dune form. Dune areas have also been de-vegetated to allow, or induce greater aeolian activity and processes to take place. In all these activities, and other similar actions, the aim has generally been to restore the natural aeolian functioning of the system, allow natural surfzone–beach–foredune interactions to occur, or create an environment where native plant species can survive and form natural dunes.
2. *Vegetation*: most of the literature that we have found on coastal dune restoration is aimed at plants. Here, a recurrent issue that is addressed in restoration actions is counteracting the negative impact of dune stabilization, which results in depressed biodiversity (Remke et al. 2009; Arens and Geelen 2006; Ketner-Oostra et al. 2006) and even loss of endemic species (Kutiel et al. 2000). In similarity to that which occurs in other ecosystems, invasive species have been considered a problem to be faced with coastal dune restoration, because they eliminate native plant and animal species (Marchante et al. 2008; Mason and

French 2007; Valtonen et al. 2006; Pickart and Sawyer 1998; Pickart et al. 1998a, b).
Several mechanisms have been used for the destabilization and recovery of coastal dune dynamics, such as bulldozing, pulling by hand or grazing (ten Harkel and van der Meulen 1996). Wet slack restoration has also been a habitat that has been frequently restored (Van der Hagen et al. 2008; Bakker et al. 2005; Grootjans et al. 2002; Nienhuis et al. 2002) within the coastal dune systems.
3. *Animals*: restoration activities have also focused on animals, mainly invertebrates. Similar to plants, controlling invasive species was relevant for Lepidoptera (Valtonen et al. 2006). Invertebrates have also been used for monitoring restoration activities (Kumssa et al. 2004; Davis et al. 2002, 2003) and, interestingly, it has been observed that grazing enhances butterfly diversity (Wallis De Vries and Raemakers 2001).
4. *Human activities*: restoration activities have also been performed after the ever increasing impact of humans such as mining (Burke 2007; Kumssa et al. 2004; Davis et al. 2003) and urban sprawl (Nordstrom et al. 2000, 2009). In addition, there are a handful of studies that consider the relevance of coastal dune restoration aimed at reversing the loss of ecosystem services (Everard et al. 2010; Martínez and Lopez-Barrera 2008; Moreno-Casasola et al. 2008). There have also been numerous attempts around the world aimed at providing protection for inland regions (e.g., against hurricanes and storms, sea level rise, etc.).

1.4 Aims and Scope of the Book

In this book we have gathered together a set of restoration activities performed in different parts of the world, including both temperate and tropical latitudes, and from most of the continents. The goal was to gather information from state-of-the-art studies on the restoration of coastal sand dunes covering a wide variety of landforms and approaches. We aimed to integrate the knowledge from this vast array of experiences and develop a coherent empirical and theoretical framework that can be of assistance in future coastal dune restoration projects. Different coastal land forms are included in this set of studies, including foredunes, slacks/swales/deflation basins and plains, blowouts, parabolics, and transgressive dunefields. We are following a broad definition of "restoration," which includes any effort to restore the original functionality of the previously existing ecosystem (Hobbs et al. 2006). Thus, we will refer to restoration activities (sensu stricto) (SER), but also to rehabilitation, reclamation and even designing ecosystems in seriously altered sites (such as after mining activities). This volume does not provide the final answer on how to restore coastal dunes and sandy beaches. Rather, we hope that the diverse array of contributions will stimulate further research that will lead to a better understanding of the actions required for improving restoration strategies.

We have invited experts from different parts of the world, and have asked them to share their experiences on coastal dune restoration (however they define that, and it is clear that there is a wide variety of opinions on what "restoration" comprises or means), whether this is a great success or a blatant failure: it all provides useful information from which we can learn. The enthusiastic response from these experts is evidence of the uniqueness of the book, where so many coastal dune restoration actions are presented and analyzed critically.

The book is organized as follows. We first divided the compilation of studies into the different landforms within the dune system that are usually restored, namely: foredunes, foredune/blowout complexes, and dune fields (coastal dunes of different shapes and mobility). We chose to follow this system because the natural dynamics and disturbances (natural and anthropogenic) are very different for each landform, and, thus, the restoration actions are obviously different too. Therefore, a wide set of actions are included with a broad variety of goals for coastal dune restoration after both natural and human-related disturbances. The third section then deals with the very relevant issue of the costs of coastal dune restoration and the value of ecosystem services from which society benefits. The fourth and last section attempts to demonstrate the need to have a multidisciplinary approach to successful restoration projects. Finally, the concluding chapter integrates all the information provided in the whole book.

The first section begins with two chapters that provide a theoretical framework for foredune restoration. On the one hand, Nordtsrom and Jackson (Chap. 2) present a broad description of foredune restoration activities while focusing on urban settings. Second, Psuty and Silveira (Chap. 3) describe a conceptual model of sediment budget and its relevance in foredune restoration activities, with examples from New Jersey and New York. In this study, the premise was that it is difficult to restore the processes of coastal foredune development because the processes themselves are the dynamics of moving wind and water.

Vestergaard (Chap. 4) and Hesp and Hilton (Chap. 5) provide the reader with well documented case studies of foredune restoration actions that range from building an artificial foredune and planting native species on it in Denmark, to reshaping foredunes and using herbicides to remove exotic invasive species in New Zealand. The last chapter in the section on foredunes deals with restoration of foredunes after the impact of hurricanes on Galveston Island, USA (Feagin, Chap. 6). Because of the impact of recurring hurricanes on the Texas shore and the repeated need for restoration, Feagin concludes that foredunes along these coasts continuously face "inevitable destruction," and thus, "certain reconstruction" is always necessary.

Section 1.2 in the book focuses on dunefields and wet slacks. Here, coastal dune restoration may refer to two contrasting types of actions: de-stabilization and re-vegetation. A broad group of studies in this multinational compendium deals with the "struggle of vegetation against sand." This set of studies provides ample evidence of how diversity decreases with dune stabilization in The Netherlands (Arens et al., Chap. 7); Wales (Rhind et al., Chap. 8); Spain (Muñoz-Reinoso et al., Chap. 9); Northwestern USA (Pickart, Chap. 10) and Israel (Kutiel, Chap. 11).

These studies show that stabilization can result from two contrasting scenarios of human activities: after human intervention such as artificial plantings (Arens et al., Chap. 7; Rhind et al., Chap. 8; Muñoz-Reinoso et al., Chap. 9; Pickart, Chap. 10), or after human intervention has stopped (when trampling ceased; Kutiel, Chap. 11). Paradoxically, and contrary to what are usually considered to be common restoration activities, restoring these dune systems requires the removal of vegetation, so that dunes can be active again through the action of aeolian processes, and vegetation can then be rejuvenated (Arens et al., Chap. 7). Projects on coastal dune remobilization are usually performed on public parks, away from human infrastructure.

In contrast with the above, coastal dune re-vegetation is also a restoration action when the goal is to halt substrate mobility and increase plant cover. Here, Acosta et al. (Chap. 12) evaluated the role of the passive recovery on Italian coastal dunes following limitations to human trampling. Revegetation is also a goal following intensive open mining activities on coastal dunes in Madagascar and Namibia. In this case, dunes were built artificially after mining activities had ceased and then native vegetation was introduced from the seed bank (Lubke, Chap. 13). Finally, an example of coastal dune restoration in Mexico is described by Moreno-Casasola et al. (Chap. 14). Oftentimes, dunes are reforested with exotic and potentially invasive species such as, for example, *Casuarina equisetifolia* trees. Here Moreno-Casasola et al. analyze whether the recovery of native vegetation is possible after the establishment of exotic *Casuarina* plantations. This is relevant because *Casuarina* has been widely used in coastal dune and beach re-forestation programs in the tropics. Often, these reforestation programs are implemented because substrate mobility is a problem for human infrastructure, such as cities and harbors.

Section 1.2 ends with two case studies on the restoration of wet slacks. In the Netherlands, Grootjans et al. (Chap. 15) show how vegetation removal is also necessary in order to recover the native diversity of vegetation typical of wet slacks. In Mexico, López-Rosas et al. (Chap. 16) modified the hydroperiod in order to eliminate the exotic grasses and recover the diversity of native species.

Section 1.3 analyzes the costs of coastal dune restoration and the potential economic value of ecosystem services. Lehrer et al. (Chap. 17) performed an interesting valuation of the public's willingness to pay to control and eliminate an invasive shrub from a coastal dune nature reserve. Pérez-Maqueo et al. (Chap. 18) summarized the costs of several coastal dune restoration projects from this book, as well as the economic value of ecosystem services. It is interesting to note that most coastal dune restoration projects have taken place in natural parks, whereas rehabilitation is mostly performed in urban settings. Frequently, it is the government who pays for these actions, although some private owners (Feagin, Chap. 6), the public (Lehrer et al., Chap. 17), and plenty of volunteer workers (Pickart, Chap. 10; Arens et al., Chap. 7) have also played a key role in coastal dune restoration projects.

The last chapters are grouped in Sect. 1.4. In this section, Lithgow et al. (Chap. 19) show how a multicriteria analysis can be of assistance in implementing actions leading to coastal dune restoration. Certainly, restoration is not merely an ecological problem. On the contrary, there are many actors and points of view involved in any restoration action: ecological, but also social and economic. A multicriteria analysis can help to achieve a common goal in which coastal dunes are restored, while there are also social and economic gains. We end the book with a discussion of all the chapters. In Chap. 20, Martínez et al. summarize the most relevant findings and show how coastal dune restoration does indeed have a wide variety of meanings, goals, and methods.

In brief, the extensive information gathered in this book shows that coastal dune restoration is much more than "assisting the recovery of an ecosystem that has been degraded, damaged or destroyed" by means of recovering the vegetation (Lithgow et al., in revision). Often, coastal dune restoration can be counterintuitive, because it may mean that it is necessary to remove vegetation in order to activate the system and revitalize the dunes and the vegetation that grows in these environments. However, coastal dune restoration also means increasing vegetation cover after human intervention. Certainly, a multidisciplinary approach is very much required and the costs and benefits of restoration should be considered. Coastal dune restoration must integrate the needs and expectations of stakeholders, as well as those of decision makers and scientists. Finally, dune restoration plans should take a regional approach, incorporating landscape-scale processes (Gallego-Fernández et al. 2011).

Acknowledgments We are very grateful to all the authors of the chapters in this book who kindly contributed their chapters and shared their restoration experiences. We are also very grateful for the insightful comments of Lawrence R. Walker.

References

Acosta ATR, Jucker T, Prisco I, Santoro R (2012) Passive recovery of Mediterranean coastal dunes following limitations to human trampling. In: Martínez ML, Gallego-Fernández JB, Hesp PA (eds) Restoration of coastal dunes. Springer, New York (Chapter 12)

Arens SM, Geelen LHWT (2006) Dune landscape rejuvenation by intended destabilisation in the Amsterdam water supply dunes. J Coastal Res 22(5):1094–1107

Arens SM, Slings QL, Geelen LHWT, Van der Hagen HGJM (2012) Restoration of dune mobility in the Netherlands. In: Martínez ML, Gallego-Fernández JB, Hesp PA (eds) Restoration of coastal dunes. Springer, New York (Chapter 7)

Bakker C, de Graaf HF, Ernst WHO, van Bodegom PM (2005) Does the seed bank contribute to the restoration of species-rich vegetation in wet dune slacks? Appl Veg Sci 8(1):39–48

Burke A (2007) Recovery in naturally dynamic environments: a case study from the sperrgebiet, southern African arid succulent karoo. Environ Manage 40(4):635–648

Carter RWG (1988) Coastal environments. An introduction to the physical, ecological and cultural systems of coastlines. Academic Press, London

Davis ALV, Van Aarde RJ, Scholtz CH, Delport JH (2002) Increasing representation of localized dung beetles across a chronosequence of regenerating vegetation and natural dune forest in South Africa. Glob Ecol Biogeogr 11(3):191–209

Davis ALV, van Aarde RJ, Scholtz CH, Delport JH (2003) Convergence between dung beetle assemblages of a post-mining vegetational chronosequence and unmined dune forest. Restor Ecol 11(1):29–42

Everard M, Jones L, Watts B (2010) Have we neglected the societal importance of sand dunes? An ecosystem services perspective. Aquat Conserv Mar Freshw Ecosyst 20(4):476–487

Feagin R (2012) Foredune restoration before and after hurricanes: inevitable destruction, certain reconstruction. In: Martínez ML, Gallego-Fernández JB, Hesp PA (eds) Restoration of coastal dunes. Springer, New York (Chapter 6)

Gallego-Fernández JB, Martínez ML (2011) Environmental filtering and plant functional types on Mexican foredunes along the Gulf of Mexico. Ecoscience 18(1):52–62

Gallego-Fernández JB, Sánchez IA, Ley C (2011) Restoration of isolated and small coastal sand dunes on the rocky coast of northern Spain. Ecol Eng 37:1822–1832

Grootjans AP, Jones P, van der Meulen F, Paskoff R (eds) (1997) Ecology and restoration perspectives of soft coastal ecosystems. J Coastal Conserv Spec Feature 3:1–102

Grootjans AP, Geelen HWT, Jansen AJM, Lammerts EJ (2002) Restoration of coastal dune slacks in the Netherlands. Hydrobiologia 478(1–3):181–203

Grootjans A, Dullo BW, Kooijman A, Bekker R, Aggenbach C (2012) Restoration of dune vegetation in the Netherlands. In: Martínez ML, Gallego-Fernández JB, Hesp PA (eds) Restoration of coastal dunes. Springer, New York (Chapter 15)

Hesp PA (1991) Ecological processes and plant adaptations on coastal dunes. J Arid Env 21:165–191

Hesp PA (1999) The beach backshore and beyond. In: Short AD (ed) Handbook of beach and shoreface morphodynamics. John Wiley and Sons, New York, pp 145–169

Hesp PA (2000) Coastal dunes. Forest research (Rotorua) and NZ coastal dune vegetation network (CDVN), 28 pp

Hesp PA (2002) Foredunes and blowouts:initiation, geomorphology and dynamics. Geomorphol 48:245–268

Hesp PA (2004) Coastal dunes in the tropics and temperate regions: location, formation, morphology and vegetation processes. In: Martínez ML, Psuty NP (eds) Coastal dunes: Ecology and conservation, Ecological Studies 171. Springer, Berlin

Hesp PA, (2011) Dune Coasts. In: Wolanski E, McLusky DS (eds) Treatise on Estuarine and Coastal Science, vol 3. Academic Press , Waltham pp 193–221

Hesp PA (2012) Surfzone-beach-dune interactions. In: NCK-days 2012: Crossing borders in coastal research. Enschede, the Netherlands

Hesp PA, Hilton MJ (2012) Restoration of foredunes and transgressive dunefields—case studies from New Zealand. In: Martínez ML, Gallego-Fernández JB, Hesp PA (eds) Restoration of coastal dunes. Springer, New York (Chapter 5)

Hesp PA, Walker IJ, (2012) Aeolian environments: coastal dunes. In: Shroder J, Lancaster N, Sherman DJ, Baas ACW (Eds.) Treatise on Geomorphology, vol 11. Aeolian Geomorphology. Academic Press, San Diego, CA

Hesp PA, Martínez ML (2008) Transverse dune trailing ridges and vegetation succession. Geomorphol 99:205–213

Hesp P, Martinez ML, Miot da Silva G, Rodríguez-Revelo N, Gutierrez E, Humanes A, Laínez D, Montaño I, Palacios V, Quesada A, Storero L, González-Trilla G, Trochine C (2011) Transgressive dunefield landforms and vegetation associations. Earth Surf Process Landf 36:285–295

Hesp PA, Thom BG (1990) Geomorphology and evolution of erosional dunefields. In: Nordstrom KF, Psuty NP, Carter RWG (ed) Coastal Dunes: process and morphology. Wiley J and Sons, Chichester, pp 253–288

Hilton MJ (2006) The loss of New Zealand's active dunes and the spread of marram grass (*Ammophila arenaria*). NZ Geogr 62:105–121

Hilton MJ, Duncan M, Jul A (2005) Processes of *Ammophila arenaria* (marram grass) invasion and indigenous species displacement, Stewart Island, New Zealand. J Coast Res 21(1):175–185

Hilton MJ, Woodley D, Sweeney C, Konlechner T (2009) The development of a prograded foredune barrier following *Ammophila arenaria* eradication, Doughboy Bay, Stewart Island. J Coast Res SI 56:317–321

Hobbs R, Falk DA, Palmer M, Zedler J (2006) Foundations of restoration ecology: the science and practice of ecological restoration (the science and practice of ecological restoration series). Island Press, Washington, 384 pp

Houston JA, Edmonson SE, Rooney PJ (2001) Coastal dune management. Shared experience of European conservation practice. Liverpool University Press, Liverpool

Ketchum BH (ed) (1972) The water's edge: critical problems of the coastal zone. MIT Press, Boston

Ketner-Oostra R, van der Peijl MJ, Sykora KV (2006) Restoration of lichen diversity in grass-dominated vegetation of coastal dunes after wildfire. J Veg Sci 17(2):147–156

Kumssa DB, van Aarde RJ, Wassenaar TD (2004) The regeneration of soil micro-arthropod assemblages in a rehabilitating coastal dune forest at Richards Bay, South Africa. Afr J Ecol 42(4):346–354

Kutiel PB (2012) Restoration of coastal sand dunes for conservation of biodiversity—the Israeli experience. In: Martínez ML, Gallego-Fernández JB, Hesp PA (eds) Restoration of coastal dunes. Springer, New York (Chapter 9)

Kutiel P, Peled Y, Geffen E (2000) The effect of removing shrub cover on annual plants and small mammals in a coastal sand dune ecosystem. Biol Conserv 94:235–242

Lehrer D, Becker N, Kutiel PB (2012) The value of coastal sand dunes as a measure to plan an optimal policy for invasive plant species: the case of the *Acacia saligna* at the Nizzanim LTER coastal sand dune nature reserve. In: Martínez ML, Gallego-Fernández JB, Hesp PA (eds) Restoration of coastal dunes. Springer, New York (Chapter 17)

Ley-Vega de Seoane C, Gallego-Fernández JB, Vidal Pascual C (2007) Manual de restauración de dunas costeras. Ministerio de Medio Ambiente, Rural y Marino, Santander, España, pp 251

Lithgow D, Gallego-Fernández JB, Martínez ML (2012) Multicriteria analysis to implement actions leading to coastal dune restoration. In: Martínez ML, Gallego-Fernández JB, Hesp PA (eds) Restoration of coastal dunes. Springer, New York (Chapter 19)

Lithgow D, Martínez ML, Hesp PA, Gallego-Fernández JB, Álvarez-Molina LL, Gachuz S, Jiménez-Orocio O, Rodríguez-Revelo N (2012) Restoration of coastal dunes: different aims and different methods. Geomorphol

López Rosas H, Moreno-Casasola P, López Barrera F, Sánchez-Higueredo LE, Espejel-González VE, Vázquez J (2012) Interdune wetland restoration in central Veracruz Mexico: plant diversity recovery mediated by hydroperiod. In: Martínez ML, Gallego-Fernández JB, Hesp PA (eds) Restoration of coastal dunes. Springer, New York (Chapter 16)

Lubke RA (2012) Restoration of dune ecosystems following mining in Madagascar and Namibia: contrasting restoration approaches adopted in regions of high and low human population density. In: Martínez ML, Gallego-Fernández JB, Hesp PA (eds) Restoration of coastal dunes. Springer, New York (Chapter 13)

Marchante E, Kjoller A, Struwe S, Freitas H (2008) Short- and long-term impacts of *Acacia longifolia* invasion on the belowground processes of a Mediterranean coastal dune ecosystem. Appl Soil Ecol 40(2):210–217

Martínez ML, Lopez-Barrera F (2008) Special issue: restoring and designing ecosystems for a crowded planet. Ecoscience 15(1):1–5

Martínez ML, Intralawan A, Vázquez G, Pérez-Maqueo O, Sutton P, Landgrave R (2007) The coasts of our world: ecological, economic and social importance. Ecol Econ 63:254–272

Martínez ML, Hesp PA, Gallego-Fernández JB (2012) Coastal dune restoration: trends and perspectives. In: Martínez ML, Gallego-Fernández JB, Hesp PA (eds) Restoration of coastal dunes. Springer, New York (Chapter 20)

Mason TJ, French K (2007) Management regimes for a plant invader differentially impact resident communities. Biol Conserv 136(2):246–259

Maun MA (2009) The Biology of Coastal Sand Dunes. Oxford University Press, New York

Moreno-Casasola P, Martinez ML, Castillo-Campos G (2008) Designing ecosystems in degraded tropical coastal dunes. Ecoscience 15(1):44–52

Moreno-Casasola P, Martínez ML, Castillo-Campos G, Campos A (2012) The impacts on natural vegetation following the establishment of exotic Casuarina plantations. In: Martínez ML,

Gallego-Fernández JB, Hesp PA (eds) Restoration of coastal dunes. Springer, New York (Chapter 14)
Muñoz-Reinoso JC, Saavedra-Azqueta C, Redondo-Morales I (2012) Restoration of Andalusian coastal juniper woodlands. In: Martínez ML, Gallego-Fernández JB, Hesp PA (eds) Restoration of coastal dunes. Springer, New York (Chapter 11)
Nienhuis PH, Bakker JP, Grootjans AP, Gulati RD, de Jonge VN (2002) The state of the art of aquatic and semi-aquatic ecological restoration projects in the Netherlands. Hydrobiologia 478(1–3):219–233
Nordstrom KF (2004) Beaches and dunes of developed coasts. Cambridge University Press, Cambridge
Nordstrom KF (2008) Beach and dune restoration. Cambridge University Press, Cambridge
Nordstrom KF, Jackson NL (2012) Foredune restoration in urban settings. In: Martínez ML, Gallego-Fernández JB, Hesp PA (eds) Restoration of coastal dunes. Springer, New York (Chapter 2)
Nordstrom KF, Lampe R, Vandemark LM (2000) Reestablishing naturally functioning dunes on developed coasts. Environ Manage 25(1):37–51
Nordstrom KF, Gamper U, Fontolan G, Bezzi A, Jackson NL (2009) Characteristics of coastal dune topography and vegetation in environments recently modified using beach fill and vegetation plantings, Veneto, Italy. Environ Manage 44(6):1121–1135
Pérez-Maqueo OM, Martínez ML, Mendoza-González G, Lithgow D (2012) The coasts and their costs. In: Martínez ML, Gallego-Fernández JB, Hesp PA (eds) Restoration of coastal dunes. Springer, New York (Chapter 18)
Pickart AJ (2012) Dune restoration over two decades at the Lanphere and Ma-le'l Dunes in northern California. In: Martínez ML, Gallego-Fernández JB, Hesp PA (eds) Restoration of coastal dunes. Springer, New York (Chapter 10)
Pickart A, Sawyer JO (1998) Ecology and restoration of Northern California coastal dunes. California Native Plant Society, Sacramento
Pickart AJ, Miller LM, Duebendorfer TE (1998a) Yellow bush lupine invasion in northern California coastal dunes—I. Ecological impacts and manual restoration techniques. Restor Ecol 6(1):59–68
Pickart AJ, Theiss KC, Stauffer HB, Olsen GT (1998b) Yellow bush lupine invasion in northern California coastal dunes—II. Mechanical restoration techniques. Restor Ecol 6(1):69–74
Psuty NP, Silveira TM (2012) Restoration of coastal foredunes, a geomorphological perspective: examples from New York and from New Jersey, USA. In: Martínez ML, Gallego-Fernández JB, Hesp PA (eds) Restoration of coastal dunes. Springer, New York (Chapter 3)
Ranwell DS (1972) Ecology of salt marshes and sand dunes. Chapman & Hall, London 258 pp
Ranwell DS, Boar R (1986) Coast dune management guide. Institute of Terrestrial Ecology, NERC, Norwich
Remke E, Brouwer E, Kooijman A, Blindow I, Roelofs JGM (2009) Low atmospheric nitrogen loads lead to grass encroachment in coastal dunes, but only on acid soils. Ecosystems 12(7):1173–1188
Rhind P, Jones R, Jones L (2012) The impact of dune stabilization on the conservation status of sand dune systems in Wales. In: Martínez ML, Gallego-Fernández JB, Hesp PA (eds) Restoration of coastal dunes. Springer, New York (Chapter 8)
Ten Harkel MJ, van der Meulen F (1996) Impact of grazing and atmospheric nitrogen deposition on the vegetation of dry coastal dune grasslands. J Veg Sci 7(3):445–452
Valtonen A, Jantunen J, Saarinen K (2006) Flora and lepidoptera fauna adversely affected by invasive *Lupinus polyphyllus* along road verges. Biol Conserv 133(3):389–396
Van der Hagen HGJM, Geelen LHWT, de Vries CN (2008) Dune slack restoration in Dutch mainland coastal dunes. J Nat Conserv 16(1):1–11
van der Maarel E (ed) (1993a) Dry coastal ecosystems, part 2A. Elsevier, Amsterdam
van der Maarel E (1993b) Dry coastal ecosystems of the Polar regions and Europe in retrospect, part 2B. Elsevier, Amsterdam

Vestergaard P (2012) Natural plant diversity development on a man-made dune system. In: Martínez ML, Gallego-Fernández JB, Hesp PA (eds) Restoration of coastal dunes. Springer, New York (Chapter 4)

Wallis De Vries MF, Raemakers I (2001) Does extensive grazing benefit butterflies in coastal dunes? Restor Ecol 9(2):179–188

World Resources Institute (2005) World resources 2002–2004. Decisions for the Earth: balance, voice and power. United Nations Development Programme, United Nations Environmental Programme, The World Bank, World resources Institute, USA, 315 pp

Part I
Restoring Foredunes

Chapter 2
Foredune Restoration in Urban Settings

Karl F. Nordstrom and Nancy L. Jackson

2.1 Introduction

Dunes will form anywhere there is a proper sediment source, a wind strong enough to entrain and move sediment, and a means of reducing the speed of winds or trapping the sand being moved. These conditions are readily met on most sand beaches, including those undergoing long-term erosion. The dunes may not survive erosion during major storms, but new dunes will re-form following recovery of the beach after storms. The reason sand dunes are not found landward of many active sand beaches in developed areas is because they are mechanically removed to enhance recreational use (Nordstrom 2008); in other cases, the beach is so severely truncated by human structures that the small unvegetated dunes that re-form are quickly eroded during the next storm. Once dunes are eliminated by human activities, they are lost from the consciousness of stakeholders, and attempts by managers to build new dunes are hampered by a lack of appreciation of their value (Nordstrom et al. 2000).

Sand dunes, like many other natural environments, provide important ecosystem functions and services (Lubke and Avis 1998; Arens et al. 2001; Peterson and Lipcius 2003; Everard et al. 2010), including:

1. Protection for human structures (providing sediment, a physical barrier or resistant vegetation)

K. F. Nordstrom (✉)
Institute of Marine and Coastal Sciences, Rutgers, the State University of New Jersey, New Brunswick, NJ 08901, USA
e-mail: nordstro@marine.rutgers.edu

N. L. Jackson
Department of Chemistry and Environmental Science, New Jersey Institute of Technology, Newark, NJ 07102, USA
e-mail: jacksonn@njit.edu

M. L. Martínez et al. (eds.), *Restoration of Coastal Dunes*, Springer Series on Environmental Management, DOI: 10.1007/978-3-642-33445-0_2,

2. Sites for passive recreation (aesthetic, psychological, therapeutic opportunities)
3. Cultural/environmental heritage
4. Educational resource
5. Sand and minerals for extraction (consumptive use)
6. Filter for pollutants
7. Retention area for groundwater
8. Ecological niche for plants adapted to dynamic conditions
9. Habitable substrate for invertebrates
10. Refuge areas (e.g., for plovers, rabbits)
11. Nest or incubation sites
12. Food for primary consumers
13. Food for higher trophic levels (scavengers, predators, humans)
14. Synergistic benefits of multiple habitat types (e.g., corridors)
15. Intrinsic value.

The greatest value perceived by most managers is the protection dunes provide against flooding and erosion. Other values can be achieved through alternative designs for creating or modifying dunes, including providing niches for flora, habitable substrate and refuge areas for fauna, and nest sites. Dunes can also have site-specific value, such as sources of groundwater. In many locations, the resource base provided by the dunes has been lost through shorefront development, and human efforts are required to restore it.

The term "restoration" implies that attempts have been made to recover the form, function, and species inventory existing prior to human modification. Dunes built primarily to increase the level of protection from storm flooding and erosion are rarely restored because the landforms are built and maintained as static features in an attempt to protect against natural processes rather than respond to them. Nevertheless, most dune-building projects can accomplish some restoration goals if construction and management practices are made more compatible by accepting more natural features and greater mobility (Nordstrom et al. 2011).

Restoration efforts must be based on ecological, geomorphological and social criteria to maximize the goods and services dunes can provide. Steps in restoring dunes where they have been eliminated in developed areas involve:

1. Getting stakeholders to accept dunes as an alternative to the human-modified environments that have replaced them
2. Creating the basic landforms where they have been eliminated or increasing the size of truncated forms
3. Allowing these newly created or modified landforms to function like natural dunes by allowing some portions of them to be dynamic
4. Controlling subsequent negative human actions
5. Favoring dune evolution through time using adaptive management strategies.

Adaptive management, with continued human input, is critical where space is restricted and long-term erosion continues. Sustainability of natural features in developed areas requires that humans act as intrinsic agents of landform change.

Space and time are important to allow incipient dunes on the backshore to survive erosion by wave uprush during small storms and to increase in size to create foredune ridges large enough to survive storms of annual or greater frequency/magnitude. If space and time are not available, these constraints can be partially overcome by aiding natural processes using sand fences or vegetation planting, or by using earth-moving equipment. These human actions can be used to build and rebuild dunes in a sand-deficient environment, but inevitably, beach nourishment is required to maintain a healthy, well-vegetated dune on an eroding shore (Mendelssohn et al. 1991).

2.2 Dune-Building Practices in Urban Areas

We use the generic word "urban" to refer to both moderately and intensively developed shores (suburbs and cities). We do not distinguish between these types of shores because the largest lengths of developed dunes in some countries (e.g., the USA) occur outside cities. In many cases, it is not the intensity of development that is the issue, but the proximity of houses, protection structures and support infrastructure to the backshore, and the frequency of beach raking. The methods used to build dunes in developed areas are similar to those used in coastal parks and preserves. The differences affecting the way dunes look and evolve in developed areas are due to restricted beach widths or changes in grain size characteristics if beach fill is used. There is no fundamental difference in the ways dunes should be restored in developed areas except for the need to control some of the natural dynamism where human infrastructure is close to the water and to control subsequent degradational human activities.

2.2.1 Accommodating Natural Aeolian Processes

The biological and geomorphological processes forming new or incipient natural foredunes on undeveloped beaches are reviewed in Hesp (1989, 1991) and Kuriyama et al. (2005). On these beaches, wind-blown sand will accumulate at the seaward-most vegetation and wrack lines that form on the backshore. These dune-forming features are missing on raked beaches. On many human-altered shores that are eroding, the presence of infrastructure close to the water restricts backshore width, requiring artificial nourishment to provide the wider sediment source for aeolian transport and greater protection for newly formed dunes from wave erosion. As a result of growth by slow natural accretion, the resulting dune will have the internal stratification, topographic variability, surface cover, and root mass of a natural dune. The wide cross-shore gradient of physical processes will allow for a suite of distinctive habitats, from pioneer species on the seaward side to woody shrubs and trees on the landward side. A large-scale beach nourishment

Fig. 2.1 Evolving dune on a nourished beach at Ocean City, NJ, USA (2009), 16 years after dune-building began. The ridge to the far left, fronting the houses, is the initial protective dune that now functions as a species-rich secondary dune. The foredune in the foreground was created by natural processes without the aid of fences or vegetation planting

project can provide the necessary sand volume and space for dunes to form (Fig. 2.1), but time is required for backdune species to colonize, and maintenance nourishment of the beach is required to retain dune integrity, given subsequent wave-induced erosion.

The characteristics of beach fill and the design elevation of the berm crest influence the likelihood of aeolian transport and delivery of sand to the dune. The removal of finer sands from poorly sorted fill sediment can leave a coarser shell or gravel lag surface that resists subsequent aeolian transport (Psuty and Moreira 1992; van der Wal 1998). Storm reworking of the backshore can replenish the finer sand and initiate new cycles of increased aeolian transport, but the elevation of the nourished beach must be low enough to allow storm wave run-up to occur on the backshore. The common practice of building a beach higher than a natural beach to provide protection to human infrastructure against run-up should be avoided if natural aeolian transfers are to be favored (Jackson et al. 2010).

Under natural conditions, restoration of the morphology and vegetation assemblages of foredunes can take up to 10 years (Woodhouse et al. 1977; Maun 2004). Accordingly, it may be desirable to use earth-moving equipment, sand fences or vegetation planting to create a dune ridge for initial protection and allow a more natural dune to gradually evolve toward the sea. Creation of this initial barrier will eliminate the need for an overly nourished beach to provide flood protection. Construction of the original protective foredune in the left background of Fig. 2.1 was facilitated by initial placement of two sand fences and plantings of *Ammophila breviligulata*. Additional fences were placed on the seaward side of the dune to encourage horizontal rather than vertical growth, so that shore-front residents could retain views of the sea. Designation of nesting sites for piping plovers by the state endangered species program resulted in the prohibition of beach raking, leading to local colonization of the backshore by plants and growth of incipient dunes that survived several winter storm seasons and grew into the new foredune ridge farther seaward (foreground of Fig. 2.1).

Fig. 2.2 Dune at Carolina Beach, NC, USA (1981), created using artificial fill reshaped by bulldozers into a dune-dike

2.2.2 Direct Deposit of Fill

Dunes are frequently constructed by directly depositing sediment and reshaping it using earth-moving equipment. These landforms, often called dune-dikes, are usually built to optimize a flood-protection function and often have a flat top or planar sides of a consistent slope with little topographic diversity (Fig. 2.2). Dune-dikes may retain their artificial form through time if no subsequent aeolian accretion occurs or if they are rebuilt to the same template when they erode. They may eventually evolve to resemble more hummocky natural dunes with greater variability of microhabitats if they are not repeatedly rebuilt or revegetated.

The grain size characteristics, rates of change, and characteristic vegetation of dunes built by mechanical placement differ from dunes created by aeolian deposition (Baye 1990; van der Wal 1998; Matias et al. 2005). Bulldozed dunes may include coarser shells and gravel, but they can have well-sorted sands that resemble dune sediments if suitable borrow areas are used to build them. Bulldozed dunes can provide habitat similar to natural dunes, if actions are taken to enhance this value during and after construction. Patchiness of habitats can be increased by creating an undulating foredune crest, resulting in local differences in drainage and wind speed and converting the landward boundary from a line into a zone (Nordstrom 2008). Subsequent deposition of wind-blown sand on a bulldozed dune can create surface characteristics similar to a natural dune if no barriers to aeolian transport are created seaward of it. This deposition can be accelerated by using beach fill materials compatible with aeolian transport, employing sand-trapping fences on (but not in front of) the nourished dune, and creating a gentle seaward slope in the initial fill deposit (Matias et al. 2004, 2005).

2.2.3 Using Sand Fences

Fencing materials include canes and tree branches inserted directly into the sand and wooden slats, plastic, and jute fabric attached to fence posts. Using sand fences with different characteristics and configurations can result in considerable variety in foredune topography and vegetation. Straight fences placed parallel to the shore appear to provide the most economical method of building protective dunes (CERC 1984; Miller et al. 2001). Zig-zag fences can create wider dunes with more undulating crest lines and more gently sloping dune faces that are closer to the shapes of natural dunes (Snyder and Pinet 1981). Paired fences can create a broader based foredune with a rounded crest that can look more natural (Schwendiman 1977). Side spurs perpendicular to straight alongshore alignments can increase trapping rates in locations of strong longshore winds. Multiple lifts of fences can create a higher dune with much greater volume than single lifts (CERC 1984; Mendelssohn et al. 1991; Miller et al. 2001).

Sand accumulation efficiency and morphological changes depend on fence porosity, height, inclination, the scale and shape of openings, wind speed and direction, the number of fence rows, separation distance between fence rows, and placement relative to the existing topography. Fencing with a porosity of about 50 %, with space between open and closed areas of less than 50 mm, can fill to capacity in about a year where appreciable sand is moving, but vegetation planting may still be required to stabilize the surface (CERC 1984) and establish a more natural trajectory.

A fence placed at the seaward limit of natural vegetation or at an existing foredune line may be far enough landward to survive a wave attack during small storms and have a wide source of wind-blown sand seaward (CERC 1984). Where diagnostic vegetation is lacking, dunes built with fences appear likely to survive storms at least a 1 year apart if they are built landward of the upper-most storm wrack line. The decision about where to place fences on a beach widened by fill or storm wave overwash could be made taking into account whether it is desirable to create a dune with space devoted to low wet habitats (slacks) between landward and seaward ridges or a higher, drier, and more continuous dune with multiple ridges (Nordstrom 2008). In these cases of a widened beach, the landward fence could be at least 100 m from the active beach, as suggested by Dahl and Woodard (1977) and Miller et al. (2001).

Sand fences or vegetation plantings are often used to stabilize newly-formed bare areas, forming gaps in the existing dunes. Depending on how, when or whether gaps are sealed, a foredune could reflect the differences in topography and cyclic reversals in vegetation succession found on a more dynamic natural landform or the more uniform sequences found on a stable landform. Constructing new fences is a more traditional and conservative approach, but one that should be evaluated more carefully considering the lower restoration potential of the resulting linear dune.

2.2.4 Using Vegetation

Vegetation reduces wind speed, traps sand, stabilizes surface sediment, provides habitat, and improves aesthetic appeal (Schwendiman 1977; Hesp 1989; Mitteager et al. 2006; Nordstrom 2008). The characteristics of the vegetation have important implications for the goods and services provided by coastal dunes. Basic dune-stabilizing vegetation for use by municipal managers and residents should consist of native species that are easy to propagate, harvest, store, and transplant with a high survival rate, be commercially available at local nurseries at a relatively low cost, and be able to grow in a variety of microhabitats on a spatially restricted foredune (Feagin 2005; Mitteager et al. 2006). The species that are most useful in building foredunes rapidly react positively to sand burial (Maun 1998). When sand deposition diminishes, initial dune-building species can degenerate, but successional species can be planted where sand deposition cannot be reinitiated (van der Putten and Peters 1995). Degeneration of *Ammophila* (*arenaria* or *breviligulata*, whichever is native), for example, is not a problem where the loss is coincident with colonization by later successional species (Vestergaard and Hansen 1992).

Other species may be more suitable for planting in environments landward of the active foredune crest, e.g., *Spartina patens* in dune swales, but the use of other species may not be necessary because they will eventually opportunistically colonize the dune themselves (Feagin 2005). Stabilizing the sand surface with one species can ameliorate environmental extremes and facilitate the establishment of other species that are less adapted to stressful environments (Martínez and Garcia-Franco 2004; De Lillis et al. 2004). Endangered species can colonize restored areas if populations and seed sources are present nearby and if dispersal mechanisms (wind, water or animals) are effective (Avis 1995; Snyder and Boss 2002; Grootjans 2004), although sometimes endangered species need some assistance to colonize.

2.3 Retaining an Element of Natural Dynamism

Natural foredunes are inherently dynamic and fragmented, with portions in an incipient state and portions at a more developed stage (Ritchie and Penland 1990; Doody 2001). Some degree of natural dynamism is important to allow landforms to be diverse and self-sustaining (Nordstrom 2008). Tolerance to burial is an important cause of zonation of plant species on coastal foredunes, and burial can have a stimulating, positive effect on the growth of dune-building plants and prevent degeneration (Maun 2004). Many plant species can occur in sand dunes, but the species most dependent on dune habitat tend to be concentrated in zones with greater sand movement (Castillo and Moreno-Casasola 1996; Rhind and Jones 1999). The variety of local topographic relief and small-scale differences in sheltering and proximity to the water table enhance the variety of habitats across and along the shore, and the resulting ecosystem services they provide (Everard et al. 2010).

The limit to how much dynamism can be tolerated by shorefront stakeholders is often related to the distance between the dune crest and the nearest human infrastructure. Foredunes can be allowed to erode or be breached by storm waves if the infrastructure is not close to the foredune and the foredune is not the only protection (Arens et al. 2001; Nordstrom et al. 2007), but dunes in most developed areas are narrow and close to human facilities and may not be allowed to evolve solely by natural processes. Ongoing human efforts may be required to allow portions of the dune to be mobile and evolve through time, while the overall feature remains intact as a barrier against storm wave run-up. These efforts will involve controls on beach raking, the use of sand fences and human-use structures on the beach and in the dune, driving on the beach, managing endangered species, and using exotic vegetation as ground cover. Establishing criteria for managing dunes in developed areas as partially dynamic systems is a difficult task. Many of the methods used to create dunes where they do not exist or to increase the dynamism of stabilized dunes represent departures from past practices, so it is likely that most projects will be experimental and small-scale and require a strong public information component until the feasibility of large-scale future projects is demonstrated.

2.4 Controlling Negative Human Actions

2.4.1 Restricting Beach Raking

Wrack (natural litter) lines contain seeds, culms, rhizomes of coastal vegetation, wood debris, and nutrients that aid in the growth of new vegetation and sand accumulations that evolve into new dunes (Godfrey 1977; Ranwell and Boar 1986) The highest storm wrack line has the greatest amount of litter and is often far enough landward of wave uprush during small storms to survive long enough to provide the base for a new foredune. Raking to remove wrack eliminates the likelihood of natural cycles of growth and destruction of incipient dunes on the beach, and eliminates backshore habitats and the associated biodiversity. Raking is one of the most common environmentally damaging actions on shores developed for human use (McLachlan 1985; Nordstrom et al. 2000; Colombini and Chelazzi 2003; Dugan et al. 2003). Finding a way to retain wrack is critical to conserving or restoring beach and dune habitats in developed areas. Alternatives for wrack management include:

1. Selectively removing the cultural litter and leaving the natural litter using nonmechanical methods
2. Leaving the highest storm wrack line on the backshore while raking the beach below it
3. Restricting cleaning operations to the summer (tourist season) or after massive fish kills

4. Leaving longshore segments unraked to develop as natural enclaves (Nordstrom et al. 2000).

2.4.2 Restricting Use of Sand Fences and Other Structures

There is often greater use of sand-trapping fences than is necessary (Grafals-Soto and Nordstrom 2009). Sand fences should not be placed where a dune of adequate size already exists, where they would trap sand in unnatural configurations, or where they cannot be buried, such as in vegetated portions of the dune or too close to the water. Fences in these locations can be conspicuous intrusions in the landscape, acting as physical boundaries to movement of fauna and reminders of the artifactual nature of the dunes. In many cases, sand-trapping fences need only be used for creation of the first dune ridge that functions as the core around which the natural dune can evolve (Nordstrom et al. 2000; Grafals-Soto and Nordstrom 2009). The location of the contact between the foredune and backshore is determined by erosion of the foredune during storms and dune accretion following storms. Storm wave uprush may eliminate the seaward portion of the dune and create an erosional scarp, but post-storm beach accretion creates a new source of sand to be blown to the foredune, reestablishing the dune sediment budget.

Biodegradable fences can be used to create an initial dune ridge while avoiding the long-term hazard to fauna and diminishing the human footprint in the landscape (Miller et al. 2001). Interference with movement of fauna can also be reduced by employing fences in configurations that create cross-shore corridors, either by leaving short gaps in longer sections of fence or deploying fences as short sections oriented transverse to shore and to prevailing winds. Sand-trapping fences are often used to control visitor traffic, resulting in dunes with unnatural shapes. Symbolic fences may be used instead of sand-trapping fences for controlling pedestrian access. Symbolic fences prevent users from trampling the dune while allowing the sand blown landward to build up dune height and volume.

Reducing the physical and visual impact of all structures can create a more naturally appearing and functioning beach and dune environment. Removable or elevated structures are often authorized on the beach and dune because of the reduced threat to their damage, but these structures enhance the feeling that coastal landforms are recreational areas. Actions to enhance recreational areas, such as grading surfaces and building walkways for access, can increase the level of physical and visual disturbance. Shore-perpendicular access paths are required to allow beach use, but they do not have to be as numerous as they are in most locations.

2.4.3 Restricting Driving on Beaches and Dunes

Vehicles driven on beaches kill fauna and disperse organic matter in drift lines, thereby destroying young dune vegetation and losing nutrients (Godfrey and Godfrey 1981; Moss and McPhee 2006; Foster-Smith et al. 2007). Vehicles compact the sand, which then becomes a hard surface and less viable for seed germination or for animal burrowing. The unnatural landscape image created by vehicle tracks can undermine attempts to instill an appreciation of the shore as a natural environment. Thus, driving should be restricted to only a few segments along a beach or to portions of the beach between wrack lines and away from incipient dunes; shore-perpendicular access could be restricted to a few prescribed crossings (Nordstrom 2003; Priskin 2003). There is little reason to allow private vehicles on beaches in developed municipalities when road networks exist landward and where undeveloped enclaves are so small that beaches can be reached on foot. Even if private vehicle use is prevented, municipal vehicles used for patrolling the beach and for emptying trash receptacles may remain a problem. Instituting a haul-in/haul-out policy for garbage generated by tourists would reduce the need for trash receptacles on the beach. Vehicles used for public safety could be confined to emergency operations.

2.4.4 Re-evaluating Endangered Species Programs

Initiatives for protecting endangered fauna, such as birds and turtles, by controlling active human uses has resulted in the creation and protection of naturally functioning beaches and dunes (Nordstrom et al. 2000; Breton et al. 2000). Identification of nests on beaches can lead to suspension of raking, bulldozing, and driving on the beach during the nesting season, leading to accumulation of litter in wrack lines, colonization by plants, and growth of incipient dunes that may evolve into established foredunes (Fig. 2.1). Protection of endangered species can have negative effects when the landscape is modified specifically for them and the needs of other species are subservient to the endangered ones. Programs to enhance conditions for nesting shorebirds can lead to artificial flattening of topography and the clearing of vegetation, and can prevent the backshores and dunes from evolving to later stages. Ways must be found to make these environments function naturally and compatible with other species dependent on a coastal setting.

2.4.5 Controlling Exotic Species

Exotic vegetation was often planted in dunes in the past because it was seen to be effective in stabilizing dunes, valuable economically, or attractive. Many public agencies now recognize the adverse effects of exotics on the biodiversity, authenticity, and successional processes, and have attempted to eliminate problematic species. The problem with the invasion of exotics from nearby residential and commercial properties is a less publicized issue. The magnitude of the problem is exacerbated if the foredune occurs on private property and if regulations allow property owners to plant exotic vegetation. Landscaping choices of individual shorefront property owners and managers of hotels and condominiums are conditioned by past experiences that are often obtained in a noncoastal setting, causing them to think that exotics are the landscaping ideal. Achieving restoration goals on private lots is an important, but virtually ignored, field of inquiry that has enormous potential for improving the natural value of coastal resources (Mitteager et al. 2006).

2.5 Gaining Acceptance for Natural Landforms and Habitats

Convincing residents and municipal officials of the need to create dunes where they do not exist or allow stable dunes to function more naturally can be challenging (Nordstrom 2008). The first step in building new dunes may be to build any dune-like feature, even if low, narrow, linear, and fixed in position, as has occurred in the State of New Jersey (Mauriello and Halsey 1987; Mauriello 1989). Gaining acceptance of larger, more naturally functioning dunes requires demonstrating their value by providing examples of good management practice and implementing public education programs, e.g., the municipal program in Avalon, NJ, USA (Nordstrom et al. 2009). Local demonstration sites can be used to provide specific technical information to local managers and provide evidence that restoration options are achievable (Breton et al. 2000; Nordstrom 2003). Demonstration sites in developed municipalities should have similar space constraints and relationships to the natural and cultural features of future restoration sites. Purely natural environments would likely represent unachievable target states. Policies should include restrictions on the mechanical removal of litter, the use of vehicles on the beach, and the use of sand-trapping fences. Portions of dunes should be dynamic once they have achieved dimensions required for shore protection. Only native vegetation should be used in planting programs, and any exotic species should be removed.

Instituting change in environmental practice is difficult without a strong education component. Education efforts must be conducted via multiple means to reach different user groups, and they must be ongoing because the turnover in population can be rapid. Actions at the municipal level include newsletters, instruction in public schools, tours of demonstration sites, displays in libraries, presentations at town meetings, mailings of flood hazard information to property owners, and information signs at key field locations (Breton et al. 2000; Nordstrom et al. 2009). It is important to specify the role of human actions in both the degradation and the restoration of natural environments.

2.6 Maintaining and Evaluating Restored Environments

Monitoring and adaptive management are important in assessing:

1. Whether restoration goals have been achieved
2. What kind of follow-up actions are required
3. How plans can be modified to achieve better future projects.

Evaluations should be conducted several years after dune building begins, so that managers can appreciate the time required for a resilient species-appropriate vegetation cover to become established and have realistic expectations of restoration outcomes. Understanding the multifaceted aspects of dune change is critical. Evidence of sand movement within the dunes may be considered beneficial in terms of habitat rejuvenation and biodiversity, but problematic in terms of the potential for the inundation of human infrastructure or the loss of an existing valued species dependent on a stable surface. Thus, changes to the form and function of the restored dunes should be expected and evaluated according to multiple criteria.

2.7 Conclusions

Attempts to restore dunes in developed areas must overcome restrictions in space and time as well as negative attitudes about their value. Restrictions in space and time can be partially offset by aiding natural processes through bulldozing, fencing, and planting programs, but dunes will provide more of their potential functions and services if they are allowed to evolve naturally and not managed as static landforms solely for shore protection. Restrictions to degradational practices, such as raking or driving on beaches, using sand fences well after the dune has formed, and allowing a single species to dominate management efforts will help dunes recover and provide a broader range of values, but recovery from the results of past practices will require more attention to education and adaptive management programs.

References

Arens SM, Jungerius PD, van der Meulen F (2001) Coastal dunes. In: Warren A, French JR (eds) Habitat conservation: managing the physical environment. Wiley, London

Avis AM (1995) An evaluation of the vegetation developed after artificially stabilizing South African coastal dunes with indigenous species. J Coast Conserv 1:41–50

Baye P (1990) Ecological history of an artificial foredune ridge on a northeastern barrier spit. In: Davidson-Arnott RGD (ed) Proceedings of the symposium on coastal sand dunes. National Research Council Canada, Ottawa, pp 389–403

Breton F, Esteban P, Miralles E (2000) Rehabilitation of metropolitan beaches by local administrations in Catalonia: new trends in sustainable coastal management. J Coast Conserv 6:97–106

Castillo SA, Moreno-Casasola P (1996) Coastal sand dune vegetation: an extreme case of species invasion. J Coast Conserv 2:13–22

Coastal Engineering Research Center (CERC) (1984) Shore protection manual. U.S. Army Corps of Engineers, Ft. Belvoir

Colombini I, Chelazzi L (2003) Influence of marine allochthonous input on sandy beach communities. Oceanogr Mar Biol Annu Rev 41:115–159

Dahl BE, Woodard DW (1977) Construction of Texas coastal foredunes with sea oats (*Uniolo paniculata*) and bitter panicum (*Panicum amarum*). Int J Biometeorol 21:267–275

De Lillis M, Costanzo L, Bianco PM, Tinelli A (2004) Sustainability of sand dune restoration along the coast of the Tyrrhenian Sea. J Coast Conserv 10:93–100

Doody JP (2001) Coastal conservation and management: an ecological perspective. Kluwer, Dordrecht

Dugan JE, Hubbard DM, McCrary MD, Pierson MO (2003) The response of macrofauna communities and shorebirds to macrophyte wrack subsidies on exposed sandy beaches of southern California. Estuar Coast Shelf Sci 58S:25–40

Everard M, Jones L, Watts B (2010) Have we neglected the societal importance of sand dunes? An ecosystem services perspective. Aquatic Conserv Marine Freshw Ecosyst 20:476–487

Feagin RA (2005) Artificial dunes created to protect property on Galveston Island, Texas: the lessons learned. Ecol Restor 23:89–94

Foster-Smith J, Birchenough AC, Evans SM, Prince J (2007) Human impacts on Cable Beach, Broome (Western Australia). Coast Manag 35:181–194

Godfrey PJ (1977) Climate, plant response, and development of dunes on barrier beaches along the U.S. east coast. Int J Biometeorol 21:203–215

Godfrey PJ, Godfrey MM (1981) Ecological effects of off-road vehicles on Cape Cod. Oceanus 23:56–67

Grafals-Soto R, Nordstrom KF (2009) Sand fences in the coastal zone: intended and unintended effects. Environ Manag 44:420–429

Grootjans AP, Adema EB, Bekker RM, Lammerts EJ (2004) Why young coastal dune slacks sustain a high biodiversity. In: Martínez ML, Psuty NP (eds) Coastal dunes, ecology and conservation. Springer, Berlin, pp 85–101

Hesp PA (1989) A review of biological and geomorphological processes involved in the initiation and development of incipient foredunes. Proc R Soc Edinb 96B:181–201

Hesp PA (1991) Ecological processes and plant adaptations on coastal dunes. J Arid Environ 21:165–191

Jackson NL, Nordstrom KF, Saini S, Smith DR (2010) Effects of nourishment on the form and function of an estuarine beach. Ecol Eng 36:1709–1718

Kuriyama Y, Mochizuki N, Nakashima T (2005) Influence of vegetation on Aeolian sand transport rate from a backshore to a foredune at Hasaki, Japan. Sedimentology 52:1123–1132

Lubke RA, Avis AM (1998) A review of the concepts and application of rehabilitation following heavy mineral dune mining. Mar Pollut Bull 37:546–557

Martínez ML, García-Franco JG (2004) Plant-plant interactions in coastal dunes. In: Martínez ML, Psuty NP (eds) Coastal dunes, ecology and conservation. Springer, Berlin, pp 205–220

Matias A, Ferreira Ó, Mendes I, Dias JA, Vila-Consejo A (2004) Development of indices for the evaluation of dune recovery techniques. Coast Eng 51:261–276

Matias A, Ferreira Ó, Mendes I, Dias JA, Vila-Consejo A (2005) Artificial construction of dunes in the south of Portugal. J Coastal Res 21:472–481

Maun MA (1998) Adaptation of plants to burial in coastal sand dunes. Can J Bot 76:713–738

Maun MA (2004) Burial of plants as a selective force in sand dunes. In: Martínez ML, Psuty NP (eds) Coastal dunes, ecology and conservation. Springer, Berlin, pp 119–135

Mauriello MN (1989) Dune maintenance and enhancement: a New Jersey example. Coastal Zone 89. American Society of Civil Engineers, New York, pp 1023–1037

Mauriello MN, Halsey SD (1987) Dune building on a developed coast. Coastal Zone 87. American Society of Civil Engineers, New York, pp 1313–1327
McLachlan A (1985) The biomass of macro- and interstitial fauna on clean and wrack-covered beaches in Western Australia. Estuar Coast Shelf Sci 21:587–599
Mendelssohn IA, Hester MW, Monteferrante FJ, Talbot F (1991) Experimental dune building and vegetative stabilization in a sand-deficient barrier island setting on the Louisiana coast, USA. J Coast Res 7:137–149
Miller DL, Thetford M, Yager L (2001) Evaluating sand fence and vegetation for dune building following overwash by Hurricane Opal on Santa Rosa Island, Florida. J Coast Res 17:936–948
Mitteager WA, Burke A, Nordstrom KF (2006) Landscape features and restoration potential on private shorefront lots in New Jersey, USA. J Coast Res SI39:890–897
Moss D, McPhee DP (2006) The impacts of recreational four-wheel driving on the abundance of the ghost crab (*Ocypode cordimanus*) on a subtropical sandy beach in SE Queensland. Coast Manag 34:133–140
Nordstrom KF (2003) Restoring naturally functioning beaches and dunes on developed coasts using compromise management solutions: an agenda for action. In: Dallmeyer D (ed) Values at sea: ethics for the marine environment. University of Georgia Press, Athens, pp 204–229
Nordstrom KF (2008) Beach and dune restoration. Cambridge University Press, Cambridge, p 187
Nordstrom KF, Lampe R, Vandemark LM (2000) Re-establishing naturally-functioning dunes on developed coasts. Environ Manag 25:37–51
Nordstrom KF, Jackson NL, Hartman JM, Wong M (2007) Aeolian sediment transport on a human-altered foredune. Earth Surf Proc Land 32:102–115
Nordstrom KF, Jackson NL, DeButts HA (2009) A proactive programme for managing beaches and dunes on a developed coast: a case study of Avalon, New Jersey, USA. In: Williams A, Micallef A (eds) Beach management: principles and practice. Earth Scan, London, pp 307–316
Nordstrom KF, Jackson NL, Kraus NC, Kana TW, Bearce R, Bocamazo LM, Young DR, DeButts HA (2011) Enhancing geomorphic and biologic functions and values on backshores and dunes of developed shores: a review of opportunities and constraints. Environ Conserv 38:288–302
Peterson CH, Lipcius RN (2003) Conceptual progress towards predicting quantitative ecosystem benefits of ecological restorations. Mar Ecol Prog Ser 264:297–307
Priskin J (2003) Physical impacts of four-wheel drive related tourism and recreation in a semi-arid, natural coastal environment. Ocean Coast Manag 46:127–155
Psuty NP, Moreira MESA (1992) Characteristics and longevity of beach nourishment at Praja da Rocha, Portugal. J Coast Res SI8:660–676
Ranwell DS, Boar R (1986) Coast dune management guide. Institute of Terrestrial Ecology, NERC, Huntingdon
Rhind PM, Jones PS (1999) The floristics and conservation status of sand-dune communities in Wales. J Coast Conserv 5:31–42
Ritchie W, Penland S (1990) Aeolian sand bodies of the south Louisiana coast. In: Nordstrom KF, Psuty NP, Carter RWG (eds) Coastal dunes: form and process. Wiley, Chichester, pp 105–127
Schwendiman JL (1977) Coastal sand dune stabilization in the Pacific Northwest. Int J Biometeorol 21:281–289
Snyder RA, Boss CL (2002) Recovery and stability in barrier island plant communities. J Coast Res 18:530–536
Snyder MR, Pinet PR (1981) Dune construction using two multiple sand-fence configurations: implications regarding protection of eastern Long Islands south shore. Northeast Geol 3: 225–229
van der Putten WH, Peters BAM (1995) Possibilities for management of coastal foredunes with deteriorated stands of *Ammophila arenaria* (marram grass). J Coast Conserv 1:29–39
van der Wal D (1998) The impact of the grain-size distribution of nourishment sand on aeolian sand transport. J Coast Res 14:620–631

Vestergaard P, Hansen K (1992) Changes in morphology and vegetation of a man-made beach-dune system by natural processes. In: Carter RWG, Curtis TGF, Sheehy-Skeffington MJ (eds) Coastal dunes: geomorphology, ecology and management for conservation. Balkema, Rotterdam, pp 165–176

Woodhouse WW Jr, Seneca ED, Broome SW (1977) Effect of species on dune grass growth. Int J Biometeorol 21:256–266

Chapter 3
Restoration of Coastal Foredunes, a Geomorphological Perspective: Examples from New York and from New Jersey, USA

Norbert P. Psuty and Tanya M. Silveira

3.1 Coastal Dune Restoration

Coasts are in dynamic evolution because drivers such as sea-level rise, storms, and sediment supply alter the conditions of the natural system. Further geomorphological changes occur as the natural system interacts with human modifications of the existing features, the alterations of the avenues of sediment transport, and the spatial interference of the ambient processes. As a result, many aspects of the ambient conditions at the coast are being modified and the landforms are responding. At a site, changing exposure or changing position in a spatial–temporal sequence may affect the relationship between sediment supply and energetics and thus the subsequent sediment budget at that coastal location. Geomorphologically, the outcome of such a change may result in displacement of the shoreline and its attendant features. Under most of the modern-day scenarios of rising sea level and diminished sediment supply, both natural and human-induced, the result is an eroding shoreline, a narrowing beach, and an attenuation of coastal dunes on the beach profile (Bird 1985; Nicholls et al. 2007).

Whereas the concept of restoration may apply to any component of the evolving coastal system as it is responding to the impetus of the environmental drivers, the focus of this paper is the variety of natural and cultural actions that affect restoration of the coastal foredune (Fig. 3.1). The foredune is the primary dune feature in the beach–dune profile and is in active exchange of sediment with the beach. Other dune forms that occur in the coastal system, such as parabolic dunes, longitudinal dunes, and barchans, may evolve from the foredune or foredune position.

N. P. Psuty (✉) · T. M. Silveira
Rutgers, The State University of New Jersey, Sandy Hook, NJ 07732, USA
e-mail: psuty@marine.rutgers.edu

T. M. Silveira
e-mail: mendes@marine.rutgers.edu

M. L. Martínez et al. (eds.), *Restoration of Coastal Dunes*, Springer Series on Environmental Management, DOI: 10.1007/978-3-642-33445-0_3,

Fig. 3.1 The foredune develops at the inland margin of the bare sand beach, where the transported sand interacts with pioneer vegetation and stabilizing vegetative cover to create an accumulative feature

These other forms may or may not be actively exchanging sediment with the beach. Although these dune forms may be the subject of restoration, they are not the focus of this paper.

3.2 Traces in the Literature

In coastal geomorphology, the issue of coastal dune restoration is at least as old as the classic publication by Marsh (1864) on the interaction of people with physical geography. Marsh described programs to re-create coastal dunes where they would have a positive effect on the land use of the hinterland. A number of decades ago, in the extensive review tome on *Man's role in changing the face of the earth* (Thomas 1956), Davis (1956) produced a seminal paper that described a variety of situations where humans had either altered the existing coastal topography or influenced the processes to produce change. He suggested that humans should play a more active role in re-establishing some of the natural features. More recently, Nordstrom (2008) has reviewed a range of actions that humans have taken to alter, to modify, as well as to re-create beach and dune topography. The product of these and many other publications on a similar theme is the description of a modified landscape and an earth-science system that is not always functioning to produce the natural outcomes of physical processes operating on sediment supply to create morphological responses. The thrust of this literature is that many features on the earth's surface are altered in some of their characteristics because of human influence and that a reduction of human influence would generate geomorphological features that would be more consistent with the natural system. Or, from a restoration perspective, a reduction of the human influence would serve to rebuild the products of the natural system to some degree.

The foundation of coastal foredune restoration in coastal geomorphology involves a focus on the active beach–dune profile and the exchange of sediment between the beach and the dune (Fig. 3.1). As described by Sherman and Bauer (1993) and Psuty (2004), the foredune is the primary dune that responds to the ambient processes and sediments to develop and evolve in the shoreline environment. Foredune restoration, therefore, should enable the input of sediment to the foredune location and provide for the opportunity of the foredune to develop in response to the dimensions of sediment supply and budget. Similar to the concept of passive and active restoration espoused by Kauffman et al. (1997), the geomorphological thrusts of foredune restoration may consist of taking action to encourage the availability and function of sediment pathways or the more passive approach of allowing natural foredune migration as well as in situ growth or attenuation.

3.3 Conceptual Basis

In the grand geomorphological scheme, there is a conceptual model of processes acting upon sediments to create geomorphological responses, thoroughly presented for the coastal system in the synthesis volumes by Trenhaile (1997) and Davidson-Arnott (2010). Restoration, therefore, is a condition that takes place within this model and it treats some aspect of the process–response sequence.

At the outset, it must be conceded that it is difficult to restore the processes of coastal foredune development because the processes are the dynamics of moving wind and water. The processes exist, although they may be altered by local interference. Likewise, it is really not within the general model to restore the forms because they are the outcomes of the action of the processes interacting with sediments and plants. The forms may be rebuilt or re-created, but that is a type of rehabilitation of an existing product, and not a restoration of the conditions that produce a product. Thus, it is logical that the concept of restoration of the coastal foredune should be directed at the restoration of the sediment supply that supports the creation, locus, and evolution of the coastal foredune (Hesp 1989; Psuty 2004). That is, given the presence of ambient processes that will mobilize sediment inland from the beach to accumulate at the site of the feature and produce the form (in association with vegetation), the constraining variable is the availability of sediment. The end product of the process–sediment response model is the foredune, and it has dimension, form, and location dependent on the presence of the processes acting upon the ambient sediment supply.

Therefore, in terms of restoration of a foredune or its return to a previous condition, it is the restoration of the available sediment supply to produce the resulting form that is the conceptual basis for the continuation of foredune development and evolution. And, importantly, there is no equilibrium form in this restoration goal because the natural evolution is dependent on the sediment budget, the net condition between gain and loss, and the vector that may lead to the natural enhancement, continuance, or to destabilization and destruction of the foredune.

Part of this geomorphological approach is related to scale. Sediment budget is measured in years to decades to centuries. It is the net measure of accumulation and release of sediment that produces the resulting landform in some specified time. It is also related to spatial association, whereby there may be alongshore sequences and stages of development. Therefore, the balances of accumulation and/or loss may vary in a geospatial arrangement of foredunes that have an array of differing dimensions in the active beach–dune profile.

There is also the concept of temporal variability. Both Marsh (1864) and Davis (1956) discuss historical accounts in Europe of coastal dune mobility and stability, and pose questions about the human role in that dynamic on a century scale. Within a somewhat longer timeframe, Arbogast et al. (2002) describe evolving natural conditions of foredune growth, stability, mobility, etc., associated with changing lake levels along Lake Michigan and point to changing sediment supply as the driver for the sequences of natural mobilization and re-establishment of coastal dunes in the Great Lakes area and on the oceanic coasts. Thus, there is an inherent temporal variability in stage of development, and restoration needs to acknowledge that the foredune feature is probably not definable in absolute terms, but as a geotemporal stage in the process–sediment response model [see further discussion in Psuty (2004)]. Indeed, some situations associated with the current sea-level rise scenarios may cause transfers of sediment in an alongshore context to encourage localized natural restoration of coastal foredunes (Psuty and Silveira 2010).

3.3.1 Geomorphological Objectives in Foredune Restoration

The geomorphological objective is to define and prescribe the restoration of a process–sediment response model as the goal for landscape evolution. Importantly, the model does not have a desired dimensional morphological outcome. It seeks to achieve maintenance of the processes that are operative at a site and to allow the vectors of change to occur. In so doing, geomorphological restoration is directed toward the continuity of the natural change/evolution. There is no final end product in the natural system; there is continual evolution, as determined by the variable dimensions of physical processes, the supply of sediment, vegetation colonization processes, species presence, and the spatial–temporal sequence of landscape change (van Zoest 1992). Therefore, the goal is to restore the heterogeneity of features in their geotemporal context. Ehrenfeld (2000) echoes that theme in stating that restoration ecology should recognize diversity as an outcome and seek more flexible goals. This is accomplished by allowing the ambient processes to interact with the natural sediment supply and create the complex of coastal foredune features. There is a general pattern and sequence that is created by the sediment budget and there is a more detailed form that is site-specific and non-predictable. It is not chaos, but it is not totally delimited because the conditions vary spatially as well as through time. Restoration, in this scenario, applies to the disencumbering of the ambient processes and the re-establishment of the spatial mobility of the sediment supply, rather than the creation of a specific ridge of sand.

In the literature, the application of restoration in the context of the coastal foredune is seen to incorporate several themes and several environmental settings.

Often, there is a contrast between restoration and repair, between restoration of the formational processes and reconstruction of the form in a designated place. In some cases, the thrust is to construct a dune form of a desirable dimension. For example, a recent interest in vulnerability to storm impacts (Pendleton et al. 2004) is largely related to the protective function of coastal dunes relative to the cultural development and infrastructure located inland of the foredune at lower elevations. Inherent in this theme are the identification of the low portions in the coastal dunes and the locations of vulnerability for storm surge penetration. And, restoration is directed toward elevating the low areas to decrease the potential for storm surge damage.

Other themes consider the repair of eroded dunes to some desirable form or assemblage of characteristics (slope, height, breadth, mass, habitat). Whereas some approaches incorporate a modest re-arranging of sediment and planting of beach grass, others create barriers in the form of sand fences and earthen barriers, or import vast quantities of sediment to build a dune form (De Lillis et al. 2004; Borowka and Rotnicki 2001). Arguably, none of the approaches that manipulate the form or its location constitute restoration. They are a type of repair or renovation at best, and may be completely out of place or ineffective at worst.

3.4 Classification of Foredune Restoration/Manipulation

A classification of foredune restoration/manipulation is applied herein to the range of conditions that extends from encouraging foredunes in their natural setting to totally creating a sand ridge in less than a natural setting (Fig. 3.2). Further, there are degrees of manipulation and positions of manipulation relative to the natural setting. And, of course, there are gradations within many of the categories. This classification is an attempt to bring some order to the gamut of conditions and to provide a basis for distinguishing logical groupings of themes and conditions.

Application of the classification is derived from examples occurring along Fire Island, where there is an opportunity to compare the outcomes in a variety of jurisdictions, with different coastal management perspectives and objectives. Fire Island is a 50-km-long barrier island situated on the southern shore of Long Island in the northeastern part of the United States (Fig. 3.3). This barrier island is the site of 17 communities, a county park, a state park, and Fire Island National Seashore of the US National Park Service, a unit that incorporates the "Otis Pike Wilderness Area". Foredune management on the barrier island ranges from attempted total control in some of the communities and local park jurisdictions to a total absence of controls in the Otis Pike Wilderness Area. Similar processes affect most of the island, although modified by the existence of jettied inlets at either terminus and a pair of stone groins located about midway through the western communities. Sediment supply is variable because of the sequestration effects of the jetties, the groins, and some small beach nourishment projects in the communities. Further, the inherited form is a variable because the foredune had a spate of dimensions and stages of accumulation and erosion. Thus, the process–sediment response model functioning at the foredune is

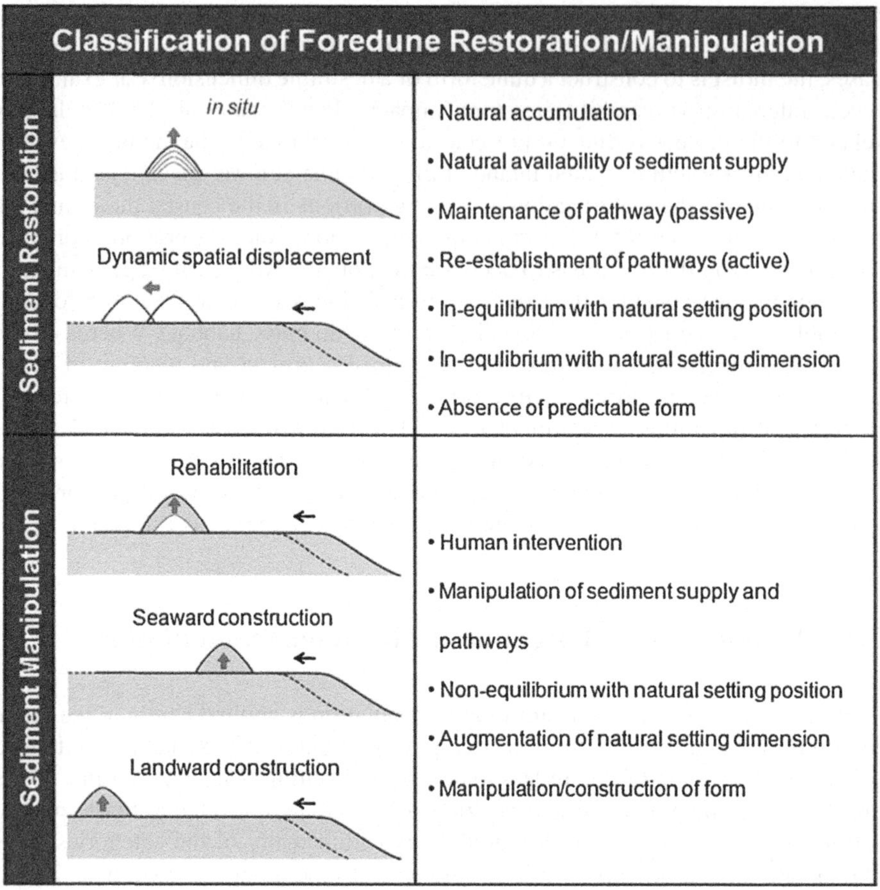

Fig. 3.2 Scenarios of foredune restoration/manipulation determined by spatial association of sediment accumulation on the erosional beach–dune profile under natural and under culturally constrained conditions. The *arrow* over the beach and the foredune refers to inland migration of the shoreline and displacement of the beach–dune profile. The *vertical arrow* at the foredune refers to sediment accumulation at the landform

under some constraints imposed by the impacts of the jetties and the existing morphologies. Yet, there is an opportunity to consider the responses of the foredune features amongst the variety of conditions and to put the morphological outcomes in a restoration/rehabilitation/construction perspective.

In general, the US National Park Service has a policy of supporting continuity of the natural system, natural processes, and the resulting landscape. In the coastal areas, this is usually translated into a reluctance to use hard structures to stabilize the shoreline, a minimum of artificial emplacement of sediment to balance the sediment budget, and an acceptance of coastal erosion and inland displacement of the beach–dune profile. In the coastal parks, restoration is viewed as restoring the

Fig. 3.3 Jurisdictions of the National Park Service, County Park and State Park, and residential communities on Fire Island. Otis Pike Wilderness Area is a component of the Fire Island National Seashore

opportunity for sediment transport into and out of the system; to support the evolution of the morphologies by minimizing the interference of sediment pathways. This is in accordance with the general geomorphological model of ambient processes working on the available sediment supply to create landforms and to allow the vectors of landform evolution to occur as far as possible.

3.5 Characteristics of the Types of Foredune Restoration/Manipulation

The variety of conditions present along Fire Island provide examples of the full range of foredune situations, from completely natural to completely manipulated in form and position. Although drawn largely from this one geographical area, the sequence of types of foredune restoration/manipulation applies to a broad system of foredune conditions. The accompanying characteristics in each category of restoration/manipulation describe the relationship among the sediment pathways, the resulting form, and the location of that form relative to its natural position on the beach–dune profile.

3.5.1 Restoration: In Situ (No Human Intervention)

Portions of the Otis Pike Wilderness Area have been overwashed in the past several decades and have a history of foredune restoration. In 2006, coastal erosion

Fig. 3.4 Location of naturally-restored foredune at the site of the 2006 washover, Otis Pike Wilderness Area

Fig. 3.5 Formation of the foredune on the washover fan, Otis Pike Wilderness Area, 26 March 2008

had removed the foredune and produced a large washover fan that extended inland (Fig. 3.4). Over time, small hillocks formed on the washover feature and pioneer vegetation extending onto the washover fan from the ends of the breached foredune created an incipient foredune that extended for much of the breadth of the fan (Fig. 3.5). The creation of this foredune is currently a bit farther inland than the former foredune, and it is being established in response to the ambient processes, sediment availability, and vegetation stabilization. The foredune form will likely shift and change dimension through time as it responds to the ambient processes. There is no equilibrium dimension to this foredune; only its natural presence on the profile and its morphological response to the environmental drivers.

3.5.2 *Restoration: Dynamic Spatial Displacement (No Human Intervention)*

Another type of natural restoration occurs where the general displacement of the beach and dune cause a translocation of the site of sediment accumulation and foredune development. An example of this dynamic restoration of a foredune occurred at Talisman, a part of the island under the jurisdiction of the National Park Service and where the foredune is functioning in a mainly natural state. In 1994, a large circulation cell caused major erosion of the beach and the existing foredune (Psuty et al. 2005). The foredune crest was lowered by about 50 % and the volume of sand above the beach surface lost about 35 % of its original mass. Following 1994, natural restoration of the foredune ensued to recreate a foredune ridge that had similar height dimensions and a similar cross-section area (Fig. 3.6). However, the foredune crestline was displaced inland about 25 m in the process of recovery. Restoration in this case was created by the available sediment supplied from the beach to accumulate in the upper beach and in the re-vegetated dune line. The process–sediment response model served to bring the pieces together in the spatial niche where the foredune was a viable element of the landscape. The foredune generating system was now positioned at a site unrestricted by sediment supply and at a location for the sediment to accumulate. Thus, the geomorphological feature was naturally re-established or restored in its equilibrium setting at this time with no human intervention.

Similar dynamic restoration occurs in other portions of Fire Island where the foredune is being reformed as it migrates inland. One such location is in the town of Point O'Woods, where the management philosophy was to allow the system to

Fig. 3.6 Natural restoration of the foredune crestline at Talisman. Through 1994–2003, the foredune was displaced inland about 25 m. The cross-section area above the beach berm surface initially decreased, recovered slowly, but was spatially displaced, 3 June 2005

Fig. 3.7 Migration of the foredune, Point O'Woods. Despite attempts to trap sediment and stabilize the foredune, the processes of sediment transport are continuing to support the foredune feature as it is transgressing inland. The house has subsequently been moved

function without attempts to restrict displacement or to create barriers to sediment transport to or across the foredune. This is a type of dynamic restoration that supports the translocation of sediment and the displacement of the foredune to a new natural accumulation setting (Fig. 3.7). The foredune feature is being restored as sediment is transported inland to compose some component of the alongshore variable foredune morphology. It persists in its spatial association with the ambient processes and sediment availability.

3.5.3 Manipulation: Rehabilitation Augmentation (Human Intervention)

Foredune rehabilitation applies to an existing foredune that is being maintained in its non-equilibrium location by small adjustments to its sediment budget that augment the feature that would normally be mobilized and displaced. At Davis Park (Fig. 3.8), the foredune has not migrated much in recent years because of repeated augmentation of the profile seaward of the foredune crestline. Sediment placed at the seaward margin of the foredune by bulldozer, by beach scraping, by transporting from off of the island, plus a constancy of fence lines have attempted to create a foredune feature seaward of its natural location (Fig. 3.9). Over time, the emplaced sediments have not been retained, primarily because the beach is relatively narrow and storm waves have reached the base of the accumulations and scoured the face of the foredune.

Fig. 3.8 Dune rehabilitation of the foredune positioned under the first row of houses. Sand is intermittently placed on the foredune, fences are erected, and various methods are applied to retain the sand ridge in place, Davis Park

Fig. 3.9 Latest accumulation of sediment at the base of the foredune, Davis Park, 26 March 2008

3.5.4 Manipulation: Constructed Foredune, Seaward (Human Intervention)

The natural foredune at Kismet has been displaced inland and completely eroded in portions of the community. However, a reconstructed foredune has been created to replace it far seaward of its former position (Fig. 3.10). Through the past three decades, the trend of the position of the natural foredune crestline at this location

Fig. 3.10 Natural foredune crestline extending under or near the seaward margin of houses at Kismet. The constructed foredune is wiped out at intervals. The newly constructed foredune line has been sited farther seaward at the second house

has migrated inland about 20 m. However, through reconstruction, the built foredune has shifted seaward about 11 m. The constructed foredune is intermittently scarped and eroded, to be further tended to with emplaced sediment, a variety of fence lines, and planted vegetation.

3.5.5 Manipulation: Constructed Foredune, Landward (Human Intervention)

There are no examples of a foredune constructed landward of its optimal position on Fire Island. All foredune construction on Fire Island served to maximize the space available for housing and infrastructure. However, in other locations there are instances of "foredune" construction considerably inland of the natural

Fig. 3.11 The constructed foredune is maintained inland of its optimal position, Wildwood, NJ, USA

foredune location (Fig. 3.11). These are usually in areas of high touristic development and related to beach use. In these cases, the artificial foredune is established inland and various means are employed to remove pioneer vegetation and any sandy hillocks that may be forming. The foredune position is manipulated constantly to maximize the open beach environment.

3.6 Summary and Conclusion

The foredune exists at the inner margin of the beach and migrates spatially as the beach either accretes or erodes. The foredune has a sediment budget, accreting or eroding, that is related to the beach budget, but that can be different in magnitude and direction. Geomorphologically, restoration of the sediment budget in the foredune is the key to the vector of change in this site. Action taken to ensure the availability of sediment reaching the location of pioneer vegetation so as to accumulate and create the heterogeneity of the natural topographical features is the essence of restoration in foredune systems. Manipulation or artificial placement of sediment on the beach profile, as well as construction of fences or other barriers to the inland transport of sediment, are aspects of foredune rehabilitation if the feature occupies the natural site of the foredune. If the emplacement of sediment or barriers are beyond the sites of natural foredune location (usually seaward), the constructed dune ridge completely contradicts the concept of foredune restoration. As seen along many coastlines, foredunes are an essential component of the natural coastal morphologies and they provide a unique set of habitats for plants and animals. Their viability is tied to the spatial and temporal supply of sediment to maintain the foredune system in situ or to maintain it as it shifts under scenarios of diminished sediment supply and sea-level rise. Restoration of foredunes consists of the active support of the geomorphological feature as it shifts and evolves under the geotemporal vector of sediment availability. Importantly, there is no equilibrium form or dimension to the functioning foredune. It has spatial and temporal variability. Heterogeneity, therefore, is a characteristic of complete foredune restoration.

References

Arbogast AF, Hansen EC, Van Oort MD (2002) Reconstructing the geomorphic evolution of large coastal dunes along the southeastern shore of Lake Michigan. Geomorphology 46:155–241

Bird ECF (1985) Coastline changes: a global review. Wiley, New York, p 219

Borowka M, Rotnicki K (2001) Budget of the eolian sand transport on the sandy barrier beach (a case study of the Leba Barrier, southern Baltic coast, Poland). Z Geomorphol 45:55–79

Davidson-Arnott RGD (2010) Introduction to coastal processes and geomorphology. Cambridge University Press, New York, p 442

Davis JH (1956) Influence of man upon coast lines. In: Thomas WH Jr (ed) Man's role in changing the face of the earth. University of Chicago Press, Chicago, pp 504–521

De Lillis M, Costanzo L, Bianco PM, Tinelli A (2004) Sustainability of sand dune restoration along the coast of the Tyrrhenian Sea. J Coast Conserv 10:93–100

Ehrenfeld JG (2000) Defining the limits of restoration: the need for realistic goals. Restor Ecol 8:2–9

Hesp PA (1989) A review of biological and geomorphological processes involved in the initiation and development of incipient foredunes. Proc R Soc Edinb B96:181–201

Kauffman JB, Beschta RL, Otting N, Lytjen D (1997) An ecological perspective of riparian and stream restoration in the western United States. Fisheries 22(5):12–24

Marsh GP (1864) Man and nature: or, physical geography as modified by human action. Scribner, New York

Nicholls RJ, Wong PP, Burkett VR, Codignotto JO, Hay JE, McLean RF, Ragoonaden S, Woodroffe CD (2007) Coastal systems and low-lying areas. In: Parry ML, Canziani OF, Palutikof JP, van der Linden PJ, Hanson CE (eds) Climate change 2007: impacts, adaptation and vulnerability. Contribution of working group II to the fourth assessment report of the intergovernmental panel on climate change. Cambridge University Press, Cambridge, pp 315–356

Nordstrom KF (2008) Beach and dune restoration. Cambridge University Press, New York, p 187

Pendleton EA, Williams SJ, Thieler ER (2004) Coastal vulnerability assessment of Fire Island National Seashore to sea-level rise. U.S. Geological Survey Open File Report 03-439. http://pubs.usgs.gov/of/2003/of03-439/index.html

Psuty NP (2004) The coastal foredune: a morphological basis for regional coastal dune development. In: Martinez ML, Psuty NP (eds) Coastal dunes: ecology and conservation. Springer, New York, pp 10–27

Psuty NP, Silveira TM (2010) Global climate change: an opportunity for coastal dunes. J Coast Conserv 14:153–160

Psuty NP, Pace JP, Allen JR (2005) Coastal foredune displacement and recovery, Barrett Beach-Talisman, Fire Island, New York. Z Geomorphol Suppl 141:153–168

Sherman DJ, Bauer BO (1993) Dynamics of beach–dune systems. Prog Phys Geogr 17:413–447

Thomas WH Jr (ed) (1956) Man's role in changing the face of the earth. University of Chicago Press, Chicago, p 1193

Trenhaile A (1997) Coastal dynamics and landforms. Oxford University Press, New York, p 366
van Zoest J (1992) Gambling with nature? a new paradigm of nature and its consequences for nature management strategy. In: Carter RWG, Curtis TGF, Sheehy-Skeffington MJ (eds) Coastal dunes: geomorphology, ecology, and management. Balkema, Rotterdam, pp 503–516

Chapter 4
Natural Plant Diversity Development on a Man-Made Dune System

Peter Vestergaard

4.1 Introduction

Natural formation of coastal foredunes involves interaction among sand, sea, wind, and plant growth (Carter 1988; Klijn 1990; Nordstrom et al. 1990; Martinez and Psuty 2004). The subsequent development of the dunes implies a gradual increase in species richness as a result of the combined effect of biotic and abiotic conditions and processes, related to dune age and distance to the sea (Carter 1988; McLachlan 1991; Lichter 1998; Hesp 1999).

Natural dunes make up a valuable natural resource because of their relatively high biodiversity and the occurrence of many specialized species of plants and animals. Coastal dunes, however, may include areas that are not purely naturally formed. Thus, many dune systems play an important role in coastal protection or may serve as an important resource for recreation or the catchment of drinking water (van der Maarel 1979; van der Meulen and Jungerius 1989). Because of the erosion, restoration measures are often needed in such dune systems (van der Meulen and van der Maarel 1989; De Lillis et al. 2004; Kiehl and Isermann (2007).

In Denmark, large dune areas occur along the North Sea coast of Jutland (Doody 1991). However, dunes also occur along the coastlines of the Kattegat and as part of the barrier coasts along the Baltic. As they are important resources for recreation and coastal protection, many Danish dunes, however, are currently restored and maintained by the local mechanical leveling of sharp dune edges and the planting of *Ammophila* to prevent sand drift, or by replenishment by means of sand nourishment (Jensen 1994, 2008).

P. Vestergaard (✉)
Department of Biology, University of Copenhagen, Universitetsparken
15, 2100 Copenhagen Ø, Denmark
e-mail: peterv@bio.ku.dk

M. L. Martínez et al. (eds.), *Restoration of Coastal Dunes*, Springer Series on Environmental Management, DOI: 10.1007/978-3-642-33445-0_4,

The general pattern of dune succession from embryonic to fixed dunes is also documented by several studies in Danish dunes (e.g., Warming 1909; Heykena 1965; Vestergaard 1991; Frederiksen et al. 2006). However, succession studies in relation to restoration measures in dunes have not been carried out.

This chapter reports a 24-year study of the development of a beach–dune system in Denmark, which in the 1970s was designed and established to meet local recreational and coastal protection purposes (Nielsen 1990). Even if the establishment and development of this system is not an example of restoration according to the narrow definition of the term—to move a degraded system to a stable target state (Hobbs et al. 2007)—the results and observations gained from the study may be relevant in a restoration context, because they indicate how a human-modified dune might be restored by natural processes.

On a global scale, structural and adaptational characteristics of beach and dune vegetation are similar, even if the taxonomic spectrum differs (Doing 1985). Therefore, the use of plant functional types, i.e., sets of species sharing similar structural and functional attributes (Box 1996), may permit ecological comparisons among coastal dunes on a more general scale than would be possible by using a taxonomic approach (Shao et al. 1996; Garcia-Mora et al. 1999). A widely used functional type system is the life form classification of Raunkiær (1907, 1934), which has been applied in several studies on sand dune zonation and succession (e.g., Ranwell 1959; Hundt 1985; Olff et al. 1993). Other functional traits, important for understanding plant colonization and succession processes in coastal dunes, are mechanisms of the dispersal of diaspores (Andersen 1993; Davy and Figueroa 1993) and tolerance to burial by sand and salinity (Hesp 1991; Garcia-Mora et al. 1999; Gallego-Fernández and Martinez 2011).

The aim of the present study was to investigate the relationship between plant diversity and dune development processes in the human-initiated dune system, and to discuss whether those processes are similar to those known to be characteristic of natural dune succession.

4.2 Locality and Methods

4.2.1 Initial Morphology of the Man-Made Beach-Dune System

The study site is located in the Køge Bay Seaside Park at the Baltic coast of Zealand, Denmark, about 15 km southwest of Copenhagen (55°36′N, 12°22′E). The seaside park consists of an artificial barrier, 8 km long, about 300 m wide, established in 1977–1978, comprising a sandy beach, an artificial foredune and a grassland area landward of the dune (Fig. 4.1). The system was created from sediment, supplied to the beach by pumping calcareous marine sand upon previously existing, natural low sandy bars (Fig. 4.1; Thougaard 1980; Nielsen 1990). During the planning of the barrier, the structure of a natural barrier island, the

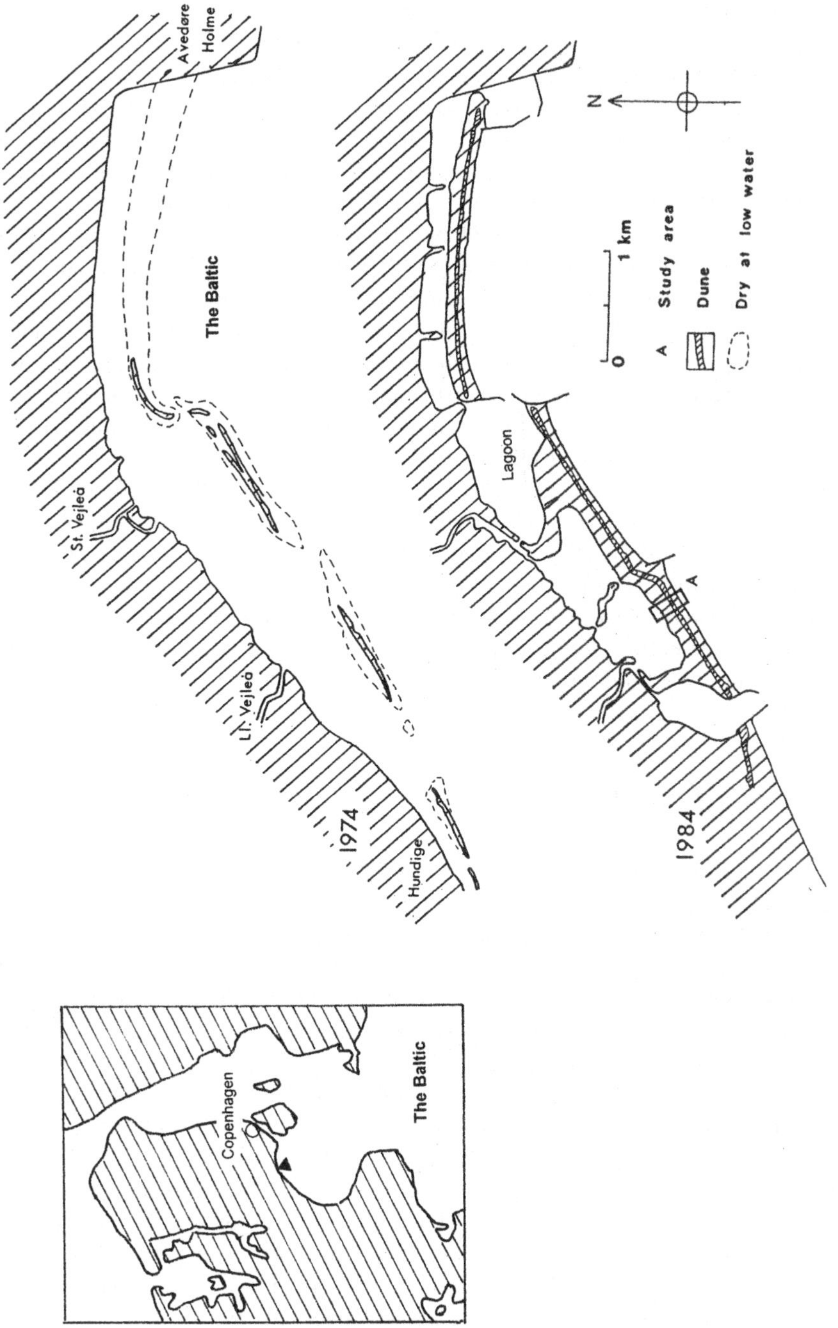

Fig. 4.1 Sketch maps based on aerial photographs from 1974 and 1984, showing the natural sandy bars existing before the construction of the Køge Bay Seaside Park, and the situation after the establishment of the artificial barrier. The *arrow* on the inset map indicates the position of the seaside park

Ølsemagle Revle, further south in the bay of Køge Bugt, which was studied by Gravesen and Vestergaard (1969), served as a model (Nielsen 1990).

The total size of the seaside park is about 500 ha. The park is managed by the I/S Køge Bugt Strandpark. The total price for the establishment of the seaside park amounted to approximately 91 million Danish krone (about 13 million US dollars). The annual management amounts to around 5 million crowns (about 0.7 million US dollars) (Valgren et al. 1986).

The initial relief of the artificial beach and dune is shown in Fig. 4.2. The beach, about 45 m wide, was built up to +2.3 m DNN (Danish Mean Sea Level). The initial dune consisted of a horizontal plateau at +3 m DNN, approximately 20 m wide, delimited by a seaward and a landward slope. The dune plateau and the seaward slope were planted with *Ammophila arenaria.* The planting was carried out according to the traditional Danish method (Jensen 2008). At about ten regularly distributed points per square meter, 2–4 *Ammophila* shoots, cut at a depth of 10–20 cm below the sand surface in established *Ammophila* dune elsewhere, were planted. After the planting, the area was fertilized with NPK fertilizer. The area landward of the dune was sown with a mixture of perennial grasses, mostly *Festuca rubra,* and locally planted with low shrubs: the native *Hippophaë rhamnoides,* which is rather widely distributed along Danish coasts, and the non-native *Rosa rugosa.* Even if *Rosa rugosa* is non-native and now known to be invasive, it was earlier commonly planted along Danish coasts for its ornamental value and its sand-binding ability (Weidema et al. 2007).

4.2.2 Data Collection

In 1979, a study area across the barrier, 100 m wide, was selected ("A" in Fig. 4.1). Along a line transect, perpendicular to the shore, six permanent plots were established. This chapter is based on four of these plots, each 5×10 m^2: two plots from the inner and outer part of the dune plateau ("dune 1a" and "dune 1b") and two from the inner and outer parts of the beach ("dune 2" and "dune 3"; Fig. 4.2). Because of later gradual seaward widening of the beach, an additional plot was established in 1993 ("dune 4").

The species composition and species diversity of the permanent plots were recorded in July or August on 11 occasions between 1979 and 2002. The distance to the sea of the plots was recorded, and from 1980 the plots were leveled in relation to DNN. A quantitative estimate of the species composition was attempted using a fixed system of 20 sampling circles of 0.1 m^2. In the circles rooted species of vascular plants were recorded, and the cover of each species was estimated using the Hult–Sernander 1–5 cover scale (1: cover <6 %, 2: cover 6–12.5 %, 3: cover 12.5–25 %, 4: cover 25–50 %, 5: cover > 50 %) (Malmer 1974;Hansen 1981, nomenclature). From these data, frequency (%) and mean cover of the species in each plot were calculated. In total 43 relevées (plot-years) were analyzed.

Fig. 4.2 Structure and change of the profile of the dune system between 1978 and 2002, with localization of the permanent plots. The year 1978 represents the initial profile. The *arrows* indicate the limit between the beach (backshore) and the dune

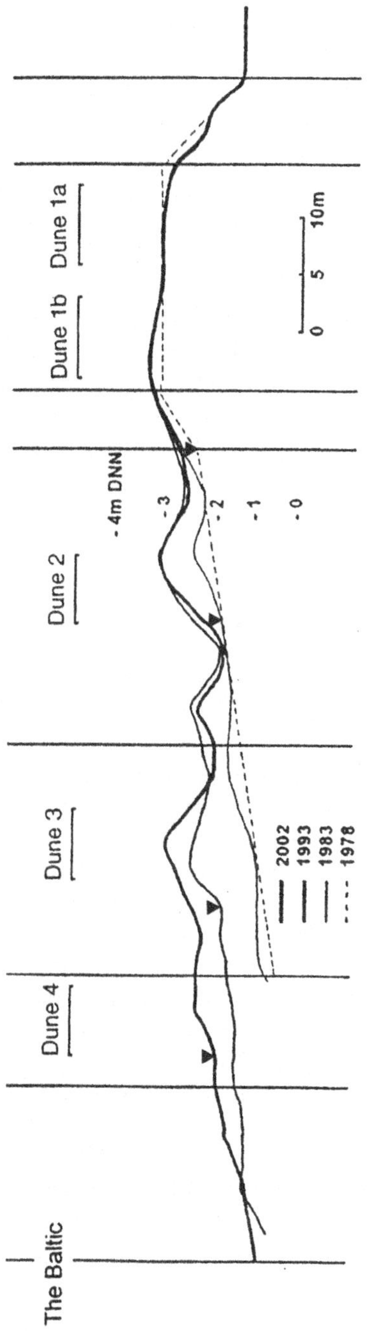

4.2.3 Plant Data Analysis

The species diversity in the permanent plots was expressed by the number of species as well as by the Shannon–Wiener diversity index. These measures were found to be highly significantly positively correlated (Fig. 4.3). Therefore, the discussion of the trends in plant diversity development presented here will be based on the number of species per plot.

For each of the dunes, the correlation between the number of species per plot and the following dune attributes was investigated: number of years since establishment, dune height, distance to the sea, and total plant cover. The plant species were classified according to their life forms (Raunkiær 1907, 1934), and diaspore dispersal strategy, following Andersen (1993).

The frequency and Hult–Sernander cover data were combined into a 1–9 cover–abundance scale (Vestergaard 2006). Based on this scale, the species composition of the relevées was subjected to TWINSPAN classification analysis, cut levels 0–2–4–6–8, using the program PC-ORD (McCune and Mefford 1999).

4.3 Results

4.3.1 Change in Dune Morphology 1978–2002

The morphology of the landward part of the initial man-made dune remained fairly unchanged during the study period, while in other parts, the morphology changed substantially (Fig. 4.2).

In the seaward part of the initial dune, a net accumulation of up to approximately 0.4–0.5 m of windblown sand was recorded. On the beach, new foredune ridges, parallel to the initial dune, gradually developed, supported by *A. arenaria*, which spread from the planting in the initial dune, as well as by other dune-forming grasses. Thus, in the inner part of the beach, the dune, represented by "dune 2," accreted rapidly (0.15 m $year^{-1}$) to about the same height as the top of the initial dune, +3 m DNN (Figs. 4.4, and 4.5). After 1990, this dune stabilized, and the height did not change significantly. On the outer part of the beach, another new dune, represented by "dune 3," developed, albeit a little more slowly (0.1 m $year^{-1}$). In 2002 this dune was about 0.5 m lower than "dune 2."

The total width of the beach–dune system increased from approximately 75 m in 1978 to 100 m in 2002, owing to the accretion of additional beach formed by less calcareous marine sand (Vestergaard 2006). On this beach, a low dune, represented by "dune 4," had developed in 2002 (Figs. 4.2, and 4.4). This propagation and development of new foredunes is similar to those examined by for example Hesp (1989) and McLean and Shen (2006).

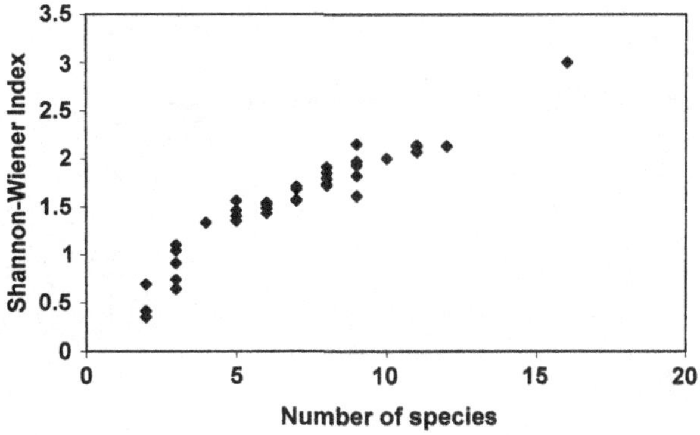

Fig. 4.3 Correlation between number of species and the Shannon–Wiener diversity index in the permanent plots (r-value for the linear trend line: $r = 0.948$, $p < 0.001$, $n = 43$)

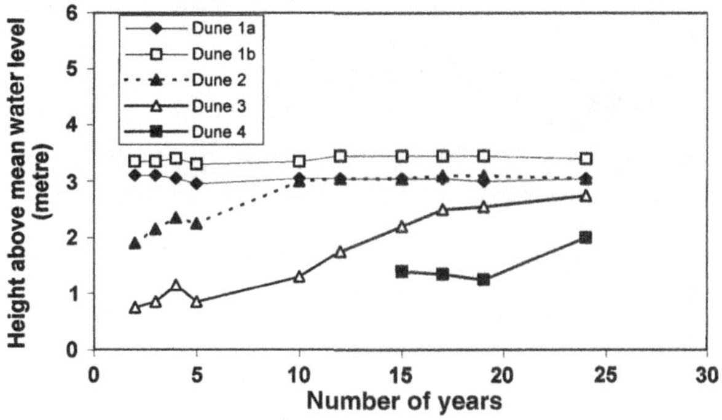

Fig. 4.4 Growth of the dunes from 1980 to 2002, expressed by the increasing height of the permanent plots above the Danish Mean Sea Level

4.3.2 Development in Species Diversity

The planted *A. arenaria* was the only species present in the beach–dune system from the very beginning. In 1979, seven species were recorded in the permanent plots, and from 1979 to 2002, the total number of species in the permanent plots gradually increased to 26 after some fluctuations, especially during the first years (Fig. 4.6a). The proportion of coast species, i.e., species exclusively or predominantly growing in coastal habitats, also changed, from 71 % in 1979 to 38 % in 2002.

The species richness in the plots changed over time, with changes in relief and distance to the sea and with total plant cover, but the changes differed among the dunes (Fig. 4.6, Table 4.1). In the relatively stable landward part of the man-made dune, "dune 1a," the number of species varied significantly only with total plant cover. In the less stable part of the man-made dune, "dune 1b," the number of species also increased significantly over time and over distance to the sea. In "dune 2" and "dune 3," which developed from beach to foredunes during the study period, compare Figs. 4.2 and 4.5: the number of species increased significantly with all the factors involved in the dune development, recorded in the present study: dune age, distance to the sea, dune height, as well

Fig. 4.5 a–d Development of the permanent plot, representing "dune 2" from 1979 to 2002, seen in a northeasterly direction. Compare with Fig. 4.2. On the right of the photos the development of "dune 3" can also be followed. *Photos* P. Vestergaard. **a** 1979, **b** 1983 **c** 1993, **d** 2002

as total plant cover. In the plot of the youngest dune, "dune 4," the number of species also increased, albeit not significantly, owing to the low number of observations.

4.3.3 Species Composition Dynamics and Functional Traits

Based on the TWINSPAN analysis, six groups of plot-year relevées ("plant communities") could be defined (Tables 4.2, and 4.3). The distribution pattern of the groups in the dune system could be related to a combination of the factors time and distance to the sea (Table 4.4).

Group A (8 relevées) represents the species composition of "dune 1a" and "dune 1b" during the first years, as well as the first year of the plot of the most landward of the new foredunes, "dune 2." The cover of the planted *A. arenaria* was mostly high, but there were also many beach and inland annuals (62 %). Most species were dispersed by wind (54 %).

Group B1 (14 relevées) represents the first years of development of "dune 3" and "dune 4" and also some years of "dune 1b." The relevées are characterized by strong sand movement and sand accretion and by dune-building rhizome

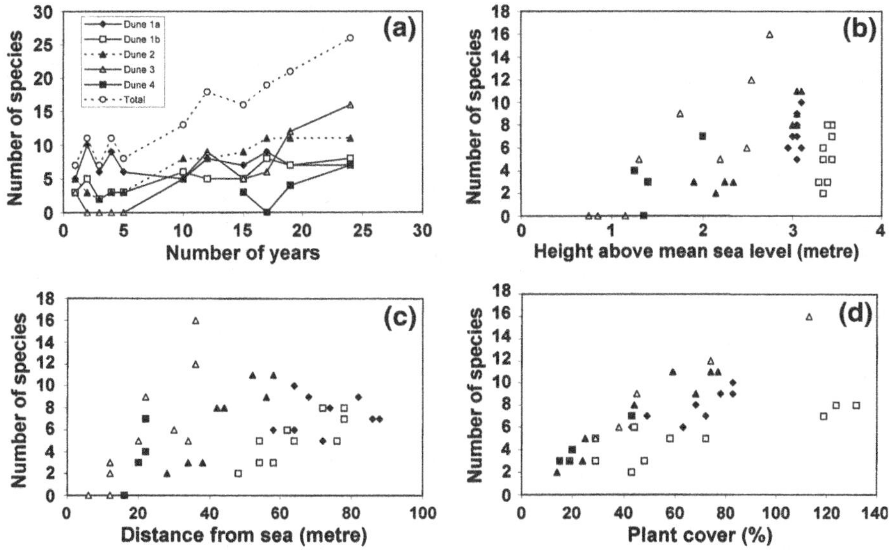

Fig. 4.6 **a** Development in the number of species in the permanent plots over time (1979–2002), with **b** dune height above the mean sea level, and **c** the distance to the sea and **d** total plant cover. For the correlation, see Table 4.1

Table 4.1 Correlation (r-values for linear trend lines) between the number of plant species in the permanent plots, and age, height above mean sea level, distance to the sea, and total plant cover of the plots

	Time	Height	Distance	Plant cover
Dune 1a	$r = 0.074$ ns	$r = 0.354$ ns	$r = 0.138$ ns	$r = 0.808$**
Dune 1b	$r = 0.867$**	$r = 0.539$ ns	$r = 0.838$**	$r = 0.845$**
Dune 2	$r = 0.928$**	$r = 0.937$**	$r = 0.935$**	$r = 0.958$**
Dune 3	$r = 0.902$**	$r = 0.871$**	$r = 0.833$**	$r = 0.986$**
Dune 4	$r = 0.792$ ns	$r = 0.749$ ns	$r = 0.898$ ns	$r = 0.997$*

Number of observations: dune 1a–3: $n = 7$–11; dune 4: $n = 3$–4. Levels of significance: *ns* $p > 0.05$; *$p < 0.05$; **$p < 0.01$)

geophytes, *A. arenaria*, *Leymus arenarius*, and *Elytrigia junceiforme*. The proportion of annuals was still high (46 %). Because of the relatively low elevation above sea level of most relevées, about 1 m DNN (Fig. 4.4), water was the most important agent of diaspore dispersal (46 %), and most species are salt tolerant.

Group B2 (6 relevées) represents the next seral stage of "dune 2" and "dune 3" after these dunes had built up to about the same height as that of the initial dune (Fig. 4.4). The rhizome geophytes, which dominated group B1, were still dominating, but they were accompanied by *Ammophila* x *baltica*. The number of species increased, and the hemicryptophytes, e.g., *F. rubra*, became the most important life form, even if their cover was still low. In this group, as well as in the following groups, wind was the dominating agent for diaspore dispersal.

Table 4.2 Important vascular plant species of the TWINSPAN groups

TWINSPAN group	A	B1	B2	C1	C2	C3
Total number of species	13	13	21	21	12	12
Cakile maritima	X	o	o			
Senecio vulgaris	X			o		
Chenopodium album	X					
Festuca rubra	X	o	X	X	X	X
Ammophila arenaria	X	X	X	X	X	X
Leymus arenarius		X	X	o		
Elytrigia junceiforme		X	X	o	o	
Salsola kali	o	X				
Ammophila x *baltica*			X	o		
Taraxacum sp.			X	X	X	X
Erigeron acre			X	o		o
Cirsium arvense			X	o		
Conyza canadensis	o		X	o		
Senecio vernalis	o	o	X	X	o	X
Epilobium montanum			o	X		
Lathyrus japonicus		o		X	X	o
Cirsium vulgare	o		o	X		o
Anthyllis vulneraria						X
Trifolium arvense				o		X
Cerastium semidecandrum			o		X	X
Rosa rugosa				o	X	
Chamaenerion angustifolium			o	o	X	o

X species present in >40 % of the relevées; o important species also present in the group in question, but in <40 % of the relevées

Group C1 (7 relevées) represents the seral situation of "dune 1a" and "dune 1b" between years 5 and 12. It also represents the situation of "dune 2" between years 19 and 24, starting about 10 years after the stabilization of the dune. The vegetation is characterized by a high number of species and increasing *F. rubra*. In "dune 1a" and "dune 1b,' *A. arenaria* declined, while *A. arenaria* and *A.* x *baltica* have kept their position in "dune 2" so far.

Group C2 (4 relevées) represents the seral situation of "dune 1b" between years 12 and 24. Hemicryptophytes were the dominating life form. *F. rubra* dominated, but *A. arenaria* and *Lathyrus japonicus* were also important, indicating that some sand movement still occurred. The invasive phanerophyte *Rosa rugosa* became established in year 17 and increased further to year 24.

Group C3 (4 relevées) represents the seral situation of "dune 1a" between years 12 and 24. The surface was stable, and the vegetation had become similar to northwest European and Baltic grassland-like fixed dune with a high percentage of hemicryptophytes, dominated by *F. rubra*, and with, for example, *Trifolium arvense* and *Anthyllis vulneraria*, but still with *A. arenaria* present (e.g., Böcher 1954; Påhlsson 1995; Thorell et al. 2001).

Table 4.3 Life form and diaspore dispersal strategy spectra of the TWINSPAN groups

TWINSPAN group	A	B1	B2	C1	C2	C3
Life form, % of species						
Rhizome geophytes	8	31	24	19	17	8
Root geophytes	0	0	10	14	8	8
Hemicryptophytes	31	23	43	33	42	50
Annuals	62	46	24	29	25	33
Phanerophytes	0	0	0	5	8	0
Dispersal strategy, % of species						
By wind	54	23	57	67	42	67
By seawater	8	46	10	10	8	8
By wind shaking	8	8	5	0	8	0
Epizoic	0	0	5	0	0	0
Endozoic	0	0	0	5	8	0
No special strategy	31	23	24	19	33	25

Table 4.4 The distribution of the TWINSPAN groups in the permanent plots from 1979 to 2002

	Dune 1a	Dune 1b	Dune 2	Dune 3	Dune 4
1979	A	A	A	B1	
1980	A	A	B1		
1981	A	A	B1		
1982	A	B1	B1		
1983	C1	B1	B1		
1988	C1	C1	B2	B1	
1990	C1	C1	B2	B1	
1993	C3	C2	B2	B1	B1
1995	C3	C2	B2	B1	
1997	C3	C2	C1	B2	B1
2002	C3	C2	C1	B2	B1

4.4 Discussion

4.4.1 Beach and Dune Dynamics

The initial relief of the beach–dune system in the seaside park differed from the relief of the natural barrier island, Ølsemagle Revle, further south in the bay of Køge Bugt, especially with regard to the amount of beach sand (Gravesen and Vestergaard 1969).

The natural beach was 10–20 m wide and had a height of up to about 1.5 m above DNN. The man-made beach was much wider and higher (Fig. 4.2). The large amount of sand contained in the man-made beach had important implications for the later development of the whole system. Several foredunes were formed by natural processes—sand movement and plant growth—on the man-made beach within 24 years. These dunes were formed by two components of sand: the initial nourishment sand and the sand subsequently deposited by the sea by natural coastal processes (Nielsen 1990). This gradually increased the width of the beach by 25 m in 24 years. The sand deposited by the sea was quite similar to the beach sand of the natural barrier in its pH and exchangeable calcium, but differed from the nourishment sand (Gravesen and Vestergaard 1969; Vestergaard 2006).

The system studied here seems to have behaved like the "high sediment supply system" described by Carter (1990), which is characterized by progressive seaward formation of several foredunes, each of which develops until the sand supply from the beach is cut off by the formation of a new foredune ridge. Thus, an important consequence of the successive seaward formation of the natural dunes on the beach in the seaside park has been a strongly reduced dynamic of the landward dunes, which instantly or within a few years stabilized at a dune height of 3 m DNN. This is typical of the prograding foredune systems such as those described by among others Carter (Carter 1990; Hesp 1999, 2002).

4.4.2 Immigration of Plant Species

The increase in species richness and development of vegetation in the dunes studied here includes the sequence of processes characteristic of primary succession in a new area (Bradshaw 1993). Among these processes, dispersal of diaspores and distance to species sources are essential for the rate and course of the succession (van der Maarel 1997; del Moral 1998; Walker and del Moral 2003).

In the nourishment sand, no seed bank was present (Hansen and Vestergaard 1986). Thus, apart from the initial planting of *A. arenaria*, the subsequent development of the vegetation in the permanent plots was due to natural influx of diaspores.

The minimum number of species dispersed into the beach and dune system can be estimated from the number of species that have been recorded in the permanent plots during the years, i.e., 38 species. Nothing indicates, however, that the influx of species to the plots has stopped. Thus, the number of species established in the beach and dune within the total study area (Fig. 4.1) was 55 (Vestergaard 2006), and in the natural Ølsemagle Revle barrier, 85 species were recorded within similar beach and dune habitats after about 37 years of development (Gravesen and Vestergaard 1969), and 102 species after a further 40 years (2008). Based on the above, it is thus expected that the species number will continue increasing, if the same trends are maintained during the coming years.

During the study period, the natural habitat of an increasing proportion of the species recorded is relatively dry, nutrient-poor inland grasslands and other open inland habitats (Hansen 1981). The source of these species is rather diffuse, but the species in question are all common and more or less widely distributed in Denmark (Vestergaard and Hansen 1989). The remaining species are found in coastal habitats. The source of these species was probably the nearby coasts along the bay of Køge Bugt and the island of Amager (Fig. 4.1). The relative increase in inland species is supposed to be due to the reduced influence of the sea upon an increasing proportion of the dune area, caused by the seaward development of new foredunes.

A large proportion (54 %) of the species recorded in 2002 specializes in seed dispersal by wind. This indicates that the dune system after 24 years was still at an early stage of succession (Fenner 1985; Prach and Pysek 1999). However, for many species it is not possible to demonstrate evident morphological specialization for long-distance dispersal. Andersen (1993) points to humans as important dispersal agents for such species. This could well be the case in the seaside park, considering the strong recreational pressure in the area (Nielsen 1990; Andersen 1995). Many species with no special spreading device can, however, be dispersed by wind because of the small diaspore size (<2 mm; Andersen 1993). Thus, *F. rubra*, which dominated the later stages of succession in the dune, has small and light diaspores (Grime et al. 1988), which could easily have been blown into the dunes from the sown grassland just landward to the initial, man-made dune by the frequent westerly winds.

4.4.3 Species and Life Form Dynamics

The increase in species richness of the dune system from 1979 to 2002 implied a change in the proportion of the plant life forms in a spatial–temporal pattern also observed during primary succession in other beach and dune systems, e.g., Hundt (1985).

In the original, man-made dune, the *A. arenaria* that had been planted declined because of a decreasing supply of fresh, wind-borne sand (van der Putten 1989; van der Putten and Peters 1995; De Rooij-van der Goes 1995; van der Stoel et al. 2002), while *F. rubra* as well as inland annuals increased. *A. arenaria* expanded seaward, however, into the area of mobile sand and was the primary colonizer initiating and forming new foredunes on the beach.

In the new dunes, beach annuals, e.g., *Cakile maritima* and *Salsola kali*, which are typically limited to open, salty and unstable beach sand (Davy and Figueroa 1993; Packham and Willis 1997), were present during the first years of dune development, when the plant cover was sparse and the substrate was fresh marine sand. During the following years the beach annuals were replaced in the sequence “dune 2” → “dune 3” → “dune 4” by the dune-forming rhizome geophytes *A. arenaria, A.* x *baltica, L. arenarius, Elytrigia junceiforme*, which are tolerant to or even favored by deposition of fresh sand (Ranwell 1972; Packham and Willis 1997). During the last years of the study period hemicryptophytes, especially *F. rubra*, began to play a role in the sequence “dune 2” → “dune 3.”

Besides the plant groups mentioned, some additional plant groups colonized the dunes during the latter part of the study period: root geophytes, e.g., *Chamaenerion angustifolium*; hemicryptophytic dicots, e.g., *Erigeron acre*, Fabaceae, e.g., *L. japonicus, Anthyllis vulneraria, Trifolium arvense*; shrubs, e.g., *Rosa rugosa, Hippophaë rhamnoides*.

4.4.4 Species Composition Dynamics

The groups of relevées that resulted from the TWINSPAN analysis appear to be meaningful in several ways.

The gradual increase in species richness of the dune system from year 1 to year 24 was reflected in the number of TWINSPAN groups represented, which increased from two to five, indicating an increasing habitat variation within the dune system.

Second, the grouping clarifies, together with the changes in morphology, the developmental trends in the dune system in the seaside park up to 2002. The spatial position of the active dune formation, as indicated by the relatively high speed of dune growth, gradually moved seaward from 1979 to 2002, leaving behind an increasing area of stabilized dunes. This dynamic was reflected in the distribution of the TWINSPAN groups (Table 4.4). The initial, man-made dune during the first years was covered by group A vegetation, which was mostly based on the *Ammophila* that had been planted. After a few years this vegetation changed into group B1 or C1 vegetation, which after 14 years stabilized into C2 vegetation or dune grassland (C3). In the oldest of the new, naturally formed foredunes seaward to the man-made dune, the successional development followed the sequence (A →) B1 → B2 → C1, which reflects the increasing stability and species richness. The younger, more seaward dune in 2002 had only reached group B2 in this sequence, while the development of the youngest dune by 2002 had occurred within the variation of group B1.

Third, the sequence of the TWINSPAN groups largely matches the succession/zonations previously described from the younger parts of coastal dunes elsewhere at the Baltic and in northwest Europe (e.g., Warming 1909; Ranwell 1972; Olsson 1974; Hundt 1985; Vestergaard 1991; Packham and Willis 1997; Frederiksen et al. 2006). This indicates that the dune vegetation in the seaside park developed according to the same successional and ecological processes as in natural coastal dunes. This result agrees with the experience of De Lillis et al. (2004) with man-made dunes on the Mediterranean coast.

4.5 Conclusions

For 24 years, the initially man-made dune in the Køge Bay Seaside Park stabilized quite quickly, while the beach expanded seaward because of the natural accretion of marine sand. On the beach, three rows of coast-parallel foredunes developed successively. The gradual change in plant diversity of the dune system, expressed by a net increase in the species richness of the permanent plots from only one planted species in 1978 to 26 species in 2002, could be related to the combined effect of the following dune succession processes:

1. Increasing the number and height of the new foredunes, seaward of the man-made dune, based on the initial nourishment, calcareous marine sand, as well as subsequent accumulation of less calcareous marine sand.
2. Increasing the distance of the man-made dune, as well as that of the new dunes, from the sea, causing a decline in the relative influence of the sea and increasing stabilization.
3. Successive immigration of species from local and regional species pools.
4. Change in the dominating dispersal agent—the seawater—during the first years of dune building and the wind during the later stages.
5. Change from a high proportion of coastal plant species to a high proportion of inland plant species.
6. Replacement of beach annuals and dune-building rhizome geophytes by hemicryptophytes and inland annuals.
7. Increasing habitat differentiation in time and space.

The increasing habitat differentiation of the initially man-made dune system was indicated by the increasing number of TWINSPAN groups with increasing dune age and distance from the coastline. This differentiation matches well with dune succession found elsewhere in younger parts of natural dune systems. After 24 years of development there is no indication so far—neither from soil processes like leaching and acidification (Vestergaard 2006) nor from the presence of species typical of stabilized, acid dunes such as *Corynephorus canescens*—of continued succession toward plant communities found in the inner, older parts of natural Baltic and NW European dune systems (e.g., Olsson 1974; Hundt 1985). It can be expected, however, that the landward-most part of the dune system will in due course be invaded by *Rosa rugosa* scrub, as observed in other parts of the seaside park. Such a development will probably reduce the species richness of the dunes (Isermann 2008).

Acknowledgments Most of the field work was carried out in cooperation with the late Professor Kjeld Hansen. I wish to thank Hans Henrik Bruun, Marisa Martinez, Patrick Hesp, and Juan B. Gallego-Fernandez for constructive comments on the manuscript.

References

Andersen UV (1993) Dispersal strategies of Danish seashore plants. Ecography 16:289–298
Andersen UV (1995) The influence of human trampling on the vegetation on artificial dune and coastal grassland in Denmark. In: Salman AHPM, Berends H, Bonazountas M (eds) Coastal management and habitat conservation, vol 1. EUCC, Leiden, pp 427–438
Böcher TW (1954) Studies in European calcareous fixed dune communities. Plant Ecol 5–6:562–570
Box EO (1996) Plant functional types and climate at the global scale. J Veg Sci 9:309–320
Bradshaw AD (1993) Introduction: understanding the fundamentals of succession. In: Miles J, Walton DWH (eds) Primary succession on land. Blackwell, Oxford, pp 1–3
Carter RWG (1988) Coastal environments. Academic Press, London

Carter RWG (1990) The geomorphology of coastal dunes in Ireland. In: Bakker TW, Jungerius PD, Klijn JA (eds) Dunes of the European coast. Geomorphology, hydrology, soils. Catena Suppl 18:31–40

Davy AJ, Figueroa ME (1993) The colonization of strandlines. In: Miles J, Walton DWH (eds) Primary succession on land. Blackwell, Oxford, pp 113–131

De Lillis M, Costanzo L, Bianco PM, Tinelli A (2004) Sustainability of sand dune restoration along the coast of the Tyrrhenian Sea. J Coastal Conservat 10:93–100

De Rooij-van der Goes PECM (1995) The role of plant-parasitic nematodes and soil-borne fungi in the decline of *Ammophila arenaria* (L.) Link. New Phytol 129:661-669

Del Moral R (1998) Early succession on lahars spawned by Mount St. Helens Am J Bot 85:820–828

Doing H (1985) Coastal fore-dune zonation and succession in various parts of the world. Vegetatio 61:65–75

Doody JP (ed) (1991) Sand dunes inventory of Europe. Joint Nature Conservation Committee, Peterborough

Fenner M (1985) Seed ecology. Chapman & Hall, London

Frederiksen L, Kollmann J, Vestergaard P, Bruun HH (2006) A multivariate approach to plant community distribution in the coastal dune zonation of NW Denmark. Phytocoenologia 36:321–342

Gallego-Fernández JB, Martinez ML (2011) Environmental filtering and plant functional types on Mexican foredunes along the Gulf of Mexico. Ecoscience 18(1):52–62

Garcia-Mora R, Gallego-Fernández JB, Garcia-Novo F (1999) Plant functional types in coastal foredunes in relation to environmental stress and disturbance. J Veg Sci 10:27–34

Gravesen P, Vestergaard P (1969) Vegetation of a Danish off-shore barrier island. Bot Tidskr 65:44–99

Grime JP, Hodgson JG, Hunt R (1988) Comparative plant ecology. A functional approach to common British species. Chapman & Hall, London

Hansen K (1981) Dansk feltflora. Gyldendal, Copenhagen

Hansen K, Vestergaard P (1986) Initial establishment of vegetation in a man-made coastal area in Denmark. Nord J Bot 6:479–495

Hesp PA (1989) A review of biological and geomorphological processes involved in the initiation and development of incipient foredunes. Proc Royal Soc Edinb 96B:181–201

Hesp PA (1991) Ecological processes and plant adaptations on coastal dunes. J Arid Environ 21:165–191

Hesp PA (1999) The beach backshore and beyond. In: Short AD (ed) Handbook of beach and shoreface morphodynamics. Wiley, New York, pp 145–169

Hesp PA (2002) Foredunes and blowou*ts*: initiation, geomorphology and dynamics. Geomorphology 48:245–268

Heykena A (1965) Vegetationstüpen der Küstendünen an der östlichen und südlichen Nordsee. Mittl d Arb.gem f Floristik in Schleswig-Holstein u Hamburg 13:1–135

Hobbs RJ, Walker LR, Walker J (2007) Integrating restoration and succession. In: Walker LR, Walker J, Hobbs RJ (eds) Linking restoration and ecological succession. Springer, New York, pp 168–179

Hundt R (1985) Phytosociological and ecological aspects of the dunes on the isle of Rügen, Baltic Sea. Vegetatio 61:97–103

Isermann M (2008) Effects of *Rosa rugosa* invasion in different coastal dune vegetation types. In: Tokarska-Guzik B, Brock JH, Brundu G, Child L, Daehler CC, Pysek P (eds) Plant invasions: Human perception, ecological impacts and management. Backhuys Publishers, Leiden, pp 289–306

Jensen F (1994) Dune management in Denmark: application of the Nature Protection Act of 1992. J Coastal Res 10:262–269

Jensen F (2008) Sandflugt og klitfredning—erfaringer og status 2008. The Ministry of the Environment

Kiehl K, Isermann M (2007) Restoration of coastal ecosystems—an introduction. In: Isermann, M, Kiehl, K (eds) Restoration of coastal ecosystems. Coastline 2007, 7. pp 1–4

Klijn JA (1990) Dune forming factors in a geographical context. In: Bakker TW, Jungerius PD, Klijn JA (eds) Dunes of the European coast. Geomorphology, hydrology, soils. Catena Suppl 18:1–13
Lichter J (1998) Primary succession and forest development on coastal Lake Michigan sand dunes. Ecol Monogr 68:487–510
Malmer N (1974) Scandinavian approach to the vegetation science. Medd f avd f ekologisk botanik, Lunds Universitet 2(1):1–32
Martinez ML, Psuty N (eds) (2004) Coastal dunes: ecology and conservation. Springer, Berlin
McCune B, Mefford MJ (1999) Multivariate analysis of ecological data. Version 4.14. MjM Software, Gleneden Beach, Oregon, USA
McLachlan A (1991) Ecology of coastal dune fauna. J Arid Environ 21:229–243
McLean R, Shen J-S (2006) From foreshore to foredune: foredune development over the last 30 years at Moruya Beach, New South Wales. Austr J Coast Res 22(1):28–36
Nielsen N (1990) Construction of a recreational beach using the original coastal morphology, Koege Bugt, Denmark. In: Fabbri P (ed) Recreational use of coastal areas. Kluwer Academic Publishers, Dordrecht, pp 177–189
Nordstrom KF, Psuty N, Carter RWG (eds) (1990) Coastal dunes. Wiley, Chichester
Olff H, Huisman J, van Tooren BF (1993) Species dynamics and nutrient accumulation during early primary succession in coastal sand dunes. J Ecol 81:693–706
Olsson H (1974) Studies on South Swedish sand vegetation. Acta Phytogeogr Suec 60:1–170
Packham JR, Willis AJ (1997) Ecology of dunes, salt marshes and shingle. Chapman & Hall, London
Påhlsson L (ed) (1995) Vegetationstyper I Norden. TemaNord 1994:665
Prach K, Pysek P (1999) How do species dominating in succession differ from others? J Veg Sci 10:383–392
Ranwell DS (1959) Newborough Wadden, Anglesey. I. The dune system and dune slack habitat. J Ecol 47:571–601
Ranwell DS (1972) Ecology of salt marshes and sand dunes. Chapman & Hall, London
Raunkiær C (1907) Planterigets livsformer og deres betydning for geografien. Gyldendal, Copenhagen
Raunkiær C (1934) The life forms of plants and statistical plant geography. Clarendon Press, Oxford
Shao G, Shugart HH, Hayden BP (1996) Functional classification of coastal barrier island vegetation. J Veg Sci 7:391–396
Thorell L, Keynäs K, Baldursson T, AaT Ekker, Vestergaard P (2001) Kustbiotoper i Norden. Hotade och representativa biotoper. TemaNord 2001:536
Thougaard N (1980) Køge Bugt Strandpark. Landskap 7:1–7
Valgren E, Front PO, Eilstrup P (1986) Køge Bugt Strandpark. I/S Køge Bugt Strandpark, Glostrup
Van der Maarel E (1979) Environmental management of coastal dunes in the Netherlands. In: Jefferies RL, Davy AJ (eds) Ecological processes in coastal environments. Blackwell, Oxford, pp 543–570
Van der Maarel E (1997) Coastal dune: pattern and process, zonation and succession. In: van der Maarel E (ed) Dry coastal ecosystems. General aspects. Ecosystems of the world 2C. Elsevier, Amsterdam, pp 505–517
Van der Meulen F, Jungerius PD (1989) The decision environment of dynamic dune management. In: van der Meulen F, Jungerius PD, Visser JH (eds) Perspectives in coastal dune management. SPB Academic Publishing, The Hague, pp 133–140
Van der Meulen F, van der Maarel E (1989) Coastal defense alternatives and nature development perspectives. In: van der Meulen F, Jungerius PD, Visser JH (eds) Perspectives in coastal dune management. SPB Academic Publishing, The Hague, pp 183–195
Van der Putten WH (1989) Establishment, growth and degeneration of Ammophila arenaria in coastal sand dunes. Dissertation, Agricultural University of Wageningen

Van der Putten WH, Peters BAM (1995) Possibilities for management of coastal foredunes with deteriorated stands of *Ammophila arenaria* (marram grass). J Coast Cons 1:29–39

Van der Stoel CD, van der Putten WH, Duyts H (2002) Development of a negative plant-soil feedback in the expansion zone of the clonal grass *Ammophila arenaria* following root formation and nematode colonisation. J Ecol 90:978–988

Vestergaard P (1991) Morphology and vegetation of a dune system in SE Denmark in relation to climate change and sea level rise. Landsc Ecol 6:77–87

Vestergaard P (2006) Temporal development of vegetation and geomorphology in a man-made beach-dune system by natural processes. Nord J Bot 24:309–326

Vestergaard P, Hansen K (eds) (1989) Distribution of vascular plants in Denmark. Opera Bot 96:1–163

Walker LR, del Moral R (2003) Primary succession and ecosystem rehabilitation. Cambridge University Press, Cambridge

Warming E (1909) Dansk Plantevækst. 2. Klitterne. Gyldendalske Boghandel, Nordisk Forlag, Copenhagen

Weidema I, Ravn HP, Vestergaard P, Johnsen I, Svart HE (eds) (2007) Rynket Rose (*Rosa rugosa*) i Danmark. University of Copenhagen, The Ministry of the Environment, Copenhagen

Chapter 5
Restoration of Foredunes and Transgressive Dunefields: Case Studies from New Zealand

Patrick A. Hesp and Michael J. Hilton

5.1 Introduction

Most New Zealand dune systems have been modified in the past 100 years by grazing; by the planting or introduction (deliberate or accidental) of exotic species; by reshaping and covering with exotic materials (e.g., clays); by wholesale change or degradation associated with the development of dunes for recreational activities (e.g., golf courses; playgrounds; reserves [areas set aside for specific purposes such as recreation, research, conservation, etc.]); and change, removal and/or degradation associated with urban, rural and beach housing and tourist developments. The area of active dunes in New Zealand declined from 129,000 ha in the early 1900s to about 39,000 ha in 2000; a reduction of 70 % (Hilton 2006) at least in part because of significant planting of exotic species. The New Zealand Biodiversity Strategy recognizes dunes and wetlands as the country's two most threatened environments.

Following the widespread destruction of natural vegetation cover on dunes by cattle and sheep grazing in the 1800–1940s, most active, mobile dune systems in the North Island of New Zealand were planted with introduced species. Marram grass (*Ammophila arenaria* (L.) Link) was planted widely from the late 1880s to protect infrastructure and to stabilize transgressive dunes, particularly those in the vicinity of road and rail links, and as a precursor to exotic plantation forestry

P. A. Hesp (✉)
School of the Environment, Faculty of Science and Engineering, Flinders University, GPO Box 2100 Adelaide, SA 5001, Australia
e-mail: Patrick.hesp@flinders.edu.au

M. J. Hilton
Department of Geography (Te Ihowhenua), University of Otago (Te Whare Wananga o Otago), PO Box 56, Dunedin, New Zealand
e-mail: mjh@geography.otago.ac.nz

M. L. Martínez et al. (eds.), *Restoration of Coastal Dunes*, Springer Series on Environmental Management, DOI: 10.1007/978-3-642-33445-0_5,

(McKelvey 1999). However, marram became naturalized and subsequently spread to dune systems of high conservation value. Marram can establish from marine-dispersed rhizome and is capable of colonizing remote dune systems tens to hundreds of kilometers from source (Konlechner and Hilton 2009). Consequently, most dunes of national conservation value now contain marram grass (Hilton et al. 2000), along with tree lupin (*Lupinus arboreus*) and a wide range of other weed species.

Ammophila species, particularly *Ammophila arenaria*, along with many other species (e.g., *Acacia, Pinus*) were introduced into coastal environments in many countries to stabilize dunes (e.g., New Zealand, USA, Australia, South Africa; McKelvey 1999; Marchante et al. 2003; Milton 2004). The introduction of marram grass has led to the development of foredunes and dune types that were previously either poorly developed or non-existent, changes in the morphology of dunes, the development of mono-specific stands and exclusion of native plant species, various ecological effects including the decline of faunal habitats and fauna, and spread to and invasion of other coastal sites (e.g., Carlton 1989; Aptekar and Rejmánek 2000; Hertling and Lubke 2000; Vega et al. 2000; Kirkpatrick 2001; Hilton et al. 2006; Konlechner and Hilton 2009; Pickart 2012). Thus, there is significant interest in how to remove marram grass, restrict invasions, and restore physical and ecological functioning to a state similar to that prevailing prior to planting and invasion.

Foredunes, as the foremost dune at the rear of the beach, have always been a desirable place from which to sit and watch the sea, and from which to conduct marine activities. From very early on in the history of building towns at the coast, foredunes were commonly utilized as sites for constructing various kinds of infrastructure, particularly houses, roads, and industrial structures. When the Victorians began to popularize visits to the beach in the 1800s, foredunes were converted to promenades, and altered/modified/destroyed for urban development. Thus, we now find that foredunes around the world have been altered, modified, or destroyed for industrial and urban development, parks and easier beach access and views (e.g., Ferreira and Dias 1992; Veloso-Gomes and Taveira-Pinto 1995; Rakodi and Treloar 1997; Nordstrom and Arens 1998; Batanouny 1999; Nordstrom 2004, 2008; Gomez-Pina et al. 2002; Seeliger 2003; Eppinga et al. 2006; El Banna and Frihy 2009). Now with the reduction or elimination of building on foredunes and greater foredune protection in some countries we are faced with the question: is it possible to restore, rebuild, or create a foredune following various levels of human-induced change and/or destruction?

In the following, we describe the restoration of two foredunes and a transgressive dunefield in New Zealand (Fig. 5.1), as examples of first, how the morphology and ecology of natural dunes were changed owing to human interventions (purposely or accidentally); and second, how the morpho-ecological states of the dunes have been purposely modified or restored to something approaching a natural state, and the processes and costs associated with that activity.

Restoration here has two dimensions. In the first and second cases examined, authorities aimed to re-establish the pre-disturbed dune form and indigenous vegetation cover. This work was undertaken in a metropolitan context. In the third

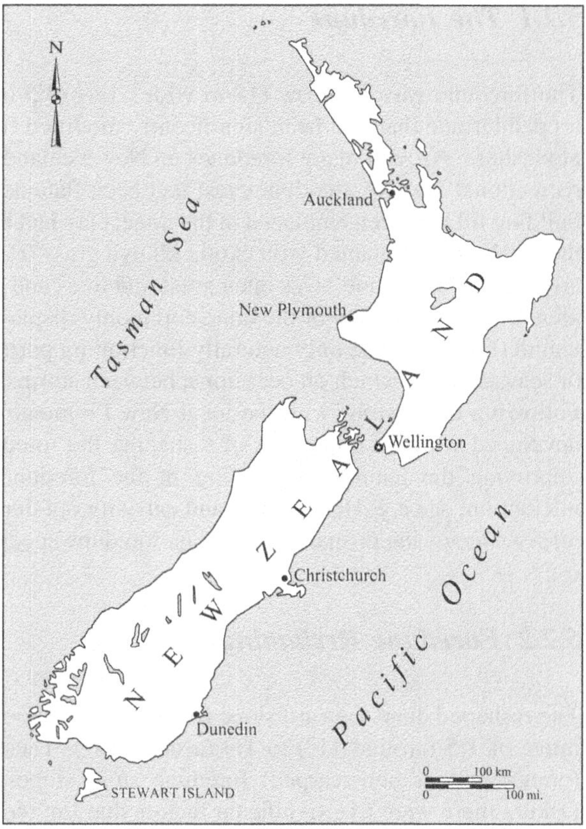

Fig. 5.1 Map of New Zealand indicating the dune restoration sites mentioned in the text

case, the goal of restoration was to restore the dynamic potential of a transgressive dune system that had been stabilized by the introduction of marram grass. Some planting of native species has occurred, but this was not the main activity or goal. The third case involves an isolated dune system in a national park setting with no history of direct human use.

5.2 Case Study 1: Foredune Restoration, Oakura Beach

Oakura Beach (New Plymouth, Taranaki Region, North Island, New Zealand) is a moderate to high wind and wave energy coastal system that has a very limited, if not zero to negative, sediment supply. It comprises a wide, rocky, predominantly sub-tidal reef that is thinly overlain by sand in the surf zone. The intertidal beach is relatively narrow at high tide and also has a low sediment volume. The entire Holocene (10,000 years to present) dune system comprises one low, narrow foredune, which indicates either that the sediment supply to this beach has always been extremely small, or that significant erosion has taken place in the past.

5.2.1 The Foredune

The foredune was a narrow (18-m wide), low (3.5 m above mean sea level) flat-topped terrace that had been significantly modified in the past. It extended 150 m alongshore. As with many foredunes on New Zealand coasts with moderate to high recreational activity, the dune crest had been flattened, some unknown amount of building fill had been emplaced in the dune, clay had been laid over the surface, and the soil had been planted with exotic kikuyu grass (*Pennisetum clandestinum*). The area behind the "dune" was open grass and trees and utilized as a recreation/picnic area. The seaward face of the dune commonly displayed an erosion scarp along its length (Fig. 5.2). The only naturally functioning portion of the dune was this stoss, or seaward face, which on occasions, between storms, would fill with aeolian sand. Following discussions with the local New Plymouth District Council, a plan was developed with the objectives of reshaping the foredune to a more natural shape, improving the natural functioning of the foredune (i.e., enabling beach–dune interaction; see e.g. Hesp 1989), and carrying out dune re-vegetation utilizing *Spinifex sericeus*, the primary indigenous foredune species in northern New Zealand.

5.2.2 Foredune Reshaping

The reshaped dune seaward slope (or stoss face) should ideally lie within the slope range of 1:5 (around 11°) to 1:3 (around 18°). These slopes approximate natural (unmodified or non-scarped) foredune stoss slopes in the region. However, at Oakura there were five significant factors that limited the dune reshaping exercise:

1. There was very limited space in which to reshape a foredune (approximately 10 m maximum from the dune scarp crest), since it was the Council's wish to retain the maximum lee dune recreation area.
2. The seaward toe of the reshaped foredune should not extend further seaward than the toe of the dune scarp and should preferentially be landward of that

Fig. 5.2 View of the Oakura Beach site prior to restoration

point. If it were seaward of the scarp toe, it would be very likely eroded in the next storm or spring high tide.

3. The foredune was quite low and should ideally be reshaped to a greater height to allow for, and mitigate against, storm erosion and potential storm overwash.
4. The foredune surface comprised clay and other materials, some of which would not be desirable for use in the reshaping and this lessened the volume of material available for dune reshaping.
5. Views from the reserve toward the beach should be maintained where possible.

Given these constraints, the very limited availability of sediment from other sources, and the potential for storm wave erosion, the reshaped dune profiles were constructed with a stoss (seaward) slope of around 18° (1:3) (Fig. 5.3). The dune height was maximized where possible to heighten the dunes a small amount while still conserving sediment. The dune lee slope was constructed with a slope of around 20° and, while it was desirable to produce a lower lee slope, dense planting and fencing ensured the stability of the slope.

5.2.3 Dune Volumes, Sand Placement, and Sand Removal from Oakura Beach

The approximate volume of sediment (per meter alongshore of the dune) required for reshaping is indicated in Fig. 5.3. The volumes required to construct the desired dune profile ranged from 1,200 to 2,700 m^3/m depending on the location alongshore. There was a greater amount of sediment available in the northern section of the area to be reshaped and a deficit in the southern area. That is, the amount of material required to produce a minimal dune shape in the southern section was less than that available from the excavated area. Since it was desirable to build the reshaped foredune as large as possible, and since there was more sediment in the northern area, some of this northern sediment was transported to the southern area to make up the deficit.

It was not clear prior to reshaping of the dune exactly how much fill of foreign materials might be in the dune, and thus, an additional 300 m^3 of sediment was potentially required for dune reshaping. This sediment was to be extracted from the beach adjacent to the Waimoku Stream (at the northern end of the beach) when stream re-alignment works took place. It was then to be transported along the beach at low tide and emplaced on the southern section.

5.2.4 Dune Planting

Prior to dune reshaping and planting the kikuyu grass was eliminated by spraying it three times with Roundup™ at 8-week intervals (Bergin et al. 2000). The dune was reshaped, restored to a relatively natural profile (Fig. 5.4), and mostly planted

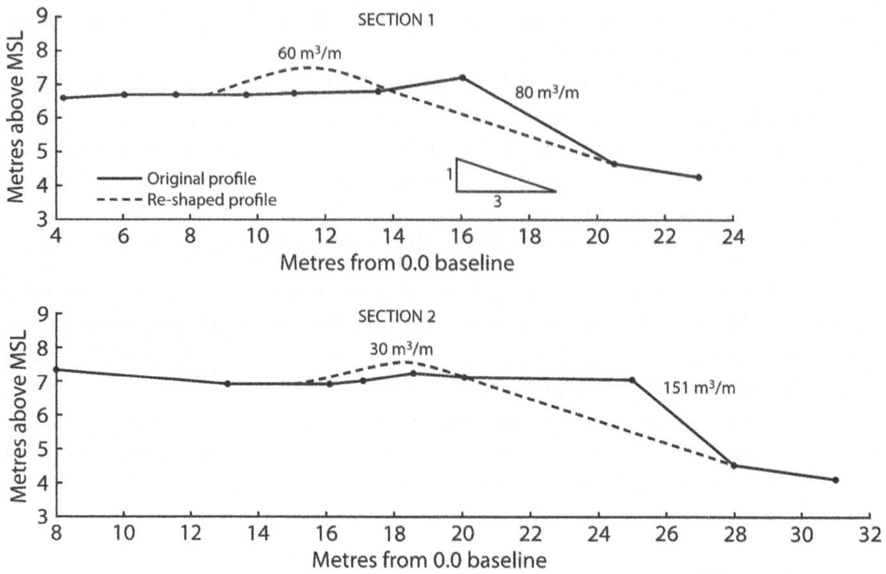

Fig. 5.3 Cross-sections of the original dune terrace, and the design foredune at two locations along the dune

Fig. 5.4 View of the heavy machinery reshaping the foredune. The former clay surface and fill have been removed, and sand from the northern end of the beach has been transported to the site to increase the dune volume

with *Spinifex sericeus* (primarily raised from seed) at a spacing of 50–70 cm. This close spacing was desirable because the operation of high-energy, west-coast, "roaring forties'" winds in this region means that there is a considerable chance of aeolian erosion occurring following planting. Some limited planting of *Carex pumila* plants was also undertaken as a trial to examine how well this native plant might survive in this environment (Bergin et al. 2000). The surface was fertilized with two slow-release NPK fertilizers at planting. A three-wire fence was erected along the front and back of the dune to restrict pedestrian access, and signs were erected to inform the public of the works. The plants grew rapidly and the dune displayed a high percentage of cover after 6 months (Fig. 5.5).

Fig. 5.5 View of the reshaped foredune in 2000, principally planted with etc., Spinifex sericeus, the native pioneer species for the region

5.2.5 Economics of the Work

The costs associated with this restoration project of a 150-m length of foredune included NZ$ 2,400 in consultancy fees, NZ$ 1,700 in heavy machinery hire, and planting and fencing at a cost of NZ$ 40 per linear meter (year 2000 prices).

5.3 Case Study 2: Foredune Restoration, East End Beach

East End Beach (New Plymouth, Taranaki Region, North Island, New Zealand) is a moderate- to high-energy coastal system that also has a very limited sediment supply. It comprises a wide, rocky, predominantly sub-tidal reef thinly overlain with sand in the surf zone and on the beach. Prior to restoration, the dune system comprised one erosional, relatively low, narrow foredune (Fig. 5.6). The foredune regularly displayed an erosion scarp along its length and had been significantly altered in the past by widening, leveling, and filling. It had been impacted by port works (to the south) and subsequent loss of sediment from the littoral drift system.

Following the short-term success of the Oakura restoration works, the New Plymouth District Council undertook to reshape and restore the foredune with the

Fig. 5.6 View of the East End foredune prior to restoration

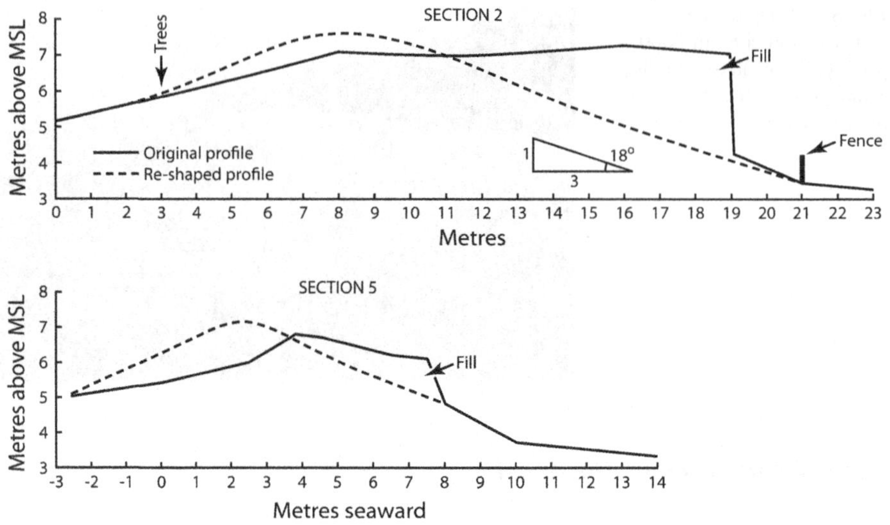

Fig. 5.7 Two cross-sections (Sect. 5.2 [southerly] and Sect. 5.5 [northerly]) of the foredune illustrating the pre- and post-restoration morphologies

objectives of restoring the dune to a more natural shape, and carrying out some dune re-vegetation utilizing native species, along a 150-m length of the dune.

5.3.1 Foredune Reshaping

The selected foredune section was situated just north of the main walkway onto East End Beach adjacent to the East End Surf Life Saving Club. Two typical topographic profiles are shown in Fig. 5.7, including the typical reshaped dune profiles. As in the case of Oakura, at East End there were four factors that limited the dune reshaping work:

1. Limited space in which to reshape a foredune (a maximum of 15–16 m (at the southern end) from the dune scarp crest to the edge of trees, which were native species and worth saving if possible).
2. The dune could not extend further seaward than the scarp toe.
3. The foredune was low.
4. The foredune contained a certain amount of fill material of unknown type.

Given these constraints, the very limited availability of sediment from other sources, and the potential for storm wave erosion, the reshaped dune profiles were constructed with a stoss slope of around 18° (1:3) to 22°. The dune height was maximized where possible to heighten the northern section of dunes a small amount, while still conserving sediment.

Fig. 5.8 The foredune at East End following restoration

The approximate volumes of sediment required to create the preferred dune profiles or morphologies were 200–400 m^3/m. The amount of material required to produce a minimum dune shape in the middle and northern sections was less than that available from the adjacent excavated areas. Since it was desirable to build the reshaped foredune at least as high as the present dune (see Fig. 5.7), some of the sediment comprising the southern section was transported to the middle and northern areas to make up the deficit in the landward parts of those sections.

Given that there was considerable uncertainty regarding the amount of fill material in the present dune, it was desirable to have available an additional 100 m^3 of sediment for dune reshaping. This sediment could be extracted from the beach near the training wall at Waiwhakaiho near the northern end of the beach. It could then be transported along the beach at low tide and placed on the 150-m section if required.

5.3.2 Dune Planting

Following the apparently "successful" trial at Oakura Beach (successful being defined here as the dune emplacement having worked according to the given specifications, it began to function normally, and the plantings survived (see Gallego-Fernández et al. 2010, paragraph 4, p. 1824); *Spinifex sericeus* was planted at 50-cm spacing across the foredune (success being measured in the short term—about 2 years—in this case). Some Pīngao (*Ficinia spiralis*), the other major pioneer species in New Zealand, was also cross-planted within the *Spinifex* (Fig. 5.8). Flax (*Phormium tenax*) and native tree species were planted along the leeward edge wherever the exotic and native existing trees had to be removed in order to obtain the desired profiles. A three-wire fence and signs were erected at planting.

5.3.3 Economics of the Work

The costs per meter of beach were approximately the same as at Oakura.

5.4 Case Study 3: Foredune Destruction, Transgressive Dunefield Restoration, Stewart Island

The impact of marram grass (*Ammophila arenaria*) on the morphology of beaches and foredunes on temperate coasts has been recognized for some time (e.g., Cooper 1958; Huiskes 1979; Carter et al. 1990; Hilton et al. 2005; Hilton 2006). Much less is known about the development of foredunes and coastal barriers following marram eradication. The potential to restore marram-dominated dunes, by eradicating the marram, is gaining recognition in New Zealand and elsewhere. There is, however, concern that marram eradication will initiate sand drift, to the detriment of down-wind infrastructure and ecosystems. An opportunity to monitor the geomorphology of a prograded bay-head barrier at Doughboy Bay, following a sustained marram eradication operation over a 12-year period (1998–2010) arose when the Department of Conservation (DoC) commenced a marram-control program in Rakiura National Park (Stewart Island, southernmost New Zealand) in 1999. Specifically, the DoC and the second author examined:

1. The response of marram to herbicide application
2. The rate of decay of marram following herbicide application
3. The nature of sedimentation following marram necrosis and decay

A parallel study of marram eradication in Mason Bay, a much larger and more complex dune system 2 km to the north of Doughboy Bay, commenced in 2000.

5.4.1 Study Area

The beach systems of Stewart Island are moderate to high wave and wind energy environments. Beach morphologies are of the intermediate type with a broad intertidal terrace comprising either sand or mixed sand and gravel. The back-beach morphology has probably narrowed and steepened following the introduction of marram grass and the development of continuous, uniform (Type I) foredunes (i.e., continuous, morphologically simple, high vegetative cover—after Hesp 2002). Hart et al. (2012) measured the wind regime at Mason Bay from June 2002 to February 2004 using a permanent Vector 3-cup anemometer at a height of 2 m located on the foredune crest. The predominant direction of sediment transport is onshore; the wind regime is dominated by strong, episodic, west to southwest (onshore) winds throughout the year. Annual average wind speed is 6.4 ms^{-1} with <1 % calm conditions. The dune systems of Doughboy Bay occur at the head of a pronounced embayment. Headlands shelter the southern and northern margins of the sandy shoreline, but the central dune system and adjoining barriers are highly exposed to onshore winds (Fig. 5.9).

Stewart Island, and Fiordland, the southern South Island beach and dune systems, retained relatively high natural values well into the twentieth century.

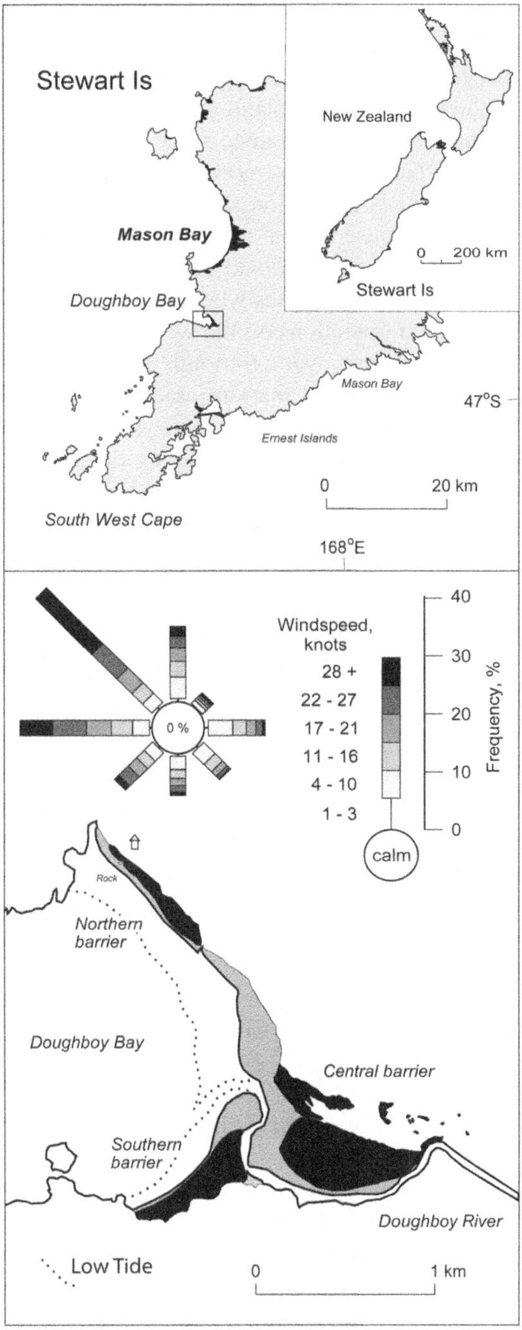

Fig. 5.9 Location of the southern barrier, Doughboy Bay, Stewart Island (Rakiura National Park). The prevailing and dominant winds are from the northwest. The wind rose is derived from data gathered at Southwest Cape, 20 km south of the study site, from 1992 to 2003

Marram has been planted at Mason Bay since the 1930s, initially at Kilbride at the southern end of the bay. Marram was probably also established from marine-dispersed rhizome. When the Department of Conservation was formed in 1987 all of the main Stewart Island dune systems contained marram grass, although many of these systems contained large areas that were free of marram. Fiordland and Rakiura National Parks now contain almost a third of New Zealand dune systems of national conservation significance (Johnson 1992; Hilton 2006). The Department of Conservation is committed to eradicating marram from both areas. Marram sites in Fiordland are sprayed annually, over just 1 or 2 days, using a helicopter to access sites where marram has been recorded. In contrast, the marram eradication operations on the west coast of Stewart Island involve the deployment of 4–5 personnel over a 6-month period, from November to May. This operation is probably the largest marram eradication program attempted.

The herbicide employed is a systemic, grass-specific chemical—native grasses are probably more vulnerable to the herbicide than marram grass. At many locations this has proved a minor issue, since marram commonly forms uniform stands and the key native grasses have already been displaced. At some sites aerial operations have been avoided to ensure that non-target species have not been affected. Precision mapping with the aid of GPS has facilitated this process. Nongrass species were monitored in a series of permanent quadrats in the Doughboy Bay dunes for 6 years, from (and just before) the first application of herbicide in 1999 till 2006. No decline in vigor was noted. A decline in species diversity did occur, however, as a result of renewed sedimentation following marram necrosis.

Marram is capable of rapid invasion of both progradational and transgressive dune systems in southern New Zealand. Marram was planted at Kilbride in Mason Bay in the 1930s and then spread north, by marine-dispersed rhizome. It became established in the foredune north of Martins Creek in the early 1950s, and then spread to the hinterland of the dune system by wind-blown seed. Some sections of the dunes were also planted by farmers in the 1960s (possibly earlier). Marram grass was also well established in Doughboy Bay by the late 1950s. By 1998 marram had occupied most available habitat in Doughboy Bay (30 ha) and was established across 676 ha (68 %) of active dunes in Mason Bay. The rate of spread was rapid—Hilton et al. (2005) estimated that areas of dense marram (>50 % cover) increased from 1.4 ha in 1958 to 74.9 ha in 1998. Without intervention, marram grass would likely have invaded most of the dune system by between 2023 and 2043 (Jul 1998).

Marram control on the southern barrier of Doughboy Bay, Stewart Island (Rakiura) was initiated by the Department of Conservation in an attempt to restore natural processes and patterns of sedimentation and dune flora and fauna. Information from the southern barrier, one of three in the bay, is presented. This barrier contains a sequence of four foredune ridges that have developed in conjunction with marram. Marram had become the dominant foredune species in Doughboy Bay by 1958 (Hilton et al. 2005; Hilton and Konlechner 2010). Prior to the establishment of marram grass the southern barrier probably comprised a simple

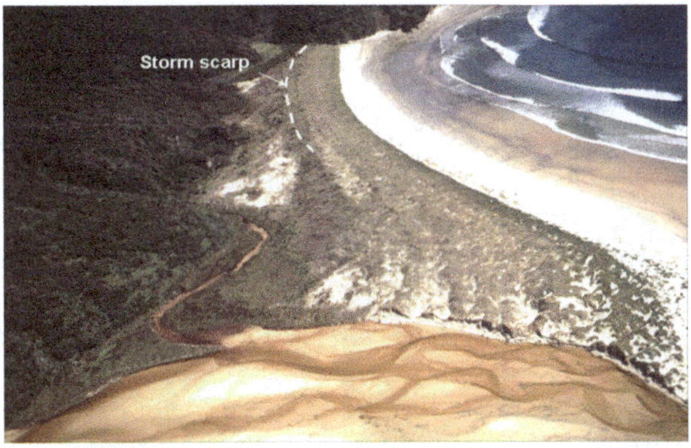

Fig. 5.10 An aerial oblique view of the southern barrier. The "scarp" separates the prograded foredune barrier associated with marram, which formed after the first aerial photograph was obtained in 1958 from the pre-marram dune system. The latter contains remnant blowouts and parabolic dunes

foredune complex, with some blowout and parabolic dune development at the northern (relatively exposed) end of the barrier. At this time the flora of the barrier would have comprised both dune-specific and cosmopolitan scrub and forest margin plants. Some species, including the endemic *Euphorbia glauca* (shore spurge) and the indigenous sand tussock, (*Poa billardierei* (Spreng.) Soreng et al. 2009; formerly *Austrofestuca littoralis)*, had probably been eliminated by introduced mammal browse and granivory before marram invasion. Pīngao (*Ficinia spiralis* (A. Rich.) Muasya and de Lange 2010; formerly *Desmoschoenus spiralis*) would have grown across much of the barrier, and moribund colonies persist where marram grass has not invaded. Only one pīngao plant was observed within the marram foredune ridge sequence in 1999. Ironically, the diversity of the marram-dominated ridges was very high—a consequence of the stability and senescence of marram on the inner ridges as the barrier prograded.

The geomorphology of the barrier was also transformed. Transgressive dunes, blowouts and, parabolic characterized the surface of the pre-marram barrier; however, these dunes are now stable and vegetated, since they are located in the lee of the sequence of marram-forced foredunes. In 1999 the seaward half of the barrier comprised a series of linear foredunes (Figs. 5.10, 5.11). Remnant blowouts and saucer-shaped parabolic dunes, now stable, occupy the landward half of the barrier. During the period from 1958 to 1998 the southern barrier prograded 60–200 seaward and 400 m to the north. This northward growth of the barrier was forced by marram colonization of the pre-marram barrier margins. Pīngao may have been established there from time to time, but both aeolian and marine sedimentation would have prevented prolonged colonization and foredune establishment.

The Doughboy case is typical of the impact of marram grass on the dune systems of southern New Zealand. The DoC intervened in the Doughboy Bay case

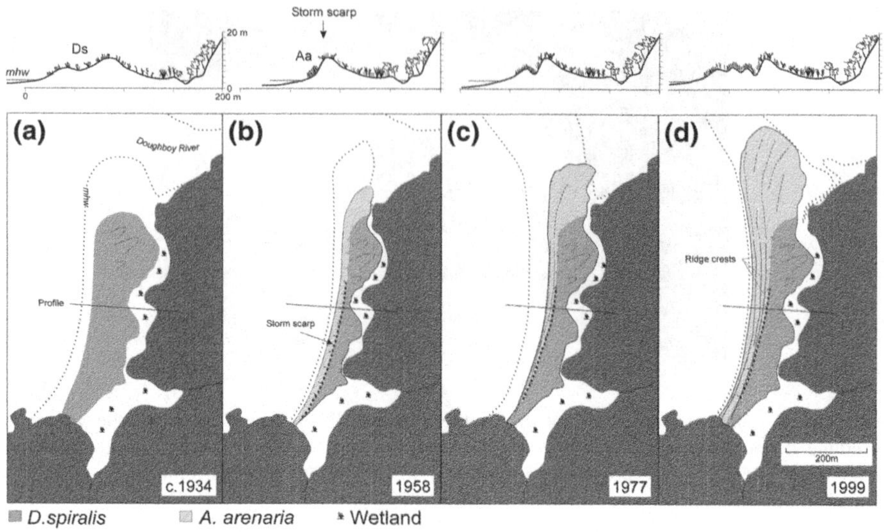

Fig. 5.11 Barrier and dune system development, 1934 to 1999 (Hilton et al. 2009)

because of initial concern that marram was adversely affecting three of the surviving six populations of *Gunnera hamiltonii* in the Bay, an endangered coastal herb. The impact was twofold. Marram was overgrowing and shading the herb, which has a low growth form. Second, marram was growing in the most exposed central section of the Bay, on the distal tip of the southern barrier. This development was forcing the mouth of the Doughboy River to the north, with the result that the northern bank of the river was eroding, with the imminent loss of the only wild female colony of *Gunnera hamiltonii*. Were it not for the conservation status of this species (nationally critical), the Marram Eradication Program (MEP) on Stewart Island might not have commenced. In 1999 there was no national overview of dune conservation in New Zealand or national program of dune restoration. Such a program has still to emerge, although the Department of Conservation is engaged in a process of national ecosystem and site prioritization.

5.4.2 Methods of Eradication

Marram grass has primarily been controlled using herbicides in New Zealand. Mechanical methods of control—excavation, burial, solarization, salt treatment—have been trialed by local authorities and conservation groups. However, unlike some jurisdictions (e.g., California and Oregon; Pickart 1997; Wiedemann and Pickart 2004), there is general acceptance that herbicides are the most effective treatment option. Some early treatments in Fiordland National Park employed Roundup^{TM}, but this caused general necrosis of nearby native species and excessive dune disturbance. Gallant^{TM} (Gallant-NF^{TM}, Gallant-Ultra^{TM}), a selective (grass-specific), systemic herbicide has been used by the Department of Conservation since the mid 1980s. Gallant can be sprayed on pīngao (or "pīkao" in southern New Zealand), a sedge, at usual concentrations and application rates, without stress or necrosis. *Spinifex sericeus*, a grass, does not grow in southern New Zealand. However, Gallant cannot be used in the presence of native grasses, including the native sand tussock, *Poa billardierei*. Gallant has two disadvantages. It is relatively expensive at around NZ$ 180/liter (in 2010). Second, the effects when used on marram grass are relatively slow to appear—in southern New Zealand (47° S Lat.) 12 months should be left between application and assessment. That is, the effectiveness of a marram eradication operation cannot be assessed until the following growth season. Complete necrosis of above-ground biomass is usually observed within 6 months of herbicide application, but regrowth can be anticipated in early spring. Other "knockdown" herbicides have been trialed, but they have been found to be less effective and more expensive over time. Gallant remains the best option.

Prior to the formation of the Department of Conservation in 1987, the New Zealand Forest Service targeted the northwest beaches of Stewart Island, which then contained small infestations of marram, of the order of 10^1–10^2 m^2. These populations had developed from rhizome that may have originated on the west coast of Stewart Island, Southland or South Westland. Knapsacks were used to apply Gallant from time to time, depending on other priorities, the availability of transport, and the weather. The beaches of northwest Stewart Island are accessed by boat or small aircraft that are capable of landing at low tide. These operations were of the utmost importance, since they prevented the establishment of large infestations at a time when there was no policy or funding for the systematic control of marram grass on Stewart Island.

The Department of Conservation commenced aerial (helicopter) operations in Doughboy Bay in February 1999. By this time marram had produced a dense cover over the northern and southern barriers and was rapidly invading the central dunes. The area of dense marram, associated with the four foredune ridges described above, an area of approximately 7 ha, was sprayed with the herbicide Gallant NF^{TM} using a Robinson-22 helicopter in February (mid-summer) 1999. Helicopters have since been used to spray large areas of marram grass, particularly marram growing in the foredune environment, primarily using Jet-Ranger aircraft (Fig. 5.12). The 1999 helicopter operations on the southern barrier at Doughboy Bay were repeated in 2000 and 2001. The initial results were spectacular—complete necrosis of leaf material was observed 4 months after the first helicopter operation. However, numerous shoots had established from surviving rhizome by late spring. The helicopter was used to spray this regrowth in February 2000. Further regrowth was sprayed during the summer of 2000/2001 using a 500 liter pump unit mounted on an ARGO amphibious vehicle (Fig. 5.13). Subsequent regrowth, primarily from in situ seeds, has been sprayed using the ARGO or with knapsacks, or hand-pulled during systematic search and destroy operations in the

Fig. 5.12 A Jet-Ranger helicopter equipped with a spray boom, engaged in marram control operations, Mason Bay, December 2010

period 2004–2010. This process is ongoing owing to the persistence of the marram seed bank (Konlechner and Hilton 2010).

There is no evidence that elements of the herbicide remain in the sand following prolonged use, consistent with the manufacturer's advice, which is that Gallant breaks down rapidly after application. This issue has not been evaluated; however, we have noticed vigorous recruitment of native plant seedlings in areas where marram has received aerial spraying for 3 years. It is likely that any residual chemical would be rapidly washed through the sand because of the high rainfall and high permeability of the sandy soils.

Observations of marram decay were made over 4 years (1999–2004) in a series of 14 permanent 2 × 1-m quadrats spaced across the southern barrier (see Woodley and Hilton 2003; Hilton et al. 2009). Five stages of decay were apparent. Permanent transects were established across the barrier to determine gross changes in barrier morphology (Fig. 5.14). Profiles were surveyed annually, from 1998 (prior to the operation) to 2007 using a dumpy level or a total station. The topography and vegetation cover of one area (approximately 40 × 140 m) was surveyed annually using a total station from 2002 to 2007. Digital terrain models were derived using a total station and topography and vegetation cover within the area was mapped and the data presented using SURFER^{TM}. The shelter afforded by different stages of Marram decay was also observed using an anemometer array and is reported elsewhere (Woodley and Hilton 2003).

5.4.3 Barrier and Dune Development Following Marram Eradication

The morphology of the southern barrier changed slowly following the aerial application of herbicide in February 1999. Substantial sedimentation did not commence until the winter of 2006 and then only north of Transect 4 (henceforth "T4"). This slow response of the barrier to de-vegetation was due to the slow rate of decay of in situ marram rhizome. In addition, it took 3 years of herbicide

Fig. 5.13 Department of Conservation rangers spraying marram grass using an ARGO all-terrain vehicle equipped with a 500 liter spray tank and two 50 m hoses

treatment to reduce marram to a density of 1 % cover or less. The leaves browned and died following the initial application of herbicide, during the winter of 1999. However, regrowth occurred from surviving rhizome during November and December 1999, particularly along the stoss face of the foredune. The rate of sand loss from the stoss face of the foredune has increased since 2004, with concomitant sand drift and deposition downwind (e.g., T2, Figs. 5.14, 5.15). By December 2006 the series of ridges was no longer visible north of T3. The surface of the barrier north of T3 now comprises an active sand-sheet. Since 2006 the distal end of the barrier, north of T3, has experienced relatively rapid sedimentation, with significant loss of sand to the Doughboy River. At the southern, relatively sheltered, end of the barrier the foredune ridges are still preserved, although the

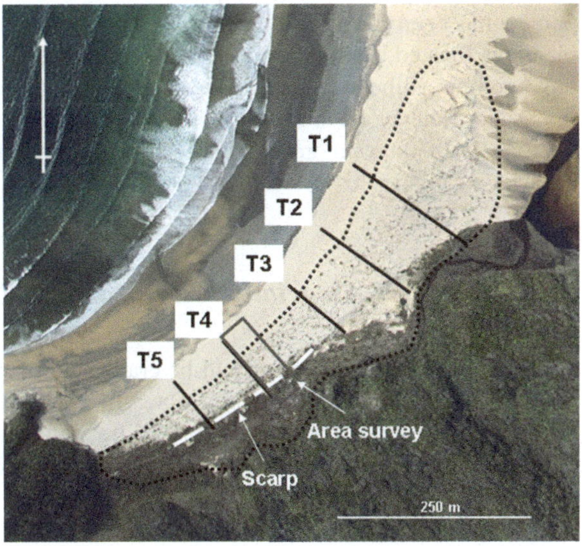

Fig. 5.14 Location of profile lines and features referred to in the text, southern barrier, Doughboy Bay

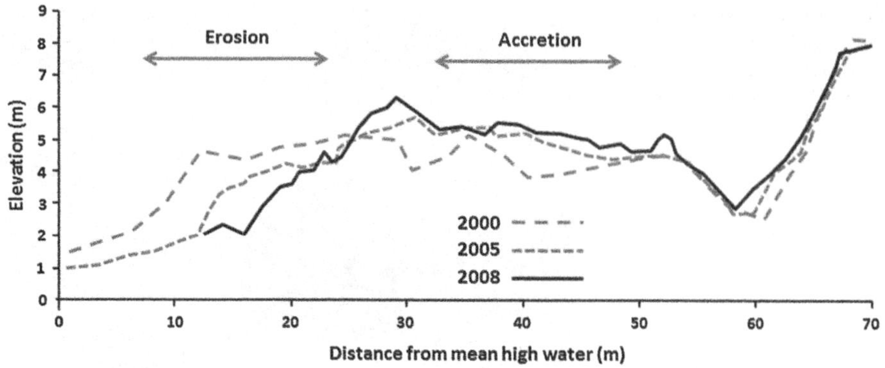

Fig. 5.15 A comparison of the first (2000), 2005 and most recent (2008) profiles surveyed across T4. Erosion is occurring seaward of 25 m. Landward of 25 m the barrier is accreting (Konlechner and Hilton 2010)

foredune has experienced some minor erosion. The zone of active sedimentation reached the relict storm scarp south of T3 (Fig. 5.16) in 2010. This sandsheet should overtop the relict scarp before 2015. In general, the dune ridges of the northern section of the barrier are eroding relatively rapidly, fueling downwind sedimentation and the development of a slip face. This feature is advancing landward (i.e., to the east).

Depositional and erosional features, aligned more or less parallel to the prevailing westerly winds, now etch the surface of the barrier. The foredune ridges are now no longer evident. Shadow dunes have formed behind surviving specimens of flax (*Phormium tenax*), and clusters of pīngao (Fig. 5.17). Pīngao was systematically planted across the barrier in 2002, 2003, and 2004, in clusters of 5–7 plants, 20–30 m apart. The subsequent plant growth and associated shadow dune development has been rapid. These dunes are now 2–3 high and 10–20 m in length. The plantings located between T2 and T3 were initially successful; however, many are now eroding as the surrounding surface of the barrier is lowered and the sides of the shadow dunes are undercut. This section of the barrier only formed in conjunction with marram grass establishment—aeolian processes now exceed the capacity of pīngao to sustain these shadow dunes.

5.4.4 Problems Associated with Marram Eradication

We anticipated that marram would be eradicated over a period of 3–5 years at the outset of the MEP in Doughboy Bay. In fact, spray operations are ongoing, although the scale of the operations will be significantly reduced by 2013. Completion of the Doughboy Bay operation has been delayed by two processes—germination of in situ seed and two marram rhizome stranding events (Hilton and Konlechner 2011). The latter is relatively trivial, but worthy of mention. A patch

Fig. 5.16 A sequence of photographs from between T3 and T4, showing the initial marram cover and progressive de-vegetation and sandsheet and nebkha development (1999–2011)

Fig. 5.17 The stoss face of the foredune after marram eradication. The arrow indicates where most marram seedlings have emerged, on an annual basis, since 2005. The shadow dunes have formed under pīngao, planted by the Department of Conservation in 2003

of marram adjacent to the Doughboy River had been left unsprayed, because of the proximity of a colony of *G. hamiltonii* to blowout, while alternative management options for the site were being considered. Two storm events, in 2005 and 2008, resulted in large quantities of marram rhizome being washed into Doughboy Bay. This rhizome was deposited above the spring high tide level on the beach fronting the three barriers. Vigorous marram growth and incipient foredune development resulted within 12 months on both occasions. This growth was subsequently sprayed, but not before some flowering occurred.

The presence of a marram seed bank has been more problematic (Konlechner and Hilton 2010). The biological literature provided little guidance as to the longevity of marram seeds, which are trapped in foredunes as they prograde and accrete. The monitoring program referred to above has involved precise annual mapping of seedlings across a DTM using a Leica total station. Seedlings started to emerge in 2004 after the decay of the marram grass leaf material and erosion of the former stoss face of the foredune (shown in Fig. 5.17). Seedlings have been emerging ever since; which indicates that marram seed banks are viable for at least 10 years. Wind-dispersed seed has been a problem in the much larger dune systems of Mason Bay, which reach 3.2 km inland of the foredune. The foredune and adjacent dunes, comprising a zone of about 800 m, is dominated by marram grass. Seed is blown from this zone into the sparsely vegetated dune hinterland in large quantities, resulting in a pattern of numerous, but widely dispersed seedlings. The implications of both in situ germination and seed dispersal by wind for the MEP have been significant. The mechanical methods used to apply spray—helicopters and the ARGO—are not suitable for the location and treatment of widely scattered plants. A system of grid search was trialed in November 2006, which employed lines of searchers, equipped with hand-held GPS units. The breadcrumb trails on the GPS screen were used to maintain direction and spacing and ensure total coverage of the search area. Every marram plant encountered was way-pointed and subsequently sprayed or hand-pulled. In November 2006, the four-person volunteer team searched 70 ha over 4 days. A total of 5,817 plants were surveyed, 4,712 of which were pulled. Effective execution of this simple method is critical to the success of the MEP on Stewart Island. Marram grass will be eradicated from Doughboy Bay over the next 3–4 years, but only if operations are conducted annually, systematically, and with precise spatial control. Eliminating the first 99 % of marram from the Doughboy site has proved relatively straightforward, compared with the task of finding and destroying seedlings. If missed, these seedlings will produce flowering plants within 24 months. The Department of Conservation operations team employed the grid search method for the first time in January 2010 and the method was honed and improved over the 2010/2011 season. Establishing smaller management units, installing the boundaries of these units on the GPS units carried by the operations team, and checking coverage on a daily basis have ensured the necessary coverage.

5.4.5 Economic Costs

The cost of the MEP on Stewart Island, primarily involving ten dune systems, eight of which are dune systems of national conservation significance, is approximately NZ$ 240,000 pa in 2010. Most of this cost, approximately NZ$ 180,000, was spent at Mason Bay. This sum comprises labor (NZ$ 86,000); helicopter spraying (NZ$ 11,000); herbicide purchase (NZ$ 12,000); transport of people and supplies by fixed wing aircraft and helicopters (NZ$ 25,000); and maintenance supplies and sustenance (approximately NZ$ 24,000). The University of Otago has an annual contract (NZ$ 13,000) to assess the effectiveness and environmental impact of the MEP at Doughboy and Mason Bay and to contribute to the development of the MEP strategy.

A technical advisory committee, the "Dune Restoration Advisory Group'" was formed in 2009 to achieve effective communication among managerial, technical, and operational staff within the Department of Conservation and between the Department and University staff. The DRAG meets twice a year. In addition, University and DoC staff participate in joint field evaluations at least twice a year. These forums have provided opportunities for important technical and strategic developments during 2010 and 2011.

The cost of annual operations at Doughboy Bay is currently just NZ$ 11,000. This sum should decline significantly over the next 2–3 years as the need for helicopter operations declines. By 2013–2015 the Department hopes to enter a surveillance phase involving the annual deployment of two rangers who will use knapsack spray equipment to spray emerging seedlings or plants growing from stranded rhizome. The annual cost of operations at Doughboy Bay should then fall to approximately NZ$ 2,000 pa.

Fig. 5.18 View of the Oakura foredune in 2008. The dune displays a prominent scarp, and has been invaded by a variety of weed species

5.5 Post-Restoration and the Future

In the case of the Oakura restoration, by 2008, the dune had been scarped by waves on numerous occasions (as expected given the very limited or negative sediment budget), but had also lost around 8 m in width. The dune had also been seriously invaded by weed species (Fig. 5.18). This may be, in part, a response to over-fertilizing, which happened at least once, but also to the fact that the spinifex senesced to some degree following dune stabilization. Like many pioneer species, spinifex requires sand deposition to drive growth. On a beach such as this, with very limited sand deposition onto the middle to upper stoss slope and crest, secondary native plants should have been introduced to mimic the natural successional tendencies in such areas. In the case of the East End restoration, the dune appears to be functioning naturally and while there is occasional wave scarping, the sediment budget on this beach is such that scarp filling and re-establishment of scarp and beach cover by native species occur.

The Stewart Island study indicates that marram-dominated dunes degrade slowly, even on high-energy windward coasts. Marram leaf, stolon, and rhizome material degrades in a predictable sequence and affords the substrate, at the early stages of decay, significant protection. Substantial loss of sand from the southern barrier in Doughboy Bay only started to occur in 2006, 7 years after the commencement of the operation. Further, this loss is only occurring from the section of the barrier in the most exposed part of the system. This section, the northern third of the landform, formed as a result of marram invasion.

The geomorphology of the study area has been fundamentally changed by the removal of marram, although the current landscape reflects landforms associated with post-marram sedimentation and landforms associated with the marram phase of progradation. The northern third of the barrier has experienced accelerating aeolian erosion since the commencement of the MEP. Re-colonization of this section of the barrier by indigenous pioneer species is now unlikely. We also anticipate ongoing erosion of the current de-vegetated foredune and a landward shift in the position of the beach. By 2030 the southern barrier should resemble the pre-marram 1950s barrier morphology (see Hilton et al. 2005). Clearly, it has been advantageous to have an understanding of the impact of marram on barrier development in setting restoration goals and interpreting landform change following the initial herbicide operations. It has been possible to account for and justify the destabilization of the barrier, the first step in re-establishing a dune system with transgressive dune elements, albeit at the expense of downwind wetlands and other ecosystems.

5.6 Discussion and Conclusion

The motivation to "restore" dunes in the cases presented was twofold. In the North Island cases, local authorities sought to re-establish a stable foredune that was foremost, a more naturally functioning foredune with native species, that was aesthetically pleasing, and that provided some protection from storm-forced episodic erosion. The scope of the restoration was constrained by the small accommodation space and sediment supply. In contrast, the Rakiura National Park, Stewart Island, operations were motivated by a desire to re-establish the natural character of dune systems dominated by transgressive dune forms, low species diversity, and high dynamism. The accommodation space is not limiting and concerns about the downwind impact of marram eradication have been addressed. At Doughboy Bay a progradational sequence of foredunes had developed in association with marram grass where none had previously existed. There is now an understanding that the operations will liberate sand and some loss of downwind habitats, in part an artifact of marram invasion, will be lost. For example, a wetland dominated by jointed wire rush (*Leptocarpus similis*) developed as the barrier prograded to the north in Doughboy Bay. This wetland will soon be inundated by sand.

There is clearly a need to adopt a long-term perspective in all such restoration projects. The Department of Conservation might have hoped that the eradication of marram grass from Doughboy Bay would take well under a decade. In fact, complete eradication may not be achieved before 2015, 15 years after the program commenced. It could take longer, depending on the longevity of buried marram grass seed. The presence of a viable marram grass seed bank, two rhizome stranding events (2005 and 2008) and the inclement weather have all delayed the completion of the project. This is hardly surprising, given that the Doughboy operation was the first, large-scale operation of its type. In the case of Oakura, the dynamics of this sediment-starved beach are having a significant effect on the stability and longevity of the reconstructed foredune, and a lack of weed maintenance and planting of intermediate species by the Council authorities has resulted in a decline in the natural ecological value of the foredune.

These three studies demonstrate the need for conservation managers to adopt a long-term management perspective, the great importance and role of monitoring, and action following monitoring. The MEP on Stewart Island will be achieved when all marram plants are destroyed and the marram seed bank is exhausted. Thereafter, the DoC will need to undertake ongoing surveillance, involving annual visits to the site and systematic searches for marram, including marine-dispersed rhizome from the mainland. In this regard, the MEP will be on-going, although the costs of these operations will be lower than at present.

Acknowledgments The research and work conducted at the Oakura and East End Beach sites in the Taranaki Region was conducted while the first author was acting as a consultant and director of Coastal and Environmental Services. Thanks to Lachlan Grant, who assisted Hesp with the consulting, and Grant Porteous, Director of Parks, and Ken Shischka of the New Plymouth District Council for their assistance and support. The research of Hilton and students on Stewart Island, Rakiura National Park is supported by the Department of Conservation. Particular thanks to Brent Beaven, Biodiversity Manager, and Al Check, Ranger-Biodiversity, Rakiura National Park; and the DoC Southland Conservancy "Dune Restoration Advisory Group" (DRAG) for invaluable support. Thanks to Marisa Martinez, Juan B. Gallego-Fernandez, and anonymous referees for their fine critiques.

References

Aptekar R, Rejmánek M (2000) The effect of sea-water submergence on rhizome bud viability of the introduced Ammophila arenaria and the native Leymus mollis in California. J Coast Conserv 6:107–111

Batanouny KH (1999) The mediterranean coastal dunes in Egypt: an endangered landscape. Estuarine Coastal Shelf Sci 49(Suppl 1):3–9

Bergin DO, Ede FJ, Kimberley MO, Davidson T, Jamieson P (2000) Establishment of indigenous sand binders on a reshaped foredune, Oakura Beach, New Plymouth. Unpublished report, Forest Research, Rotorua, New Zealand, p 19

Carlton JT (1989) Man's role in changing the face of the ocean: biological invasions and implications for conservation of near-shore environments. Conserv Biol 3:265–273

Carter R, Hesp P, Nordstrom K (1990) Erosional landforms in coastal dunes. In: Nordstrom K, Psuty N, Carter R (eds) Coastal dunes: form and process. John Wiley & Sons Ltd, Chichester, pp 217–250

Cooper WS (1958) Coastal sand dunes of Oregon and Washington (Memoir 72). Geological Society of America, Boulder, Colorado, p 85

El Banna ME and Frihy OE (2009) Human-induced changes in the geomorphology of the northeastern coast of the Nile delta, Egypt. Geomorphology 107(1–2):72–78

Eppinga MB, Rietkerk M, Dekker SC, De Ruiter PC (2006) Accumulation of local pathogens: a new hypothesis to explain exotic plant invasions. Oikos 114(1):168–176

Ferreira O, Dias JMA (1992) Dune erosion and shoreline retreat between Aveiro and Cape Mondego (Portugal). Prediction of future evolution. Proceedings of the international coastal conference, Kiel, pp 187–200

Gallego-Fernández JB, Sánchez IA, Ley C (2010) Restoration of isolated and small coastal sand dunes on the rocky coast of northern Spain. Ecol Eng 37(2011):1822–1832

Gomez-Pina G, Munoz-Perez JJ, Ramirez JL, Ley C (2002) Sand dune management problems and techniques, Spain. J Coast Res SI 36:325–332

Hart A, Hilton MJ, Wakes S, Dickinson K (2012) The impact of Ammophila foredune development on downwind aerodynamics and parabolic dune development. J Coastal Res 28(1):112–122

Hertling UM, Lubke RA (2000) Assessing the potential for biological invasion—the case of Ammophila arenaria in South Africa. S Afr J Sci 96:520–527

Hesp PA (1989) A review of biological and geomorphological processes involved in the initiation and development of incipient foredunes. In: Gimmingham CH, Ritchie W, Willetts BB, Willis AJ (eds) Coastal sand dunes, Proceedings of the royal society of Edinburgh, 96B:181–202

Hesp PA (2002) Foredunes and blowouts: initiation, geomorphology and dynamics. Geomorphology 48:245–268

Hilton M, Macauley U and Henderson R (2000) The New Zealand Active Duneland Inventory.Department of Conservation. Wellington, New Zealand, p 152

Hilton MJ (2006) The loss of New Zealand's active dunes and the spread of marram grass (Ammophila arenaria). NZ Geogr 62:105–121

Hilton MJ, Konlechner TM (2010) A review of the marram grass eradication program (1999–2009), Stewart Island, New Zealand. Proceedings, the New Zealand plant protection society inc. and the council of Australasian weed societies Inc.—17th Australasian weeds conference, 26–30 September 2010, Christchurch, New Zealand, pp 386–389

Hilton MJ, Konlechner T (2011) Incipient foredunes developed from marine-dispersed rhizome of Ammophila arenaria. J Coast Res SI 64:288–292

Hilton MJ, Duncan M, Jul A (2005) Processes of Ammophila arenaria (marram grass) invasion and indigenous species displacement. Stewart Island, New Zealand. J Coast Res 21(1): 175–185

Hilton MJ, Harvey N, Hart A, James K, Arbuckle C (2006) The impact of exotic dune-grasses on foredune development in Australia and New Zealand: a case study of Ammophila arenaria and Thinopyrum junceiforme. Aust Geogr 37:313–335

Hilton MJ, Woodley D, Sweeney C, Konlechner T (2009) The development of a prograded foredune barrier following Ammophila arenaria eradication, Doughboy Bay, Stewart Island. J Coast Res SI 56:317–321

Huiskes AHL (1979) Biological Flora of the British Isles. Ammophila arenaria (L.) Link (Psamma arenaria (L.) Roem. et Schult.; Calamagrostis arenaria (L.) Roth). J Ecol 67:363–382

Johnson PN (1992) 'The sand dune and beach vegetation inventory. II. South Island and Stewart Island'. DSIR land resources scientific report no. 16, Christchurch

Jul A (1998) Marram grass (Ammophila arenaria) invasion of Mason Bay, Stewart Island. Diploma of Wildlife thesis, University of Otago, Dunedin New Zealand, p 80

Kirkpatrick JB (2001) Ecotourism, local and indigenous people, and the conservation of the Tasmanian Wilderness World Heritage area. J Royal Soc N Z 31(4):819–829

Konlechner TM, Hilton MJ (2009) The potential for marine dispersal of Ammophila arenaria (marram grass) rhizome. J Coast Res SI 56:434–437

Konlechner TM, Hilton MJ (2010) An examination of the seed bank of Ammophila arenaria (marram grass). The New Zealand plant protection society inc. and the council of Australasian weed societies inc.—17th Australasian weeds conference, 26–30 Sept 2010, Christchurch, New Zealand, pp 390–393

McKelvey P (1999) Sand forests. Canterbury University Press, New Zealand, p 168

Marchante H, Marchante E, Freitas H (2003) Invasion of the Portuguese dune ecosystems by the invasive species Acacia longifolia Andrews Willd: effects at the community level. In: Child LE, Brock JH, Brundu G, Prach K, Pysek P, Wade PM, Williamson M (eds) Plant invasions: ecological threats and management solutions. Backhuys Publications, Leiden, Netherlands, pp 75–85

Milton SJ (2004) Grasses as invasive alien plants in South Africa. S Afr J Sci 100:69–75

Muasya AM, de Lange PJ (2010) Ficinia spiralis (Cyperaceae) a new genus and combination for Desmoschoenus spiralis. NZ J Bot 48:31–39

Nordstrom KF (2004) Beaches and dunes of developed coasts. Cambridge University Press, Cambridge, p 356

Nordstrom KF (2008) Beach and dune restoration. Cambridge University Press, Cambridge, p 187

Nordstrom KF, Arens SM (1998) The role of human actions in evolution and management of foredunes in The Netherlands and New Jersey, USA. J Coast Conserv 4:169–180

Pickart A (1997) Control of European beachgrass (Ammophila arenaria) on the west coast of North America. In: Kelly M, Wagner E, Warner P (eds) Proceedings of the California exotic pest plant council symposium, vol. 3, pp 82–90

Pickart AJ (2012) Dune restoration over two decades at the Lanphere and Ma-le'l Dunes in 543 Northern California. In: Martínez ML, Gallego-Fernández JB, Hesp PA (eds) Restoration of 544 coastal dunes. Springer, Berlin. Chap. 10

Rakodi C, Treloar D (1997) Urban development and coastal zone management: an international review. Third World Plan Rev 19(4):401–418

Seeliger U (2003) Response of Southern Brazilian coastal foredunes to natural and human-induced disturbance. J Coast Res SI 35:51–55

Soreng RJ, Gillespie LJ, Jacobs SWL (2009) Saxipoa and Sylvipoa—two new genera and a new classification for Australian Poa (Poaceae: Poinae). Aust Syst Bot 22:401–412

Vega LE, Bellagamba PJ, Fitzgerald LA (2000) Long-term effects of anthropogenic habitat disturbance on a lizard assemblage inhabiting coastal dunes in Argentina. Can J Zool 78:1653–1660

Veloso-Gomes F, Taveira-Pinto F (1995) Portuguese urban waterfronts expansion near coastal areas. In: Reis Machado J, Ahern J (eds) Environmental challenges in an expanding Urban world and the role of emerging information technologies conference, Lisbon, Portugal

Wiedemann AM, Pickart AJ (2004) Temperate zone coastal dunes. In: Martinez ML, Psuty NP (eds) Coastal dunes ecology and conservation. Springer, Heidelberg, pp 53–66

Woodley D, Hilton MJ (2003) A model of Ammophila arenaria necrosis, decay and sedimentation following herbicide application. Proceedings, coasts & ports conference, Auckland

Chapter 6
Foredune Restoration Before and After Hurricanes: Inevitable Destruction, Certain Reconstruction

Rusty Feagin

6.1 Introduction

On Gulf of Mexico coastlines, hurricanes and tropical storms disturb beach and foredune development quite frequently. For example in Texas, a good "rule of thumb" is that a tropical system makes landfall within 100 km of a given point, once in every 5 years (Morton et al. 1983). In natural areas where the sediments can move dynamically, dunes simply re-form over time, although in new spatial locations.

However, in urban or developed areas, housing and infrastructure often impede dune migration, leaving no space for dune-building processes to occur (Nordstrom 2008; Psuty 2004). In this urban context, for example on Galveston Island, Texas, there are only a few meters available for a dune in the cross-shore direction.

At many locations around the world, people often restore foredunes for ecological purposes, for example, to enhance habitat for plants or animals. However, on Galveston Island, with cross-shore erosion rates that threaten to eliminate any available width within a few years, restoration provides minimal habitat value and only briefly establishes a human-managed ecosystem. Still, dunes are consistently restored in this context on Galveston (typically, they are not restored in natural areas in Texas as there is no financial incentive in these isolated, rural areas).

Why do both private individuals and government agencies restore foredunes in locations with minimal habitat value, and in an environment where there is a certainty of loss within a few years? In such a context, dunes are restored because:

1. They effectively demarcate the public beach from private property, protecting homeowners from wandering tourists.

R. Feagin (✉)
Spatial Sciences Laboratory, Department of Ecosystem Science & Management, Texas A&M University, College Station, TX 77845, USA
e-mail: feaginr@tamu.edu

M. L. Martínez et al. (eds.), *Restoration of Coastal Dunes*, Springer Series on Environmental Management, DOI: 10.1007/978-3-642-33445-0_6,

2. They are aesthetically-pleasing, providing a sense of nature and wildness to homeowners and tourists.
3. They protect housing and infrastructure from the impacts of future storms. People on the Texas Coast routinely invest in expensive foredune restoration projects, while understanding that they are likely to be destroyed in less than a decade.

In this chapter, the incentives behind three dune restoration projects on Galveston Island, and the storms that created and/or destroyed them, are examined. Additionally, an example is presented where Light Detection and Ranging (LIDAR) laser altimetry, Global Positioning System (GPS), and Geographic Information System (GIS) were used to develop a novel technique for dune restoration.

6.2 Delineation of Public–Private Property Boundary: Pirates' Beach Project

Tropical Storm Frances (1998) destroyed foredunes in the Pirates' Beach area on Galveston Island, Texas, leaving a flat profile with a low slope. No vegetation remained between the ocean and the residential community of Pirates' Beach that was located approximately 75 m further landward.

Under Texas law, the Texas Open Beaches Act (§61.011), demarcates private from public property using the vegetation line. Thus, after Frances, the residents of Pirates' Beach were concerned that the State of Texas might appropriate their land through this Act, since the vegetation line had receded to a location landward of their houses. In the past, the State had subjectively applied the act, often allowing homeowners to remain, regardless of whether the line of 100 % vegetation coverage was landward or seaward of the housing structure. Although the State would often declare all land seaward of these structures as publicly owned, they largely avoided condemning the structures by declaring eminent domain over them, because of the expense of fighting lawsuits that were often filed by the homeowners. Typically, the State waited until the erosion had become so severe that the housing structures physically collapsed into the ocean. Most of the homeowners in Pirates' Beach were not in jeopardy of losing their homes to the State after Frances; however, they had little to no land seaward of their homes.

Residents of the Pirates' Beach community were concerned that the public easement was encroaching upon their private lands, and they could foresee that the most seaward structures would be lost in a few years. Moreover, residents expressed concern that the public was accessing the beach through their property, walking through their manicured lawns, creating the potential for crime, and resulting in a loss of privacy for their high-income community.

The community of Pirates' Beach pointed out that erosion at their location had been hastened by the existence of shoreline armoring that had been in place for decades prior (the Galveston Seawall and jetties), located upstream in the littoral current to the east. Pirates' Beach residents argued that their downstream loss of

Fig. 6.1 Foredune restoration project at the Pirates' Beach. *Left image* shows the geotube covered with sand, prior to planting vegetation. *Right image* shows the geotube after Hurricane Lilly removed the sand and vegetation in 2002. By 2003, the geotube was partially torn apart by Hurricane Claudette. Photos by Rusty Feagin

private land was partially due to the State's intervention in the natural sedimentary dynamics of the region, and that the State should extend the armoring further westward to protect their properties as well. By 1998, however, "hard" shoreline armoring was no longer allowed or approved by the State, in general. As an alternative solution, geo-textile tubes were proposed for Pirates' Beach. Geo-textile tubes, or geotubes, are essentially large textile socks filled with sand. They can be cut open and removed relatively quickly if desired.

A large foredune was thus reconstructed along a 2-km stretch of Pirates' Beach in the spring of 2000, using a geotube at its core (Fig. 6.1, see also Feagin 2005, for a full description of the project). Sand was placed on top of and around the geotube, making it look like it was natural. In June through July 2000, vegetation was planted on the reconstructed dune at 1-m spacings. *Sporobolus virginicus* was planted at its seaward base, *P. amarum* over the entirety of the dune, and *Spartina patens* at its landward base. A drought ensued, and temperatures were at all-time highs in the Galveston region for much of the late summer. By winter, it was anticipated that the majority of the plants were dead; however, by the following spring it became apparent that the survival of *P. amarum* was over 90 %.

Residents were quite happy with the reconstructed dune, and felt that it looked natural. It was aesthetically pleasing and kept the public at a distance from their homes. Their private property was relatively secure, since the vegetation line had been re-established tens of meters seaward of where it had been post-Frances. In an ironic twist, the residents' private property was thus extended seaward by the State itself, funding portions of the reconstruction project with grants.

In 2001, 2002, and 2003, a series of tropical storms and hurricanes removed all of the sand and vegetation that covered the geotube (Feagin 2005). The storms also destroyed the walkovers and bridges that stretched across the geotube, reducing public access to the beach. In some locations, the geotube had ripped open.

Residents of Pirates' Beach felt that the reconstructed dune had protected their homes from the waves and surges of these storms, and they were quite happy with its performance. However, public citizens were angry that the geotube was now uncovered, unnatural, and ugly. Several public interest groups challenged the subsequent use of geotubes at other locations, arguing that they blocked access to the public resource and violated the Texas Open Beaches Act. Over the following years, sand and vegetation were never replaced on the geotubes at Pirates' Beach, and the geotubes were not well-maintained. On 13 September 2008, Hurricane Ike completely destroyed what remained of the geotube at Pirates' Beach. The surge during this storm was approximately 4 m at this location with waves several meters high (Williams et al. 2009), and the entire community was flooded the day before Ike actually made landfall. The remnant portions of the geotube were removed by 2010. Still, the example of the dune reconstruction at Pirates' Beach elucidates the strong incentive for private landowners to build dunes, and maintain the vegetation line, for the purpose of delineating private versus public land.

6.3 Aesthetics and Natural Habitats: Galveston Island State Park Project

The image of sea oats (*Uniola paniculata*), with tall stems and majestic seed-heads, is a classic icon that graces Texas coastal literature, photos, and paintings. Coastal management literature often features images of *U. paniculata*. Sea oats seem to be synonymous with the beach and dune ecosystem in Texas.

However, there are no *U. paniculata* plants of the native Caminada variety left on Galveston Island, except in paintings and pictures on the walls of hotel rooms. This genetic variety is apparently limited to the Western Gulf (Subudhi et al. 2005) and its seeds are essentially unviable (NRCS 2005). Restoration has been done previously using this variety (NRCS 2008) and specific recommendations have been made regarding its planting, although not on Galveston Island. This variety of the species spreads primarily through rhizomatous runners and forms large clumps on the tops of the dune. These traits are not favorable on an island where a high hurricane frequency, a history of overgrazing, and modern development have wiped out any local source from which the species could spread (Feagin et al. 2005). The few Caminada variety plants that were propagated on Galveston were grown in the nursery, did poorly, and were never transplanted to natural dunes.

During one project, a primary objective of restoration was to re-introduce sea oats, *U. paniculata*, to Galveston Island (see Feagin et al. 2008 for full project description). Several sites were chosen for the re-introduction of the species to Galveston Island. Sites were chosen because they were located on public property owned by the State of Texas, and they were appropriately located to maximize the potential spread of *U. paniculata* across the island. A workshop was conducted on sand dune restoration at Galveston Island State Park, and local residents were recruited to assist with the re-introduction of *U. paniculata*.

The Caminada variety of *U. paniculata*, the variety that is native to the Western Gulf, was initially sought. The UDSA, NRCS Plant Materials Program offices in Kingsville, Texas and Galliano, Louisiana were contacted for plant material. The Kingsville office could not provide plants. The Louisiana office provided a list of growers; however, only one grower claimed to have *U. paniculata* and was requesting a very high price per plant. After consulting with this grower, who refused to discuss where the sediment for the plants came from, it was inconclusive as to whether sediment in the plant pots may already be mycorrhizal. Moreover, there was a good chance that these *U. paniculata* plants were of the Eastern Gulf variety.

For these reasons, *U. paniculata* was purchased from growers in Florida. This plant material was almost certainly the Eastern Gulf variety, as the seeds were viable. Although the Caminada variety was preferred, the Eastern Gulf variety would also allow the plants to potentially spread in the future.

On 5–6 May 2008, approximately 1,000 Eastern Gulf variety plants were planted at three restoration sites. At Galveston Island State Park, the plants were placed where the back beach graded into the dunes, in an area with low embryonic dunes that were already vegetated. This less-than-ideal location along the beach–dune gradient was chosen because the area immediately seaward was covered with beach wrack, denoting an elevation and location that was often inundated during high tides, and the area immediately inland was already covered with 100 % plant cover, which would have likely resulted in high competition for the re-introduced plants.

Individual plants were spaced at 0.5-m intervals in the long-shore direction, and 1-m intervals in the cross-shore direction (between rows). Rows were offset from each other by 0.5 m, and plot locations were offset from each other by 1.5 m.

For planting, a small hole was dug to 10 in. in depth; mycorrhizae were added and mixed with sand at the bottom of the hole. A single plant was then placed into the hole, and the hole was refilled with the original sand. Plants were not watered and no additional fertilizer was added, although the sand was damp to two inches below the surface because of recent rains.

After the restoration, a minor drought occurred and no rainfall fell for approximately 1.5 months. All plants were monitored in late July for their survival, number of stems, length of each stem, and total vegetative length (sum of all stem lengths). Owing to the minor drought that immediately followed the restoration plantings, survival was relatively low. The survival rate for all plants at Galveston Island State Park site was 22.91 %. Although such a value is typical of many inland restoration projects, it is quite low for coastal sand dune projects in Texas. Still, those plants that did survive were growing well. For those that survived only, the average number of stems per plant was 3.4, the average length of a stem was 24.8 cm, and the average total vegetative length was 84.58 cm.

On 13 September 2008, Hurricane Ike destroyed all of the three restoration sites that had been planted. At the Galveston Island State Park site, the location that had

Fig. 6.2 Hurricane Ike destroyed the Galveston Island State Park restoration site. *Left images* are before Ike, *right* are after Ike. *Top images* show the general area, *bottom images* are zoomed in. Restoration plantings were placed even with the black line in the bottom picture, which sits at the exact same location in both pictures. Imagery courtesy of the Texas Natural Resource Information System (TNRIS)

been restored sat at the waterline shortly after the storm (Fig. 6.2). At the other two sites, the dunes were completely leveled and the plants were washed away. Thus, after approximately 6 months, there was 0 % survival for all three sites.

Hurricane Ike ultimately destroyed all of the restoration sites, leaving no plants. Still, several valuable lessons were learned that can be passed on to future attempts at re-introducing *U. paniculata* (sea oats) to Galveston Island.

First, there is no seed stock available for the genetic Caminada or Western Gulf variety of this plant, nor is there any readily available plant stock, regardless of the fact that the USDA had worked to build such a stock in the past. The Western Gulf variety has sterile seeds with effectively zero viability. All of the *U. paniculata* that is currently available is actually a genetic variety from the Eastern Gulf. Although it would be ideal to obtain the Western Gulf variety for the restoration projects west of the Mississippi River, it is not feasible. The only other possibility is to take cuttings from already established Western Gulf plants growing in the field, but this is both destructive to the few plants that remain on the Upper Texas Coast and has never been proven to work. Thus, this explains why all restoration projects use the Eastern Gulf variety. Since the Eastern Gulf variety has viable seeds, the long-term outcome of this restoration practice is that *U. paniculata* will likely become a more resilient species on the Upper Texas Coast, but genetic diversity may be lost (Franks et al. 2004).

In terms of planting survival, *P. amarum* is likely to be a much better choice than *U. paniculata* for Galveston Island dunes. *U. paniculata*, if planted, needs to be consistently watered for at least the first 3 months. In previous experience, *P. amarum* has been seen to withstand much longer periods of drought than the *U. paniculata* plants withstood in this project, with little to no adverse effects on their long-term growth (Pirates' Beach project, see Feagin 2005). Thus, if *U. paniculata* re-introduction is not a priority, *P. amarum* should be used for restoration in the Galveston area. Another potential strategy is to plant a mixture of the two plants, with the majority being *P. amarum*, as recommended by the NRCS (2008).

Finally, hurricanes are a fact of life on Galveston Island, and when coupled with the low seed viability of the native Western Gulf variety of *U. paniculata* and historical usage of Galveston for cattle grazing and modern development of the landscape immediately behind most foredunes on the island, may explain their absence from Galveston Island. The plants cannot simply re-establish, and for this reason it may be good practice to introduce the Eastern Gulf variety and allow the two varieties to interbreed. Future work could attempt to breed these two varieties in a planned manner, so as to preserve as much genetic diversity as possible while also re-introducing the species to Galveston Island.

6.4 Protection of Infrastructure and Housing: Sandhill Shores Project

Hurricane Ike flattened nearly all of the dunes on Galveston Island, Texas on 13 September 2008 (Williams et al. 2009). When Ike hit the Galveston Island community known as Sandhill Shores, the surge was approximately 4 m, with waves of several m in height (Williams et al. 2009). The dune sands were pushed

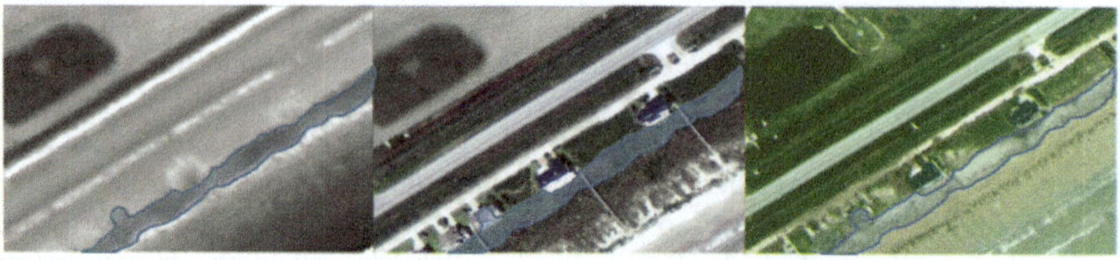

Fig. 6.3 Using a GIS–LIDAR–GPS survey approach to restoration. Hurricane Ike altered the dune and swale topography at the Sandhill Shores site. *Left image* shows pre-storm LIDAR elevation data, clearly indicating the swale and dune locations (*black* is low elevation, *white* is high). *Center image* is pre-storm aerial imagery, *right image* is the post-storm image. In all three geo-referenced images, the outlined *blue* area denotes the same swale location, for comparison. Imagery courtesy of the TNRIS

landward and deposited into swales that were formerly behind them, filling them. A small amount of remaining sand was washed under and around the houses (many houses were on stilts), and further landward onto the streets. Much of the beach sands seaward of the dunes were eroded and deposited in the nearshore.

Residents of the Sandhill Shores residential community clearly stated that the dune and swale structure protected their houses and investments from Ike's wrath. The 3-m dunes at Sandhill Shores were considerably taller than dunes on most of Galveston Island, but the residents were especially vocal that the 3-m deep swale provided the majority of the protection. According to eyewitness reports, the breached dunes were swept into the swales, and the majority of the wave energy continued to break upon and rework these areas. Although the surging waters did sweep further landward and under the houses, there was no erosion in these areas, but rather accretion via overwash deposits. The residents strongly stated that they wanted the dune–swale structure restored after Ike.

A primary objective of the Sandhill Shores project was to restore the dune–swale structure, as closely to its pre-Hurricane Ike state as possible. Residents wanted both the restored dune and swale volumes to match pre-Ike volumes, along the entirety of the community's 1-km-long shoreline. The challenge was finding a historical reference to help rebuild the dune and swale to the exact former dimensions, as the residents desired. A novel technique was developed using LIDAR, GIS, GPS, and surveying to address this challenge (Fig. 6.3).

Within a GIS, virtual cross-shore transects were created that started at the back-beach and stretched landward across the dunes, beyond the swales up to the base of the houses. These transects were distributed every 50 m alongshore direction. Several points were then placed on these transects, with particular care taken to mark the tops of the dune and its base, the depths of the swale and its interior sides.

Once the points along each transect were selected, their horizontal location was noted in latitude and longitude coordinates. Pre- and post-Ike aerial photography was utilized, to confirm locations and assist in the process of adding additional points. In the field, a GPS was used to find these points, and flags were used to mark the pre-Ike horizontal locations of the dune and swale.

Fig. 6.4 Digging out the swale and reconstructing the foredunes, Sandhill Shores project. The reconstructed locations of the dune, swale, and their heights/depths were marked on flags for the bulldozer operator to follow. The dune was reconstructed after the storm with the aid of a GIS, GPS, and survey equipment, but the dimensions and volume were defined to match pre-storm topography, as based upon a LIDAR elevation dataset and aerial photography taken before the storm. Photo by Paul White

To virtually reconstruct the vertical elevations, the pre-Ike absolute elevation of each point was found, using a 2008 pre-Ike LIDAR dataset that was acquired from the Texas Natural Resource Information System (TNRIS). The resolution of this dataset was 1 m horizontally, 0.01 m vertically. This dataset allowed the topography and structure of the dune–swale structure to be viewed, prior to Ike's rearrangement of the sands.

However, after Ike hit, the absolute elevation had completely changed in the area (the entire beach profile was approximately 0.75 m lower). A changing profile would modify the base structure upon which the dune and swale vertical locations were fixed. Thus, when using the LIDAR dataset, the absolute elevation of each of the points was adjusted relative to a nearby fixed point at the base of each house (its concrete driveway, or similar feature that presumably did not change in elevation pre- to post-storm). In the field, survey equipment was then utilized to assess the vertical difference between the same nearby fixed point and the surveyed points, so that the vertical distance could be defined that the surface would have to change to approximate the pre-Ike absolute elevation. The flags were then marked to denote this information. To double-check the calculations of depth and height, several holes were also dug at the location of the swale until the original surface had been hit. The vertical and horizontal match was quite strong, with calculated depths less than 0.05 m off from actual field depths.

Once vertical and horizontal locations were marked in the field, backhoes and tractors were used to reconstruct the pre-Ike topography. *Sporobolus virginicus* was then planted at the base of the dunes, with *P. amarum* covering the rest of the dunes. *Spartina patens* was planted in the swales. Over the next year, the beach profile also partially recovered as much sand returned from the nearshore. Accretion visibly occurred at the base of the reconstructed dunes, although it was not quantified.

The residents of the Sandhill Shores community were quite happy with the reconstructed dune–swale ecosystem (Fig. 6.4). They spent less than $100,000 (in 2009) on the restoration along the 1-km stretch of beach, and felt that the original dune and swale structure had protected their homes from a much greater expense during Hurricane Ike. Moreover, residents expressed that the reconstructed dunes enhanced the aesthetic value of their property and buffered them from public intrusions onto their property.

6.5 Conclusion

As these three projects show, several incentives exist to restore coastal sand dunes, even if hurricanes or other storm events destroy them every few years. They are continually rebuilt because:

1. They effectively demarcate the public beach from private property, which is a significant legal feature in many parts of the world.
2. They are aesthetically pleasing, providing a sense of nature and wildness.
3. They protect housing and infrastructure from the impacts of future storms.

It is gratifying when governments or individuals contribute money toward conserving or restoring natural dune ecosystems. Often, foredune restoration projects are initiated because people realize the value of a functioning dune ecosystem. However, in the context of repeated destruction of these projects, and in a world of fiscal constraints and competing interests, there must be other incentives that drive their construction. The financial benefit must outweigh the cost of such projects.

This is a powerful story that is not to be viewed cynically—dune restoration projects can find support even when there is no ecological reason to do so, and where there is no support from an environmental lobby or concerned citizens. Small pockets of the natural world can exist located within a largely urban matrix, simply because it is in people's financial interest to maintain them. In Galveston, Texas, this interest is strong enough to ensure that these small, semi-natural dune areas are diligently restored after every storm.

References

Feagin RA (2005) Artificial dunes created to protect property on Galveston Island, Texas: the lessons learned. Ecol Restor 23:89–94

Feagin RA, Sherman DJ, Grant WE (2005) Coastal erosion, global sea-level rise, and the loss of sand dune plant habitats. Front Ecol Environ 3:359–364

Feagin RA, Koske RE, Gemma JN, Williams AM (2008) Restoration of sea oats (*Uniola paniculata*) with mycorrhizae on Galveston Island. NOAA report # NA07NOS4190144, Texas General Land Office report # 08-020

Franks SJ, Richards CL, Gonzales E, Cousins JE, Hamrick JL (2004) Multi-scale genetic analysis of *Uniola paniculata* (Poaceae): a coastal species with a linear, fragmented distribution. Am J Bot 91:1345–1351

Morton RA, Pilkey OH Jr, Pilkey OH Sr, Neal WJ (1983) Living with the texas shore. Duke University Press, Durham,

Nordstrom KF (2008) Beaches and dune restoration. Cambridge University Press, New York

NRCS (2005) Caminada sea oats: *Uniola paniculata*. Golden Meadows Plant Materials Center, Galliano

NRCS (2008) Coastal and dune restoration. Plant Materials Technical Note No: TX:PM-08-01

Psuty NP (2004) The coastal foredune: a morphological basis for regional coastal dune development. In: Martinez M, Psuty NP (eds) Coastal dunes: ecology and conservation. Springer, Berlin

Subudhi PK, Parami NP, Harrison SA, Materne MD, Murphy JP, Nash D (2005) An AFLP-based survey of genetic diversity among accessions of sea oats (*Uniola paniculata*, Poaceae) from the southeastern Atlantic and Gulf coast states of the United States. Theor Appl Genet 111:1632–1641

Williams AM, Feagin RA, Smith WK, Jackson NL (2009) Ecosystem impacts of Hurricane Ike: perspectives of the Coastal Barrier Island Network (CBIN). Shore & Beach 77:71–76

Part II
Restoring Inland Coastal Dunes: Dunefields and Wetslacks

Chapter 7
Restoration of Dune Mobility in The Netherlands

Sebastiaan M. Arens, Quirinus L. Slings, Luc H. W. T. Geelen and Harrie G. J. M. Van der Hagen

7.1 Introduction

Coastal dunes in The Netherlands are part of important, multifunctional landscapes. In addition to their geomorphological values, they harbor many rare species, of both flora and fauna, protect the hinterland from flooding, provide us with drinking water, and serve as recreational space. Sustainable management of this landscape demands a thorough understanding of the system. The exact origin of many coastal dune systems in the world is unknown. In the literature, several hypotheses regarding the origin of mobile dunes are explained. One hypothesis argues the dunes are related to large-scale coastal processes, leading to a massive input of sand into the coastal system, particularly as the sea level rises in the postglacial marine transgression. In this case, the birth of coastal dunes is the result of sand supply forced by sea level rise. Another hypothesis claims that large-scale mobile dunes must be related to the remobilization of older systems due to coastal erosion and enhanced by climatic fluctuations. A third hypothesis attributes dune mobility to human activity, caused by the destruction of vegetation through wood gathering for fuel, overgrazing, and other exploiting activities.

S. M. Arens (✉)
Bureau for Beach and Dune Research, Amsterdam, The Netherlands
Civil Engineering and Geosciences, Delft University of Technology, Delft, The Netherlands
e-mail: arens@duinonderzoek.nl

Q. L. Slings
nv PWN Drinking Water Company, Velserbroek, The Netherlands

L. H. W. T. Geelen
Research and Development, Waternet, Vogelenzang, The Netherlands

H. G. J. M. Van der Hagen
Dunea, Voorburg, The Netherlands

M. L. Martínez et al. (eds.), *Restoration of Coastal Dunes*, Springer Series on Environmental Management, DOI: 10.1007/978-3-642-33445-0_7,

Most coastal dune systems in western Europe are currently in a phase of stabilization (e.g., Provoost et al. 2011). Again, the true reason is not well understood. Stabilization of these systems may occur because of stabilizing efforts of man over centuries, reduced anthropogenic pressure, like grazing and wood gathering, nitrogen input from air pollution, the absence of rabbits due to diseases or climatic change. Because of stabilization, pioneer stages are becoming rare, causing biodiversity to decrease in many dune systems. Managers try to counteract these negative effects, for example, by removing vegetation in order to set back vegetation succession, or by restoration of mobile dunes. The question here is whether stabilization can be reversed in a sustainable way, for example, by restoring natural processes. Or will managers be forced to fight stabilization "forever"? What are the key factors in dune mobility or dune stability? Which can be adapted? What are the links between nature management and coastal defense? How can the two benefit from one another? And finally, how can we manage dune landscapes in a responsible way, that justifies the important geomorphological, ecological, and other societal values.

7.2 Mobility Versus Stability: The Struggle of Vegetation Against Sand

Dunes all over the world experience alternating phases of mobility and stability (e.g., McFadgen 1985, 1994; Arbogast et al. 2002; Tsoar and Blumberg 2002; Hugenholtz and Wolfe 2005; Clemmensen and Murray 2006; Kutiel 2013). Also, within a system, mobile and stable phases can coexist (Barbosa and Dominguez 2004; Tsoar 2005; Yizhaq et al. 2007). The state of the system is governed by climatic parameters and the availability of sand, but on a smaller scale random processes may interfere.

7.2.1 Driving Forces

Three major driving forces can be recognized (Hesp and Thom 1990): wind energy, the growing capacity of vegetation, and sand availability (Fig. 7.1). Human impact (management, nitrogen input, overgrazing) exerts its influence through its effect on vegetation growth. Transitions from one phase to the other are driven by changes in climate, sand supply, human impact and other disturbing factors that might act randomly in space (e.g., rabbit digging). In essence, the problem of mobility versus stability can be reduced to the struggle of vegetation against sand: which of the two dominates?

Wind energy and the growth characteristics of vegetation are governed by climate. For example, increased storminess will increase the available wind energy

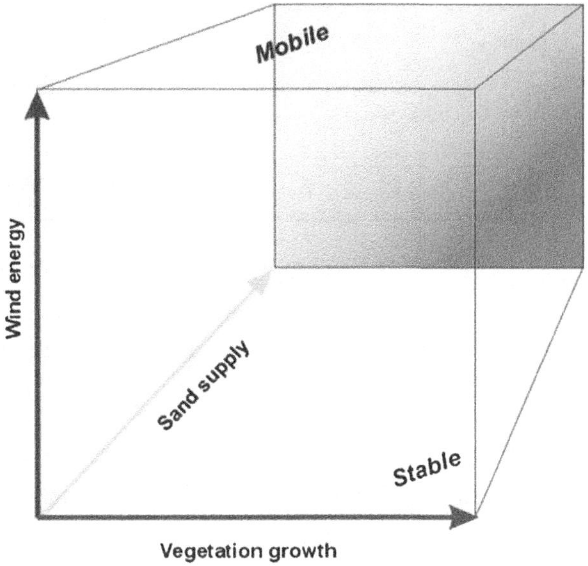

Fig. 7.1 Driving forces in dune development. After Hesp and Thom (1990)

for aeolian transport, and therefore enhance dune mobility. Increased rainfall and a rise in the average temperature will favor vegetation growth, and tend to direct the state of the system toward stabilization. An increase in biomass and consequently soil development also add to stability, making it more difficult for the wind to erode the sand (hysteresis, see Tsoar 2005). The growing capacity of plants is strongly influenced by man, being either negative or positive. Management (mowing, grazing, etc.) reduces the accumulation of biomass. Wood gathering or sod cutting simply removes vegetation from the system. On the other hand, planting activities, nitrogen input through fertilization of soils or by air pollution, and the covering of bare surfaces with branches to prevent wind erosion, all add to the growing capacity of plants.

Several studies point to increase mobility of dune systems owing to climate change, especially in areas where aridity rises (Lancaster 1997; Forman et al. 2001; Thomas et al. 2005). Few mention the probable reduction in dune mobility in areas with increasing rainfall, and the resulting loss of biodiversity (Provoost et al. 2011; Rhind and Jones 2009; Jones et al. 2010).

Sand supply is governed by coastal processes or the input of sediment by rivers. Coastal erosion may contribute to sand supply by releasing erodible sand due to the removal of vegetation and exposure of the sand in active dune cliffs. Humans may directly or indirectly influence sand supply in many ways and on several scales, with important consequences. For example, the construction of jetties to prevent erosion in one place leads to erosion downstream due to a lack of sand. The canalization of rivers or sand extraction results in a reduction in sediment recharge, which may affect coastal processes on a large scale. Fixation of the foredune by planting interferes with the sediment exchange with the beach and

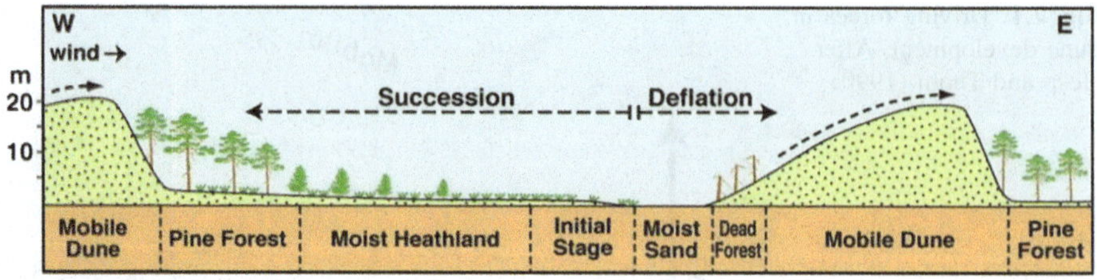

Fig. 7.2 Sequence of succession stages in a landscape with mobile dunes (after Piotrowska 1991)

shoreface; consequently, erosion of the shoreface results in a steepening of the coastal profile.

7.2.2 Scales of Mobility and Biodiversity

In the case of mobile dunes, being either parabolic or transgressive, the processes act on a large scale, and influence the whole landscape. Without restriction to mobility, mobile dunes tend to migrate inland until they naturally stabilize. In the case of a completely dynamic landscape, any part in the landscape will be affected by the mobile dunes in time, first by burial, when the dune passes, and later by deflation. As most mobile dunes tend to stabilize at some distance from the shoreline, this is a somewhat hypothetical case. However, there are examples of active mobile dunes at several tens of kilometers inland (for example, Madagascar, Australia). The result is that any part of the landscape will at some stage be overridden by a dune, and vegetation development has to start from scratch. In the lee of the dune we will find the oldest succession stages, at the windward side bare sand, followed by a series of succession stages (Fig. 7.2; Piotrowska 1991). In dune systems where deflation is governed by groundwater depth, deflation planes "follow" the changes in groundwater depth in space and time, thus assuring the continual presence of wet slack ecosystems. This is the reason why biodiversity is high in these systems, since vegetation will hardly reach the climax state, and pioneer stages are continuously replenished by the passing of new dunes. In the landscape all stages in the vegetation succession are present, and will be, as long as the mobile state is preserved.

When the system is stabilized, processes might still be active, but on a smaller scale. Random disturbances, for example, rabbit digging, followed by water erosion of the exposed sand, create small, patchy spots of bare sand, which may turn into blowouts (Jungerius and van der Meulen 1988). In the blowouts the surface is bare, only partly covered by pioneer species. Around the blowout we will find a sequence of vegetation types which need some burial by sand; the specific types depend on the rate of burial. Depending on the dimensions and the specific setting a blowout could provide its surroundings with a very thin layer of sand over tens to hundreds of meters (Van Boxel et al. 1997). There are few data on the areas influenced around blowouts, but measurements in The Netherlands indicate that one blowout can affect several hectares with its sand spray. The result is a landscape with a mosaic of younger vegetation types surrounded by older, even climax stages. The percentage of younger types depends on the number of blowouts.

In both mobile dunes and blowouts, several stages of vegetation succession are present, but their arrangement depends on the geomorphological dynamics, and is completely different for the two systems.

7.3 Dune Mobility in The Netherlands

What does the Dutch system look like? What changes have affected the coastal dunes in the past few centuries and what was their impact on dune mobility?

The Dutch coast is a sandy coast that has developed since the last Ice Age. With a rising sea level a system of beach barriers has developed. Around 5000 BC, despite a further rising sea level, the beach barriers reached their seaward-most extension (Zagwijn 1986). Small dune systems, now referred to as the Older Dunes, existed on the barriers. Since 4000 BC the barriers have gradually eroded. Around AD 800 a new phase in coastal development started, when vast quantities of sand, probably released by increased erosion rates, invaded the older barrier systems and started to form the Younger Dunes. These dunes developed in several phases of major dune mobility, roughly between AD 800 and 1600 (Jelgersma et al. 1970; Klijn 1990). The exact mechanisms are not well understood. Klijn (1990) and Jelgersma et al. (1970) mention storm activity, coastal erosion, and exploitation by man as probable causes. But the main question remains about the origin of the sand that built up the Younger Dunes: which process triggered the invasion of these massive amounts of sand? A study by Pool and van der Valk (1988) pointed out that during development of the Younger Dunes as much as 50 m^3/m/year of sand was stored in the dunes, over several hundred years and along several tens of kilometers. These amounts exceed the present day transport by an order of magnitude. Probably, the input of sand was related to processes of unknown origin on the shoreface, resulting in a steepening of the shoreface, thereby providing the beach with large quantities of sand (Zagwijn 1986). Storm surges in the tenth to sixteenth centuries caused large-scale coastal erosion and a probable remobilization of sand. Whether there is a connection to the supply of sediment by the large rivers is unclear. Climate variation may have played a role, with periods of increased storminess and severe droughts as triggers for dune mobilization. A recent study in the Dutch Delta area (Beekman 2007) relates massive phases of dune building to coastal erosion, triggered by changes in coastal inlets due to human interference.

The present morphology of the dunes reflects the extent and vigor of dune mobility in the past. Most of our dunes consist of large series of parabolic dunes, increasing in dimension from west (coastline) to east, alternating with large

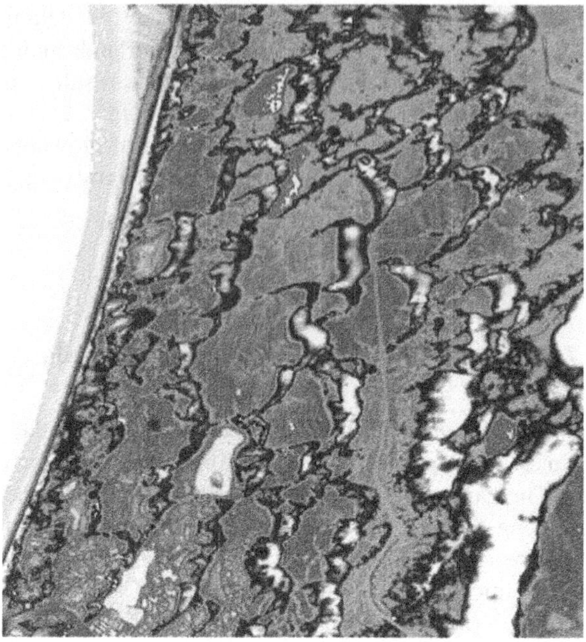

Fig. 7.3 The coastal dunes of Kennemerland. *Source data* PWN

deflation plains, and ending up in a huge precipitation ridge in the east (Fig. 7.3). The maximum width of the dune belt is approximately 4.5 km. The height of the dunes ranges between 10 and 50 m.

The specific setting of this series leads us to believe that parabolic dunes started off as blowouts in the foredunes, gradually released from the foredune when growing larger and moving inland, coalescing to larger systems and moving farther inland, and finally ending up in a precipitation ridge. The main origin of the parabolic dunes, then, is related to processes at the foredunes. However, human influence cannot be ignored. From several sources we know that at least until the nineteenth century dune mobility in many locations was related to human disturbance, mainly the gathering of wood for fuel, branches, marram grass, etc., overgrazing, and digging out by rabbits. Also, there is evidence that the size of the precipitation ridge, which locally exceeds 50 m DOD, is caused by the plantation of trees by humans who needed to stop the gradual invasion of sand, since at least around 1600. Beekman (2007) pointed out that of the two phases of parabolic dune development on the island of Schouwen (southwestern Netherlands) the last phase resulted in much higher dunes because of plantation activities. Several paintings from the seventeenth century show severely eroding coastlines, but also heavy abuse of the foredunes (Giepmans et al. 2004). Many stories tell of villages swallowed by the sea or buried under sand (Rentenaar 1977). The present coastline is situated several hundreds to thousands of meters east (landward) of its position at the start of the building of the Younger Dunes. We believe that the sand of the

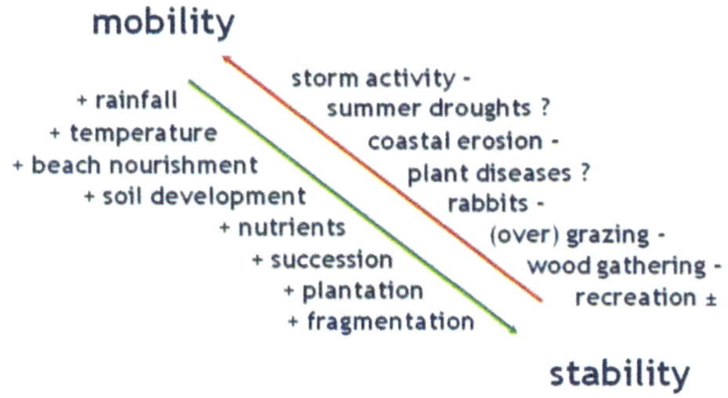

Fig. 7.4 Impact of environmental stress factors on dune mobility and stability

Younger Dunes was derived from this vast coastal erosion, while much of its present morphology is in some way related to human activity.

7.3.1 Current State of the System

Nowadays, the system has changed completely. For centuries, man has put efforts into stabilizing dunes by planting pine trees and marram grass. From the start of the nineteenth century this gradually resulted in decreasing dynamics in the dunes. At the end of the nineteenth century large areas of dunes were still mobile. By the end of the twentieth century, only a small percentage of the surface was left bare.

Figure 7.4 summarizes most environmental and climatic factors that have exerted some influence on dune mobility in The Netherlands. On the left are those factors that enhance stability; on the right, those factors that enhance mobility. Between AD 1000 and now, many of the factors on the right have decreased in importance (most of all coastal erosion and exploitation by man). Especially in the last century, many of the factors on the left have increased in importance, for several reasons. As a result, the dunes in The Netherlands have been forced into the direction of stability.

Despite predictions of increased storm activity because of global warming, The Netherlands has experienced considerably fewer storms since 1990, resulting in reduced coastal erosion and aeolian activity. Rain fall and temperatures have been steadily increasing over the last century, with positive impacts on the growing capacity of vegetation. The growing season over the last few decades has extended by approximately 1 month. Vegetation growth is further stimulated by nitrogen input from air pollution, which is probably one of the major effects of the last few decades. The exploitation of dunes by man has ceased. Because of increasing vegetation, organic matter in soils is accumulating. Meanwhile, vegetation succession continues, as time goes on. All these factors together have resulted in an incredible increase in biomass.

Because of the flourishing marram grass, seeds are everywhere in large quantities. Dispersal of marram from seeds was thought to be insignificant before 2000, but is

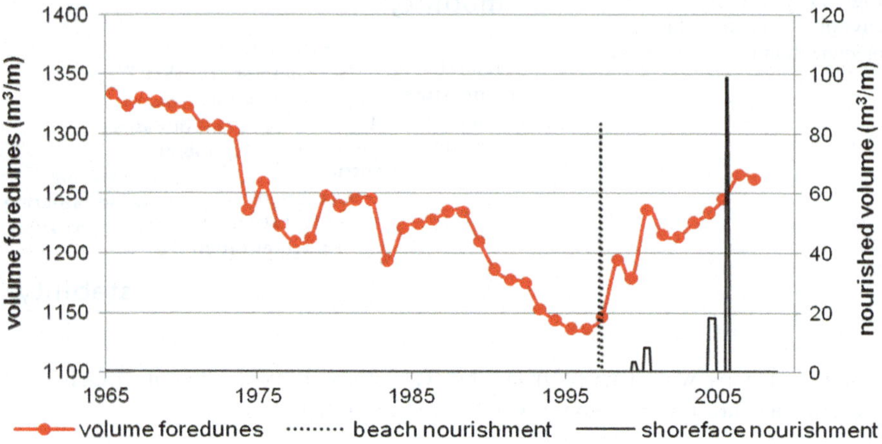

Fig. 7.5 Trend breaks in volume development due to nourishment. *Source data* Rijkswaterstaat

now responsible for the rapid colonization of bare surfaces (e.g., Provoost et al. 2011). Spreading of marram by seeds is certainly more significant now than vegetative expansion. It is not clear if this is a result of the warmer climate, increased rainfall, nitrogen in the system, a positive feedback or a combination of factors.

Although most of the landscape is stabilized, some bare sand is left through the activity of blowouts. These smaller scaled features are superimposed on the underlying landscape of parabolic dunes. In some places the former parabolics are mostly covered by blowouts, in other parts they are untouched, depending among others on the rabbit population, the interference of human use or management and vulnerability to water erosion (Jungerius and van der Meulen 1988). Since rabbit populations declined owing to myxomatosis and VHS, blowout activity has also decreased considerably in the last few decades. The percentage of dune surface covered with active blowouts is nowadays less than 1 %.

7.3.2 Changing Perspectives

Around 1990, coastal defense strategies changed. Since then, beach and shoreface nourishments are applied on a very large scale, to counteract erosion processes, and to fixate the coastline in its position. This strategy is very successful in that most of the coastline retreat has stopped. As a consequence of the input of sediment onto the beach and shoreface, the transfer of sand into the dunes has increased considerably. The state of the foredunes has changed, with an important increase in accreting foredunes, while most foredune cliffs have disappeared. Calculations for the mainland dunes indicated that the net input of sand has

Fig. 7.6 Small-scale dune building owing to beach nourishment. North of Scheveningen. *Photograph* Bas Arens

changed from about 3.5 m^3/m/year in the 1980s (de Ruig and Louisse 1991) to about 10 m^3/m/year in the last few decades (Arens et al. 2010). On the Wadden Islands, the input is even larger (Arens et al. 2010). When studying volume changes along coastal transects, it appears that for many sections trend breaks in the volume development coincide with the start of nourishments (Fig. 7.5). Compared with the total volume of nourishments (about 12 $10^6 m^3$/year), the amount of sand stored in the dunes is approximately 25 %.

The increased input of sand into the dunes has resulted in important changes in the system. In many places, incipient foredunes have been developing in front of the previous foredunes. There are signs that this inhibits the interaction among the beach, foredunes, and inner dunes, limits the input of fresh sand into the inner dunes, and causes a further fixation of the coastal dune landscape. Also stimulated by a lack of big storms, large parts of the coastline are now covered with newly built dune ridges (Fig. 7.6). Gentle sand input of around 5–15 m^3/m/year is not sufficient to kill vegetation, but on the contrary helps marram grass to grow vigorously (Willis 1989; observations by the authors in several areas).

However, because of the nourishment, there is no need to manage the foredune with the same intensity as in the past. The practice of "dynamic preservation" (Hillen and Roelse 1995) is widely applied now. There are many locations where either wind or wave erosion, or both, leads to increased dynamics in the foredunes. For example, the foredunes along the coast of North-Holland have changed drastically within a decade with the development of breaches and blowouts. Before 1998 the development of blowouts was prevented by the coastline managers. Since the application of dynamic preservation, numerous blowouts have been developing. Locally, blowouts in the foredune seem to develop into new parabolic dunes (Fig. 7.7). In other parts of the coast, with the same change in management attitude, nothing happens. It is important to find out why these differences in development occur.

Fig. 7.7 Foredune remobilization, Noordhollands Duinreservaat, mainland coast. Photograph Bas Arens

Since beach and shoreface nourishments are applied worldwide on a large scale, the impact of nourishment on dunes is a big issue, and currently intensively studied in The Netherlands. The above examples illustrate that, depending on the coastal setting, the effects might be negative (a blocking of the interaction between shoreface and dunes) or positive (a new release of sand and reactivation of dynamic processes). Furthermore, if the nourished sand differs from the original sand, e.g., in grain size, carbonate content or mineralogical compounds, there may be ecological consequences. Recent research (Stuyfzand et al. 2010) indicates that the nourished sand mainly differs in carbonate content. We need to identify the mechanisms behind the effects of nourishment to be able to optimize nourishment strategies (i.e., minimize the negative, maximize the positive effects).

Some current developments may further favor aeolian processes and give rise to new phases of dune mobility. Grazing as a management tool replaces the former activity of rabbits. Increased summer droughts are expected, and occasionally observed, as in the summer of 2006. There are observations that the combination of higher temperatures and increased rainfall intensities results in higher levels of water erosion: intense drought enhances the water repellency of the soil (Dekker and Jungerius 1990), resulting in surface run-off and considerable water erosion during heavy showers, enabling the development of new blowouts (Jungerius and van der Meulen 1988).

7.4 Restoration Projects

Currently, because of stabilization, many of the dunes tend to end up with the same climax vegetation, dominated by shrubs. On average, biodiversity is decreasing (Arens and Geelen 2006; Kutiel 2013). New ways of management are being developed to maintain biodiversity, or increase it where possible. Which factors can be manipulated in such a way that the system will be forced back into a durable state of mobility? Is this possible at all? The first restoration projects in The Netherlands were aimed at restoring pioneer stages by sod cutting in dune slacks (Grootjans et al. 2002; 2013). However, these kinds of measures are designed to restore the existing pattern. Within a matter of years, the same method has to be applied again, finally resulting in a lowering of the surface and a relative increase in groundwater levels, wetter circumstances and at the end a fundamental change of the system. Better results might be obtained if processes were restored, resulting in the development of new dune slacks. In the last decade, several experiments were started with the aim of restoring large-scale mobility. The main idea is that if we succeed in restoring dune mobility, nature itself will keep biodiversity at the same level, by creating freshly deflated valleys through deflation on the upwind side, with opportunities for pioneer species and destroying climax vegetation by burial of sand on the downwind side (Arens et al. 2004). With a certain percentage of mobile dunes within the area, permanent renewal of pioneer stages and rejuvenation of the landscape are ensured.

7.5 Restoration of Aeolian Dynamics in Dunes

Restoration of aeolian dynamics in The Netherlands started at the end of the 1980s with small-scale experiments on the island of Schiermonnikoog, in areas that would in no way interfere with sea defense. In the 1990s the first blowouts were reactivated (e.g., Van Boxel et al. 1997). Since then the scale and number of experiments have increased. Restoration of blowouts resulted in a temporary revitalization of aeolian dynamics. Only a small percentage of the reactivations remained active after a period of years. Also, the area that benefited from the interference was limited. The idea was that larger scaled restoration projects, like the reactivation of parabolic dunes, would be more successful, since in the past considerable effort was needed to stabilize them.

However, restoration of dune mobility turned out to be complicated. Monitoring of these experiments indicates that restoration is not immediately successful (Arens et al. 2004, 2005; Arens and Geelen 2006). After 15 years, with several large-scale experiments, we can conclude that the removal of vegetation and soil is not enough to restore large-scale landscape development. This type of interference leads to a sudden and dramatic increase in aeolian dynamics (Fig. 7.8), but re-stabilization from root remnants causes important problems. The formation of a desert pavement of blown out (dead) roots prevents further erosion. Thus far, most reactivated areas have gradually declined in scale. Table 7.1 shows the stabilization rates for a number of projects in The Netherlands. For example, in the Van Limburg Stirum area (Fig. 7.8) the former bare sand surface of 30 ha (1995) was transformed into a mostly stabilized area alternated with 12 active blowouts and 18 freshly deflated surfaces. In 2009, 26 % of the original area was left bare, mostly deflating. On the parabolic dune of the Verlaten Veld only six blowouts are left,

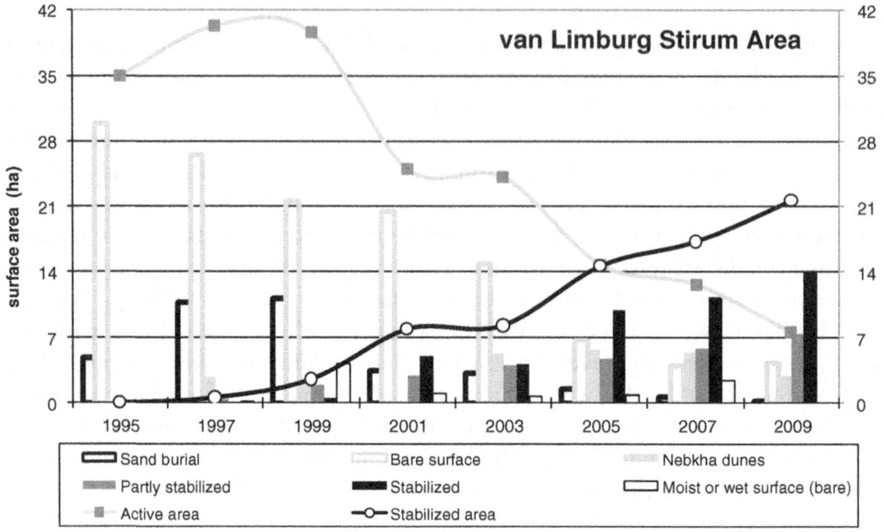

Fig. 7.8 Changes in dynamic and stabilized surfaces in the Van Limburg Stirum area

Table 7.1 Overview of interventions and stabilization rate

Area	Intervention type	Start	Start area (ha)	Area 2009 (ha)	Percentage of start	Decline rate (%/year)
Oude Golf 2	Dunes/slack	2002	4.2	5.4	128	−4.1
Bruid	Parabolic dune	2002	7.8	2.8	36	9.1
Verlaten Veld	Parabolic dune/slack	1998	12.5	2.3	19	8.1
Schoorl	Blowouts	1999	0.8	0.09	12	8.8
Huttenvlak	Deflation area	1999	7.4	1.8	25	7.5
Kerf	Foredune/slack	1997	7.7	1.4	18	6.9
Oude Golf 1	Dunes/slack	1996	4.7	3.2	67	2.5
Van Limburg Stirum	Dunes/slack	1995	30.0	7.8	26	5.3

Start area is the area of interference. The bare area in 2009 is expressed as a percentage of the start area. A percentage >100 % means that the active, bare area has extended

but the deflation slack extended by approximately 1 ha. The same happens in the Bruid van Haarlem. The Oude golf, a former golf course, was destabilized in two phases, the last in 2002. This is the only area where, after an intervention, the bare area is extending over time. Probably this is related to the removal of all roots, leaving a truly bare surface after the intervention.

With respect to the large-scale, landscape-forming processes, some conclusions can be drawn from these experiments. Dune slacks either stabilize quickly or remain bare for a longer period, if the surface is eroded by a couple of centimeters per year. The groundwater level is probably the dominant factor here. If the surface is dry for a certain period in a year, this apparently is enough to keep erosion going, while wet circumstances during the year prevent the establishment of marram. However, this shallow erosion only delivers a limited amount of sand to the surrounding dunes, too limited to bury or kill vegetation. On the slopes of the reactivated (parabolic) dunes, regrowth from marram roots is a major problem. The roots extend to several meters in the soil, therefore, complete removal is unrealistic. Despite higher wind speeds than in the dune slacks, the remaining roots prevent erosion and easily restabilize the surface. Only locally, serious erosion occurs, resulting in the development of blowouts. In contrast, removing vegetation from the crest of the dune also results in huge erosion. Apparently, near the crest the eroding capacity of the wind is high enough to kill the marram grass. However, as a consequence, the crest erodes, loses sand, and the dune form may disintegrate.

Apparently, a single restoration intervention is not sufficient to restore the large-scale, landscape-forming process. A certain form of maintenance, removal of blown out and regrowing roots, for a number of years, is necessary to get the dune rolling. New projects are started where regrowth of (mainly marram) roots is prevented by volunteers, who regularly remove all visible roots by hand. Our hypothesis is that continuous removal of roots for a number of years will result in a smooth and truly bare surface at the end, which might then be the start of autonomous mobility. Further monitoring in the coming decades will reveal if these experiments are successful in the long run.

7.6 Foredune Projects

The situation near the foredunes is quite different from that of the inner dunes. Stress factors are much larger (wind speed, exposure, salt spray, blowing sand from the beach, coastal erosion, less biomass, no soil development). Earlier experiments with foredune remobilization proved that by means of simple interventions (removal of marram grass) the foredunes turned into transgressive dunes. In the province of North-Holland removal of vegetation and the digging of some trenches in the direction of the wind caused the foredune to move inland, with measured sand transport rates of about 45 m^3/m/year, mainly derived from redistribution of foredune sand. Similar results were obtained on the Wadden Island of Terschelling (Fig. 7.9), but here the sand input is derived from the beach as well. These examples suggest that in the foredune situation, the removal of vegetation might be enough to release huge quantities of sand that blow inland. Also, regular removal of vegetation is easy owing to the accessibility of the beach for heavy equipment.

Fig. 7.9 Remobilization of foredunes after intervention, Terschelling. *Photograph* Bas Arens

7.7 Restarting the Engine? Toward a New Management Strategy

These foredune experiments and the current spontaneous development of new blowouts and parabolics in foredunes due to "dynamic preservation" show us that the large-scale processes of dune mobility can be restored when they are connected to the coastline (see Chap. 6). Exploring different systems around the world with Google Earth shows that most systems with active parabolic dunes are in some way connected to the shore. This confirms our hypothesis that large-scale dune mobility should start at the foredune. Also, it supports our conclusion that the fixation of the coastline for 150 years significantly contributed to the stabilization of the dune landscape.

Conclusions from the literature, and local developments in The Netherlands and other sites in western Europe have convinced us of the necessity to incorporate beach and foredune dynamics into restoration projects. Be it by erosion or excessive sand supply (e.g., excessive beach nourishment), the foredunes should act as a transfer system between beach and dunes, which might be the key to sustainable development. In our situation, coastal erosion, breakdown of the foredune and remobilization of sand is considered to be the engine for large-scale dune mobility. Mobility due to climatic change is unlikely to happen within the next few decades, but may happen in the long run (>2040). Mobility due to a massive supply of sand to the beach is unrealistic. Dimensions of beach nourishments might be oversized to enable transgressive dune development, but the costs will be enormous. Better results could be achieved on parts of the coast with enough space for a natural development. Allowing foredune erosion and consequently dune remobilization at these places could "restart the engine."

However, large structures like parabolic dunes take a long time to develop (>15 years). It may take ages before the released sand reaches dunes further landward. At larger distances from the foredunes (>500 m) additional measures to maintain biodiversity will remain necessary for the coming decades (if there is a continued call for increased biodiversity).

Nourishments can be adapted to meet restoration needs. For example, nourishment could be started after a few years of foredune breaching, releasing the sand, and providing possibilities for landward transfer. Then, some extra amounts of sand could be nourished to feed the dune system. In the past, coastal defense engineers considered any landward transport of sand as a loss for the defense system. Nowadays, with sea level rise, any input of sediment into our country should be regarded as a welcome aid to keep our heads above the water.

7.8 Costs of Restoration Projects

The costs of the restoration projects largely depend on the specific situation. In the Dutch coastal dunes the costs usually imply:

1. Geomorphological research and engineering.
2. Archeological research (to prevent damage to archeological heritage).
3. Removal of explosives from World War 2.
4. Removal of soil; the depth of removal depends on the depth of soil development (presence of organic matter). Usually, all soil containing organic matter is removed.
5. Removal of vegetation; the type of vegetation determines the efforts needed. The costs of removing a forest (cutting of trees, removal of the stubs) are higher than those incurred in removing a grass land (except when the wood can be sold).
6. Communication to create local support.

Often, the costs of the transportation of soil and sand are the most expensive, depending on local infrastructure, transport distance, and marketing possibilities. For example, if the sand can be sold for other purposes, transport is often free, and the project may even be completely financed by the profits. This was the case in the Verlaten Veld, where the sand was used for construction of a noise barrier around the racing circuit of Zandvoort. Another option is the use of sand for other nature restoration projects, for example, filling in former excavations, which was the case with the Bruid of Haarlem. In foredune projects, the sand is usually not removed from the site, but spread over the beach and dunefoot. Then, the costs are limited to the hiring of the equipment. Of course, the costs depend on the scale of the project. For remobilization projects the average depth of soil and sand removal is about 0.3 m, which means that per hectare about 3,000 m^3 have to be removed (price approximately 2–8 €/m^3). However, this depends on the present vegetation. In the case of the dominance of marram grass, the removal depth increases to up to 1 m, in order to get rid of most of the roots. This increases the costs enormously. An alternative is to manually remove the roots from the area for a number of years.

7.9 Conclusion and Discussion

Coastal dunes alternate between phases of mobility and stability. A conceptual model is presented in which the main factors influencing either mobility or stability and their complex interactions are described. Climatic factors impose their influence through two of the driving forces: wind energy and vegetation. Environmental factors, often influenced by man, affect vegetation growth and sand availability in several ways. Both climate- and man-induced changes can alter the state of the system. For managers, it is important to know how they can interfere with these factors, to achieve their management goals.

The discussion of dune mobility and stability, and of environmental and climatic control has brought us to the conclusion that dunes in general reflect two different settings. In desert dunes (mostly continental dunes), there is an almost 1:1 link between dune mobility and climate. Vegetation growth is fully governed by climate (and humans). With a deterioration in climate (or increased exploitation), vegetation disappears and dunes become mobile. These dunes respond directly to climatic change. In many of the coastal dunes in the world climate is not arid or semi-arid, but temperate and/or humid. Although these dunes only can develop under the guarantee of wind, the other decisive factor is sand supply. Without sand supply, these dunes finally will stabilize. Sand supply can either be from the shoreface to the beach, resulting in burial of vegetation that cannot cope with deposition rates, or the release of previously deposited sand by erosion. Under climatic conditions with limited possibilities for plant growth, small amounts of sand suffice. Under climatic conditions favorable for vegetation growth (like ours), large amounts of sand are needed. Bare foredune cliffs 15–25 m high, in combination with the right winds, can produce these amounts of sand. The response of these systems to (delicate) climatic change is much more complicated.

In The Netherlands dune mobility is related to sand supply. Under the current conditions, a huge supply is needed to overcome fixation by vegetation. Removing vegetation alone is probably not sufficient to maintain dune mobility and consequently biodiversity. Some help from the sea will sustain a more durable development. In the case of foredune erosion, significant amounts of sand may be released and transferred inland by the wind. Allowing foredune erosion followed by dune remobilization may “restart the engine.” In other cases, where the foredune cannot be incorporated, maintenance of mobility is necessary, at least until all roots are removed from the system. However, this is no guarantee of durable mobility either: dispersal of marram from seeds may under favorable conditions stabilize bare surfaces rapidly.

A smart design of nourishments could give room for coastal erosion, creating “corridors” in the foredunes where sand can be transported inland. More knowledge is needed on the transfer mechanisms. It is clear that wave erosion releases sand for subsequent aeolian transportation, and that a very large supply of sand from the beach kills vegetation and results in mobility as well (as demonstrated by Rhind and Jones 2010, in Wales). But gentle supply leads to enhanced vegetation growth and blocks the transfer. Is there some equilibrium? Can we manipulate the system in such a way that both wave erosion and (gentle) sand supply keep the transfer function intact? Can we nourish our coastline in such a way that its position is largely preserved and the sand-sharing system of the beach and dunes is stimulated at the same time? This is an important challenge for the years to come. If we succeed, we can meet both requirements for coastal defense *and* nature management.

Acknowledgments The authors thank Marisa Martinez and Patrick Hesp for their support and comments during the writing of the manuscript.

References

Arbogast AF, Hansen EC, Van Oort MD (2002) Reconstructing the geomorphic evolution of large coastal dunes along the southeastern shore of Lake Michigan. Geomorphology 46:241–255

Arens SM, Geelen LHWT (2006) Dune landscape rejuvenation by intended destabilisation in the Amsterdam Water Supply Dunes. J Coast Res 23:1094–1107

Arens SM, Slings QL, De Vries CN (2004) Mobility of a remobilised parabolic dune in Kennemerland, The Netherlands. Geomorphology 59:175–188

Arens SM, Geelen LHWT, Slings QL, Wondergem HE (2005) Restoration of dune mobility in The Netherlands. In: Herrier J-L et al (eds) Proceedings: dunes and estuaries 2005. International conference on nature restoration practices in European coastal habitats, Koksijde, Belgium, 19–23 Sept 2005. VLIZ Special Publication, vol 19, pp 129–138

Arens SM, van Puijvelde SP, Brière C (2010) Effecten van suppleties op duinontwikkeling; geomorfologie. Rapportage fase 2. Arens Bureau voor Strand- en Duinonderzoek en Deltares RAP2010.03 in opdracht van Directie Kennis, LNV, 141 pp + bijlagen

Barbosa LM, Dominguez JML (2004) Coastal dune fields at the São Francisco river strandplain, Northeastern Brazil: morphology and environmental controls. ESPL 29:443–456

Beekman F (2007) De kop van Schouwen onder het zand. Duizend jaar duinvorming en duingebruik op een Zeeuws eiland. Uitgeverij Matrijs, Utrecht

Clemmensen LB, Murray A (2006) The termination of the last major phase of aeolian sand movement, coastal dunefields, Denmark. ESPL 31:795–808

Dekker LW, Jungerius PD (1990) Water repellency in the dunes with special reference to The Netherlands. Catena Suppl 18:173–183

De Ruig JHM, Louisse CJ (1991) Sand budjet trends and changes along the holland coast. J Coast Res 7:1013–1026

Forman SL, Oglesby R, Webb RS (2001) Temporal and spatial patterns of Holocene dune activity on the Great Plains of North America: megadroughts and climate links. Global Planet Change 29:1–29

Giepmans SE, Kos A, Van 't Zelfde R (2004) Hollandse stranden in de Gouden Eeuw. Katwijks Museum, 136 pp

Grootjans AP, Geelen LHWT, Jansen AJM, Lammerts EJ (2002) Restoration of coastal dune slacks in the Netherlands. Hydrobiologia 478:181–203

Grootjans A, Dullo BW, Kooijman A, Bekker R, Aggenbach C (2013) Restoration of dune vegetation in the Netherlands. In: Martinez ML, Gallego-Fernández JB, Hesp PA (eds) Coastal dune restoration. Springer, Berlin

Hesp PA, Thom BG (1990) Geomorphology and evolution of active transgressive dunefields. In: Nordstrom KF, Psuty N, Carter RWG (eds) Coastal dunes; form and process. Wiley, New York, pp 253–288

Hillen R, Roelse P (1995) Dynamic preservation of the coastline. J Coast Conserv 1:17–28

Hugenholtz CH, Wolfe SA (2005) Biogeomorphic model of dunefield activation and stabilization on the northern Great Plains. Geomorphology 70:53–70

Jelgersma S, De Jong J, Zagwijn WH, Van Regteren Altena JF (1970) The coastal dunes of the western Netherlands; geology, vegetational history and archaeology. Mededelingen Rijks Geologische Dienst, Nieuwe Serie 21:93–167

Jones MLM, Sowerby A, Rhind PM (2010) Factors affecting vegetation establishment and development in a sand dune chronosequence at Newborough Warren, North Wales. J Coast Conserv 14:127–137

Jungerius PD, van der Meulen F (1988) Erosion processes in a dune landscape along the Dutch coast. Catena 15:217–228

Klijn JA (1990) The younger dunes in the netherlands; chronology and causation. In: Bakker ThW, Jungerius PD, Klijn JA (eds), Dunes of the European coast; geomorphology—hydrology—soils. Catena Suppl 18:89–100

Kutiel PB (2013) Restoration of coastal sand dunes for conservation of biodiversity—the Israeli experience. In: Martinez ML, Gallego-Fernández JB, Hesp PA (eds) Coastal dune restoration. Springer, Berlin

Lancaster N (1997) Response of eolian geomorphic systems to minor climate change: examples from the southern Californian deserts. Geomorphology 19:333–347

McFadgen B (1985) Late Holocene stratigraphy of coastal deposits between Auckland and Dunedin, New Zealand. J R Soc N Z 15:27–65

McFadgen B (1994) Archaeology and Holocene sand dune stratigraphy on Chatham Island. J R Soc N Z 24:17–44

Piotrowska H (1991) The development of the vegetation in the active deflation hollows of the Leba Bar (N. Poland). Fragmenta Floristica et Geobotanica 35:172–215

Pool MA, Van der Valk L (1988) Volumeberekening van het Hollandse en Zeeuwse Jonge Duinzand. Kustgenese Project, Taakgroep 1000, Rapport BP10705, RGD, Haarlem

Provoost S, Jones MLM, Edmondson SE (2011) Changes in landscape and vegetation of coastal dunes in northwest Europe: a review. J Coast Conserv 15:207–226

Rentenaar R (1977) De Nederlandse duinen in de middeleeuwse bronnen tot omstreeks 1300. Geografisch Tijdschrift 11:361–376

Rhind P, Jones R (2009) A framework for the management of sand dune systems in Wales. J Coast Conserv 13:15–23

Stuyfzand PJ, Arens SM, Oost AP (2010) Geochemische effecten van zandsuppleties langs Hollands kust. KWR-rapport KWR 2010.048, 71 pp

Thomas DSG, Knight M, Wigs GFS (2005) Remobilization of southern African desert dune systems by twenty-first century global warming. Nature 435:1218–1221

Tsoar H (2005) Sand dunes mobility and stability in relation to climate. Phys A 357:50–56

Tsoar H, Blumberg DG (2002) Formation of parabolic dunes from barchan and transverse dunes along Israel's Mediterranean coast. ESPL 27:1147–1161

Van Boxel JH, Jungerius PD, Kieffer N, Hapelé N (1997) Ecological effects of reactivation of artificially stabilized blowouts in coastal dunes. J Coast Conserv 3:57–62

Willis AJ (1989) Coastal sand dunes as biological systems. Proc R Soc Edinb 96B:17–36

Yizhaq H, Ashkenazy Y, Tsoar H (2007) Why do active and stabilized dunes coexist under the same climatic conditions? Phys Rev Lett 98:188001–188004

Zagwijn WH (1986) Nederland in het Holoceen. Staatsdrukkerij, 's-Gravenhage, 46 pp

Chapter 8
The Impact of Dune Stabilization on the Conservation Status of Sand Dune Systems in Wales

Peter Rhind, Rod Jones and Laurence Jones

8.1 Introduction

Dunes in Wales represent one of the most natural habitats, with many sites having UK or European protection status (Fig. 8.1, Table 8.1). However, over the years much has been lost to industrial, urban, and tourist developments and large areas have been planted with commercial forests (Everard et al. 2010; UKNEA 2011a, b). This loss in quantity is further compounded by a loss of habitat quality due to increasing amounts of stabilization and soil development, with most sand dunes in Wales having virtually full plant cover (Dargie 1995) (Fig. 8.2). This issue is not restricted to sand dunes in Wales, but occurs elsewhere in the UK (Radley 1994), and indeed is part of a consistent trend across north-west Europe's temperate dune systems (Provoost et al. 2011). Overstabilization becomes a problem when the balance between early successional and later, more stable successional phases is lost. The early successional phases are important for many rare plant and invertebrate species in the UK. Furthermore, overstabilized systems lack the natural dynamic mechanisms of geomorphological processes that allow them to adapt naturally to threats such as climate change and sea-level rise.

There are many factors that influence the dynamic state of dune systems (see below), but evidence from the Holocene stratigraphic record indicates that periods

P. Rhind (✉) · R. Jones
Countryside Council for Wales Bangor, Gwynedd LL57 2DN Wales, UK
e-mail: p.rhind@ccw.gov.uk

R. Jones
e-mail: rd.jones@ccw.gov.uk

L. Jones
Centre for Ecology and Hydrology, Environment Centre Wales, Deiniol Road, Bangor LL57 2UW Wales, UK
e-mail: LJ@ceh.ac.uk

M. L. Martínez et al. (eds.), *Restoration of Coastal Dunes*, Springer Series on Environmental Management, DOI: 10.1007/978-3-642-33445-0_8,

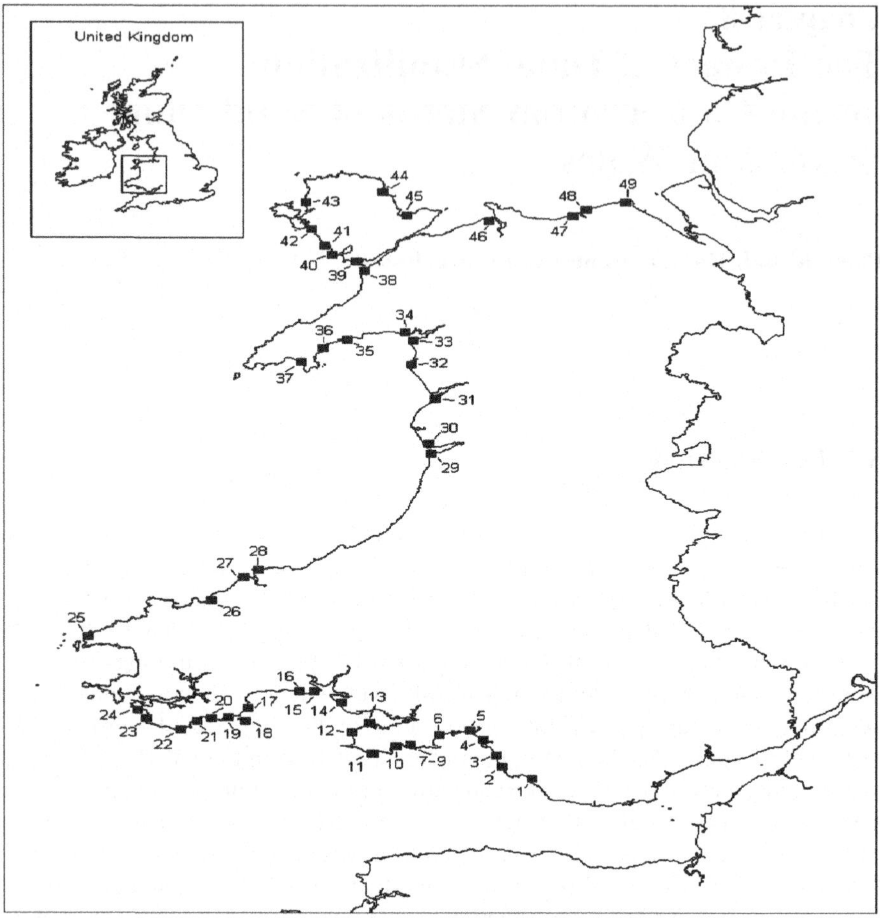

Fig. 8.1 Locations of sand dune sites in Wales. Site names are given in Table 8.1. OS base map reproduced with permission of HMSO. Crown copyright reserved. CCW license No. 100018813 (Scale 1: 1253585.5)

of warmer, wetter climate resulted in widespread dune stabilization and soil development, while relatively cold periods with lower levels of precipitation are associated with increased mobility (e.g., Wilson et al. 2001). The so-called Little Ice Age (1450–1850), for example, is well documented as a time of widespread dune mobility in northwest Europe (Pye 2001), although this period is also associated with widespread cutting of forests, over-use of dunes, and population expansion (e.g., Knight and Burningham 2011). However, today's dunes are characterized by exceptionally high levels of stability, which seems to be due to the fact that climatic and geomorphological factors have been enhanced by widespread artificial stabilization measures. On the other hand, it could be argued that the geomorphological maturity of dune systems is not reflected in the maturity

Table 8.1 Dune systems in Wales and their designations in terms of conservation status

Site Number	Site	Area (ha)	Welsh region	SSSI	NNR	SAC	LNR	AONB	HER	NP	BIO	RAM	GCR
1	Merthyr Mawr	342	Bridgend	x	x	x			x				
2	Kenfig Burrows	602	Bridgend	x	x	x							
3	Margam Burrows	101	Bridgend										
4	Baglan Bay	78	Bridgend										
5	Crymlyn Burrows	118	Swansea	x									
6	Black Pill to Bryn Mill	16	Swansea										
7–9	Pennard, Penmaen and Nicholaston	87	Swansea	x				x	x				
10	Oxwich Burrows	93	Swansea	x	x			x	x				x
11	Port Eynon to Horton	19	Swansea										
12	Hillend to Hills Tor Burrows	224	Swansea										
13	Whiteford Burrows	142	Swansea	x	x	x		x	x				x
14	Pembrey Burrows	591	Carmarthenshire	x									x
15	Laugharne Burrows	431	Carmarthenshire	x		x							x
16	Pendine Burrows	173	Carmarthenshire	x		x							x
17	Tenby Burrows	92	Pembrokeshire	x					x	x			
18	Caldey Island	3	Pembrokeshire										
19	Lydstep Haven	23	Pembrokeshire										
20	Manorbier and Swanlake Bay	10	Pembrokeshire										
21	Freshwater Bay East	17	Pembrokeshire										
22	Stackpole Warren	179	Pembrokeshire	x	x	x			x	x			
23	Brownslade and Linney Burrows	253	Pembrokeshire	x		x							
24	Broomhill Burrows	183	Pembrokeshire	x		x			x	x			
25	Whitesand Bay	28	Pembrokeshire										
26	The Bennett	20	Pembrokeshire										
27	Poppit Sands	11	Pembrokeshire										
28	Towyn Warren	30	Ceredigion	x						x			

(continued)

Table 8.1 (continued)

Site Number	Site	Area (ha)	Welsh region	SSSI	NNR	SAC	LNR	AONB	HER	NP	BIO	RAM	GCR
29	Ynylas	68	Ceredigion	x	x						x	x	x
30	Tywyn to Aberdovey	111	Gwynedd	x						x			
31	Fairbourne	15	Gwynedd										
32	Morfa Dyffryn	313	Gwynedd	x	x	x				x			x
33	Morfa Harlech	341	Gwynedd	x	x	x				x			x
34	Morfa Bychan	169	Gwynedd	x			x						
35	Pwllheli to Pen y Chain	44	Gwynedd	x									
36	Traeth Crugan	22	Gwynedd										
37	Tai Morfa	20	Gwynedd										
38	Morfa Dinlle	66.6	Gwynedd	x		x							x
39	Newborough Warren	529	Gwynedd	x	x	x		x					x
40	Penhrhnoedd—Llangadwaladr	25	Gwynedd	x									
41	Aberffraw	248	Gwynedd	x		x		x					x
42	Valley Airfield	192	Gwynedd										
43	Tywyn Gwyn	17	Gwynedd	x									
44	Traeth Dulas and Traeth Lligwy	7	Gwynedd										
45	Red Wharf Bay	6	Gwynedd										
46	Conwy and Deganwy Dunes	75	Gwynedd										
47	Kinmel Bay	11	Gwynedd										
48	Rhyl to Prestatyn	53	Gwynedd										
49	Gronant to Talacre	190	Flintshire	x		x							

SSSI site of special scientific interest, *NNR* national nature reserve, *SAC* Special Area of conservation, *LNR* local nature reserve, *AONB* area of outstanding natural beauty, *HER* heritage coast, *NP* national park, *BIO* biosphere reserve, *RAM* Ramsar site, *GCR* geological review sites

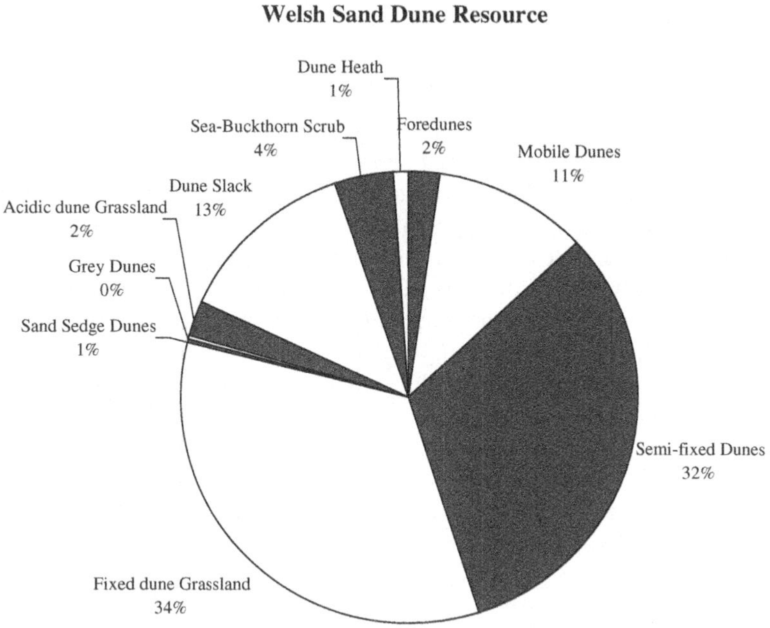

Fig. 8.2 Proportions of the major sand dune vegetation categories in Wales based on data collected in the 1980s and 1990s

of the vegetation. Grazing and other influences have held most coastal dunes systems to a sub-climax or plagio-climax, with a generally limited extent of shrub and tree cover. In all likelihood, in the absence of this interference, large parts of many dune systems may have been developing into mature woodland by now.

However, in terms of conservation we are still facing the dilemma that if we do not reverse this trend we will continue to witness the loss of suitable habitat for rare and uncommon obligate or semi-obligate psammophytes such as *Eryngium maritimum*, *Euphorbia paralias*, *Matthiola sinuata*, *Mibora minima*, *Phleum arenarium*, and *Vulpia fasciculata,* and dune slack hygrophytes, such as *Epipactis leptochila* var. *dunensis, Liparis loeselli* var. *ovata*, and *Pyrola rotundifolia,* together with many dune fungus and invertebrate species (e.g., Howe et al. 2010).

In order to try to further clarify issues relating to stabilization, and to have better information on how to deal with this problem, a detailed look at the changes to the vegetation at Newborough Warren has been compiled, together with accounts of how stabilization is affected by external factors such as sediment supply, nutrient enrichment, climate change, and predicted sea-level rise, and internal factors such as soil development, grazing, and scrub development.

8.2 Case Study Showing Evidence of Change: Newborough Warren

Rhind et al. (2001) showed that over a 50-year period (1950–1990s), the vegetation at Newborough Warren changed almost beyond recognition. Nearly 75 % of the total dune area in the 1950s consisted of mobile dunes and embryonic dune slacks with open vegetation (Fig. 8.3), but by 1991 only about 6 % of the site could be classed as mobile and open, and embryonic dune slacks were more or less non-existent. Ranwell (1955) estimated the average rate of landward dune migration at that time to be in the order of 0.3–0.6 m per year. Today, there is little evidence of any major landward migration of dunes.

At that time the main sand binding grass, *Ammophila arenaria*, was mainly confined to dune crests, the base of lee slopes and sheltered hollows. In other, more level, areas with high sand mobility, a community described as an open *Salix repens* association occurred (Fig. 8.4), often with no other associated species. Nowadays, there are no remains of this *Salix* community at Newborough Warren, and there is no recognized equivalent within the UK's National Vegetation

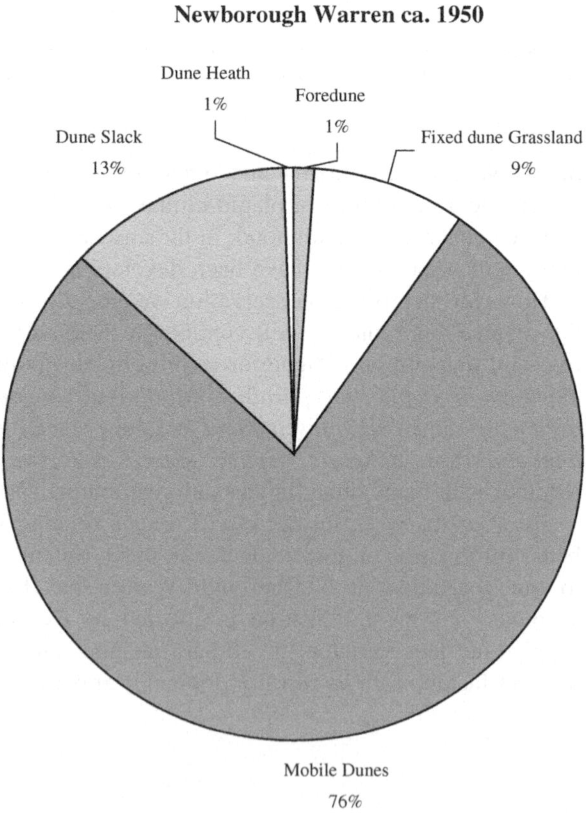

Fig. 8.3 Estimated relative proportions of the major sand dune vegetation categories at Newborough Warren around 1950, based on data from Ranwell (1960a)

Fig. 8.4 Photograph of Newborough Warren taken by Derek Ranwell in the 1950s. This type of vegetation, described as a mobile *Salix* community, no longer occurs in the UK

Classification (NVC) (Rodwell 2000). Furthermore, most of the modern mobile dune vegetation, which is still mainly dominated by *A. arenaria,* is restricted to the coastal fringe, whereas in the 1950s, mobile dune vegetation extended inland in some areas for over 1 km.

Shortly after the onset of myxomatosis (a rabbit disease caused by the deliberate introduction of the *Myxoma* virus) in 1954, virtually all of the rabbits at Newborough Warren were eliminated (Ranwell 1959, 1960b). The resulting reduction in grazing pressure gave rise to conditions much more suitable for scrub invasion and by 1981 scrub had spread over much of the Warren (Hodgkin 1984). Hawthorn (*Crataegus monogyna*) and to a certain extent birch (*Betula pendula*) were the most frequent species. No attempt to control hawthorn has been undertaken, but stands of sea buckthorn (*Hippophae rhamnoides*) and some birch have been eradicated.

8.3 External Factors Influencing Stabilization

8.3.1 Sediment Supply

Much of the sand present on coastal dunes in Britain accumulated during the Holocene, often as transgressive beaches rolling inland over Pleistocene land surfaces as the sea level rose with the melting of the ice-sheets. However, today we are facing reduced levels of primary accretion and growth. This has been ascribed to the ending of a long phase of available quantities of sand-sized sediment in shallow coastal waters at the conclusion of the post-glacial and there is evidence that sand dune geomorphology in the UK is now entering a phase of senility (Ritchie 2001).

Sediment today is mainly supplied from the offshore zone, but this is frequently supplemented by longshore transport although much of this has now been disrupted by sea defense structures. The majority of dune sites in Wales have experienced an increase in dune area over the last 100–120 years. This pattern is consistent with a more widely observed trend for net sediment accumulation within many Welsh estuaries over the same time period (Pye and Saye 2005). However, at least 27 of the 49 dune systems in Wales are subject to net erosion (Dargie 1995).

8.3.2 Nitrogen Deposition

The stabilization of dunes is thought to be at least in part attributable to the input of artificially high levels of nitrogen from atmosphere sources (Jones et al. 2004; Ketner-Oostra et al. 2006; Kooijman 2004; Remke et al. 2009). Since the onset of the industrial revolution, the levels of reactive nitrogen in the atmosphere have increased twofold (Fowler et al. 2004) and this has had a negative impact on many semi-natural habitats that are adapted to very low levels of nitrogen deposition (Bobbink et al. 1998). Dune systems are no exception, although empirical evidence of negative effects is sparse in the literature compared with other habitats. In UK dune systems, Jones et al. (2004) showed that elevated nitrogen deposition was correlated with increased above ground biomass in early successional and fixed dune habitats, and with decreased species richness in fixed dune grasslands. Soil parameters, such as the C:N ratio and available nitrogen, showed significant correlations with nitrogen deposition and soil seed banks in dune slacks may also be negatively affected (Plassmann et al. 2008). The increase in biomass has important consequences in that it increases the supply of organic matter to the soil and may be one mechanism by which nitrogen deposition can accelerate the early stages of soil development. In the Netherlands, mesocosm experiments showed that the nitrophilic grass *Calamagrostis epigejos* could quickly become the dominant species with increasing nitrogen deposition (van den Berg et al. 2005). Other studies suggest that the invasive spread of *Brachypodium pinnatum* is also linked to the deposition of atmospheric nitrogen (Bobbink and Willems 1987). This species has now become problematic on certain dune systems in Wales (Rhind 1993).

The critical load range for nitrogen on sand dune vegetation had been established at 10–20 kg N ha^{-1} $year^{-1}$ (Achermann and Bobbink 2003). On the basis of this, most dune systems in Wales approach threshold levels for atmospheric inputs alone (Jones et al. 2002a, 2002b). More recently, the potential for relatively small additional levels of nitrogen above the critical load to affect dune ecosystem processes has been shown in an experimental "nitrogen-addition" study at Newborough Warren (Plassmann et al. 2009). In fact, significant increases in moss biomass and ecosystem retention of nitrogen were observed at treatment levels within the critical load range (total deposition of 18 kg N ha^{-1} $year^{-1}$ including background deposition). Recent work in the Baltic shows that negative effects of nitrogen can be observed on acidic sites at deposition levels as low as 8 kg N ha^{-1} $year^{-1}$ (Remke et al. 2009), suggesting that critical load ranges may have to be modified downward for dune habitats. Under current revisions of the critical loads for nutrient nitrogen in European habitats, a new critical load range of 8–15 kg N ha^{-1} $year^{-1}$ is proposed for fixed dune grassland, with the lower end of the range applied to acidic sites (UNECE 2010).

In addition, background nitrogen deposition at the lower end of the critical load makes a site very vulnerable to other local sources of nitrogen, e.g., local atmospheric point sources such as poultry farms (Jones 2010), or from ground water (Jones et al. 2006; van Dijk 1989; van der Meulen et al. 2004). Additional nitrate inputs from ground water at Merthyr Mawr in South Wales, where the atmospheric inputs of nitrogen are 10.1 kg N ha^{-1} $year^{-1}$, just within the lower end of the established critical load range (Jones et al. 2005), have the potential to push parts of this site well above the critical load. In one area, nitrogen inputs in surface flooding caused by upwelling of ground water input amounted to 51.9 kg N per day. On this basis just 1 day's supply of nutrients in the ground water, if added to the annual atmospheric loading, could cause potential exceedance over an area of 4 ha. These levels of nutrient input are likely to drastically enhance rates of soil development and to accelerate vegetation succession (e.g., Ketner-Oostra 2006).

8.3.3 Sea-Level Rise

Sea level rise is an important issue in sand dune management not least because of the flood defense role it plays. Pye and Saye (2005), using an assumed sea level rise of 0.41 m by 2100 [the median value based on Intergovernmental Panel on Climate Change (IPCC) predictions at the time], have predicted that several sites in Wales are likely to experience a significant net loss of dune area/habitat over the course of the next century. However, in the light of recently revised estimates for sea level rise produced by the UK Department of the Environment, Farming, and Rural Affairs (Defra), the values used by Pye and Saye may underestimate medium- to long-term changes, suggesting even higher levels of habitat loss in the longer term. In fact, rates of sea level rise are predicted to increase to values approximating to 1.4 cm per year by 2100. Such rates have not been experienced on the Welsh coastline for some 6,000 years or so. It is therefore difficult to predict exactly how the coast will respond, but a key aim is to try and conserve the body of sand. In other words, to allow it to translate landward, but this process will be fraught with difficulties and may potentially be controversial since assets, such as golf courses, housing, farmland, and caravan parks, are often located to the rear of dune systems. On the other hand, the sea level rise will tend to increase the erosional state of the dunes and this may lead to an increase in mobile dunes over time (Carter 1991; Davidson-Arnott 2005).

8.4 Climate Change

In addition to potential sea level rise, other climate change factors that need to be considered based on recent updates from the UK Climate Impact Programme (UKCIP) are as follows:

1. Annual average temperature predicted to rise by 1–3 °C or more during the present century.
2. Warming likely to be greater in autumn and winter than in spring and summer.
3. Winter minimum temperatures predicted to rise more rapidly than maximum temperatures, reducing the diurnal range, while summer maximum temperatures predicted to rise more rapidly than minimum temperatures, increasing the diurnal range.
4. Variability of the winter temperature between years is likely to decrease, with cold winters becoming rare.
5. Variability of summer temperatures likely to increase, with very hot summers becoming more common.
6. Annual precipitation probably increasing by 3–5 % by 2050, with greater increases in winter and autumn, but with no change or a decrease in summer.
7. Year-to-year variability of seasonal precipitation likely to show changes such that the frequency of dry summers will double and wet winters treble by 2080.
8. More of the increased precipitation is likely to occur in intense storm events than at present, especially in winter.
9. Evapotranspiration is likely to increase year round, but particularly in autumn and summer.

The likely results of these interacting factors in terms of stabilization are difficult to predict, but there is clear evidence of the impact of climate on rates of stabilization at Newborough. Faster rates of vegetation colonization since the 1940s, assessed using the aerial photographic record, were associated with lower values of Talbot's Mobility Index ($M < 0.3$)—a function of windspeed, temperature, and rainfall (Talbot 1984). The rate of vegetation colonization slowed down in the late 1960 and 1970s when values of M exceeded 0.3, then increased again as the Mobility Index declined (Jones et al. 2010a). The predicted rise in temperature is unlikely to have a major impact on the distribution of individual dune species, since many in Wales, particularly plants and invertebrates, have a southern distribution and tend to be thermophilic. However, there are exceptions, such as the dune grass *Leymus arenarius,* which is close to its southern limit in Wales. Also, most dune plant species in this part of the world use the C3 photosynthetic pathway enabling them to better utilize any increase in CO_2 levels. One might expect, therefore, if all else is equal, that dune grasses will grow more rapidly as global warming proceeds (Carter 1991). Changing climatic conditions are also likely to affect soil processes and community-level vegetation development. Warmer, wetter conditions as demonstrated from the Holocene stratigraphic record of dunes and recent chronosequence evidence from Newborough are likely to favor further stabilization and soil development, and possibly increased rates of succession (Jones et al. 2007, 2008), but this may be offset or even reversed by a higher incidence of summer droughts and intense storm events.

8.5 Internal Factors Influencing Stabilization

8.5.1 Soil Development

Studies by Jones et al. (2007, 2008) trace the progress of soil development at two major dune systems in Wales (Newborough Warren and Merthyr Mawr). Soil development rates at the two sites were similar. Organic matter content (measured as % loss on ignition) was around 0.2 % in the bare sand of the mobile dunes, increasing to about 1 % when vegetation cover reached 100 %. This initial phase of soil development may take up to 40 years. There was then a steeper increase in the rate of soil development and by 60 years after the onset of stabilization, soils in dry dune habitats reached an organic matter content of around 4 %, becoming comparable to the much older soils. In dune slacks, soil development was much faster, and reached higher levels of organic matter. On average, wet dune habitats took only about 12 years to become fully vegetated and some soils attained organic matter contents of up to 5 % in under 60 years. The fact that soil development progresses faster under full vegetation cover has important implications for management and suggests that soil development can be retarded if open areas are maintained. Jones et al. (2008) suggest that these rates appear to be faster than rates for comparable habitats reported in the Netherlands and those from the UK at the beginning of the twentieth century. The influences of physico-chemical factors affecting rates of soil development were also assessed using the chronosequences developed for dry and wet dune habitats at Newborough. Large-scale factors such as temperature and atmospheric nitrogen deposition showed positive correlations with changing rates of soil development over time, while site-specific factors such as soil pH, microclimate, slope angle, and slope aspect did not have any significant effect.

8.5.2 Human Activity and Grazing

Sand dunes have been subject to centuries of human exploitation, with much of this activity being instrumental in causing extensive destabilization. One example was the practice of *A. arenaria* (marram) harvesting for thatching or for making mats, ropes, and baskets. The practice started in the sixteenth century and eventually became a thriving domestic industry. However, it led to extensive sand drift and so legislation promoting dune stabilization came into force in 1779 and 1792, but the industry went into complete decline after the Enclosure Act of 1815 prevented free access to dunes.

Much more extensive changes to dune vegetation have been caused by the grazing of domestic stock and rabbits (e.g., Ranwell 1960b; Plassmann et al. 2010). In terms of biodiversity, the levels of grazing intensity are critical. Undergrazing can lead to rapid loss of species-rich grassland, while overgrazing can eliminate sensitive, palatable species. Rabbit grazing, in particular, has played a major role in shaping the sand dune community structure since they were introduced, probably during the Roman Period some 2,000 years ago. The introduced disease myxomatosis decimated their numbers in the 1950s and this seems to have been one of the factors that initiated dune stabilization. Changes in land use, mainly abandonment of agricultural practices, have further exacerbated the problem (Provoost et al. 2011). On the other hand, there is every possibility that dune systems would be even more stable today in the absence of past human activity (e.g., Hesp and Thom 1990).

8.5.3 Scrub Development

Scrub represents part of the natural succession on European dunes, often including blackthorn (*Prunus spinosa*) and wild privet (*Ligustrum vulgare*), and these can be important for certain bird species. However, there is now major concern regarding the invasive spread of certain species, particularly sea-buckthorn (*H. rhamnoides*). This species is not native to Wales, but has now become well established on a number of dune systems, and has become particularly problematic at Merthyr Mawr in South Wales. Over 52 ha have now been cleared at substantial cost. Much of it was grubbed out using an excavator fitted with a weed rake, while herbicide foliar spraying was used on smaller specimens. Several other sites have now been identified as requiring intensive scrub control. Extensive scrub development can severely restrict dune mobility.

8.6 Conservation Management and Restoration Strategies

8.6.1 Destabilization

A fully functioning dune ecosystem would normally include a range of successional stages from pioneer, mobile dune vegetation and embryonic dune slacks through to fixed dune grassland and possibly scrub and woodland on the landward perimeters. This balance has now been lost on many dune systems with much of the early successional vegetation confined to the seaward fringes. To restore this balance it is now deemed necessary to introduce a degree of managed remobilization.

In addition to maintaining the early successional stages of dune vegetation, where many of our obligate dune species reside, our invertebrate specialists advise that up to 10 or 20 % bare sand is needed to maintain invertebrate interests. In addition to open conditions, they also require warm substrates and a range of habitats and habitat structures (Howe et al. 2010). At present we are far from achieving this target, but the future focus will be on delivering greater mobility and dynamism. However, such management can only be taken forward where it does not conflict with the flood defense role of dune systems (Everard et al. 2010). This means that efforts to increase mobility are likely to be concentrated on just a few larger sites. Historically, there has been a tendency toward "patch management" rather than having a more integrated approach. There has also been a focus on preventing both natural and man-induced erosion, which has no doubt aggravated the problems we see today.

8.6.2 Techniques for Countering Stabilization

A range of options to counter stabilization is available. Small-scale, local disturbance can be promoted by increasing the levels of grazing pressure through encouraging rabbits, or by the use of domestic stock such as cattle, sheep or ponies. Domestic stock, particularly cattle, can also have a destabilizing influence with their poaching activities. Heavy grazing pressure by domestic animals on dunes on San Miguel Island, California, for example, resulted in severe destabilization (Erlandson et al. 2005).

Destabilization can also be generated through the reactivation of stabilized blowouts. Studies (van Boxel et al. 1997) showed that for these to remain active the area of destabilization needs to be in excess of 50 m^2. Reactivations of blowouts smaller than this tend to rapidly restabilize, but there is also now some evidence (Arens and Geelen 2006) that even extensively destabilized areas (tens of hectares) are likely to restabilize within a few decades and that new measures to reduce stabilization may be required every 10 or 20 years (see Arens et al., this volume).

Larger scale, more invasive options include mechanical disturbance, such as topsoil stripping and deep ploughing (Fig. 8.5). The latter, which inverts the soil profile, burying any surface nutrients and unwanted seeds while exposing low fertility subsoil, has been trialed at Talacre Warren in North Wales (Jones et al. 2010b). Soil cores prior to deep ploughing showed that the organic layer extended to a depth of 25–40 cm with high organic matter content (4.7 %) at the surface. After deep ploughing, the soil profile was successfully inverted with the organic layer buried under 45 cm of pure mineral sand, and soil pH in the surface layers was raised by around 0.5 pH unit, closer to that of the mineral sand from depth (Fig. 8.6). These soil conditions should facilitate the establishment of typical dune species and hinder the re-invasion of nitrophiles. Recent results show a degree of success in providing conditions conducive to natural sand movement, and bare sand cover remained at >70 % for the first 15 months. However, as studies in the

Fig. 8.5 Deep ploughing operation at Talacre Warren dune system in North Wales

Netherlands (Arens and Geelen 2006) have also shown, regeneration from deep roots or rhizomes of perennials can be a problem, and winnowing of sand from the ploughed surface brings the buried soil layers closer to the surface. Subsequent vegetation growth has showed a dominance of *Equisetum arvense* and *Rubus caesius*, as well as naturally recolonizing dune species such as *Erodium maritimum* and *Carex arenaria*. Depending on the aims of restoration, it may be desirable to immediately plant with *A. arenaria* (marram grass) following topsoil inversion to stabilize the bare sand surface for vegetative colonization by early successional species. The organization Landlife, which specializes in topsoil restoration, also recommends spraying with herbicide 14 days prior to ploughing; then again 28 days after ploughing to control deep-rooted perennials and prevent regeneration from root fragments or rhizomes (Landlife 2008).

Beach nourishment is another solution and, although expensive, its value has been clearly demonstrated at Talacre Warren. Here in 2003, 150,000 m^3 of dredged sand was pumped on to the foreshore along a section of eroding dunes. The material was made available from navigation dredging operations in the nearby Dee Estuary, thereby reducing the cost. New foredunes are now developing and blowouts are now channeling the sand into hind dune areas, creating new mobile dunes. However, the lack of sand supply per se does not appear to be the primary reason for stabilization, since a number of sites that still have positive sediment budgets have undergone stabilization.

Another possible technique is the use of controlled burning. This has the advantage of potentially removing above ground biomass and its associated nutrients without the use of heavy machinery. An accidental fire at Newborough Warren in North Wales resulted in a marked improvement in the structure and composition of the affected vegetation (Rhind and Sandison 1999).

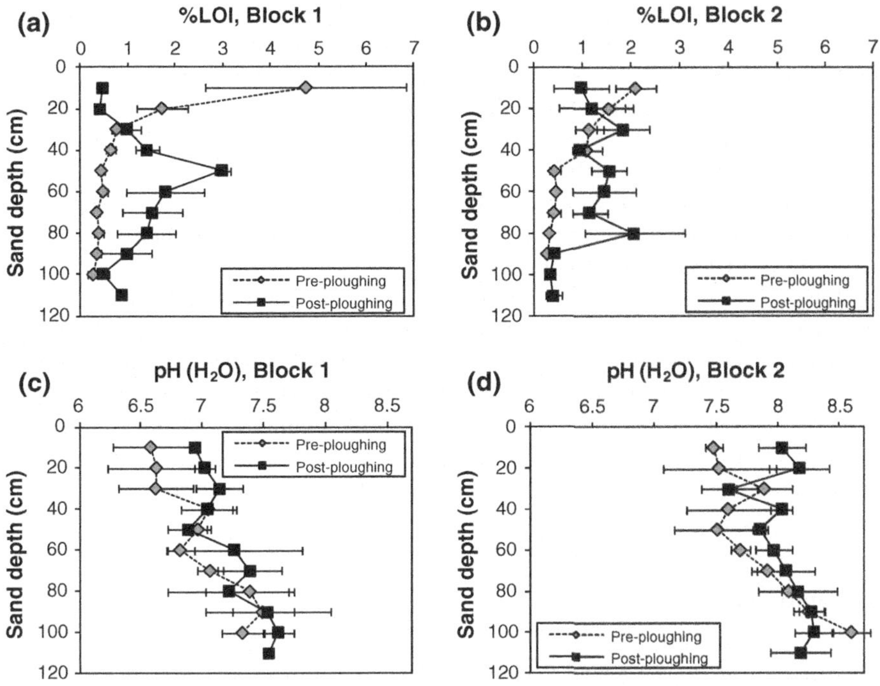

Fig. 8.6 Soil parameters with depth pre-ploughing (*gray diamonds*) and 1 month post-ploughing (*black squares*), showing Block 1 on *left-hand* side, Block 2 on the *right-hand* side respectively for: **a**, **b** Organic matter content (%LOI), **c**, **d** soil pH (measured in de-ionized water). *Bars* show ±1 s.e. Adapted from Jones et al. (2010b)

8.7 Conclusions

Dune systems in Wales are clearly on a path of increasing stability and current management prescriptions, such as recently introduced grazing, are failing to arrest this change. It is assumed that some of the key drivers of this change, such as climate change, raised levels of N deposition and raised CO_2 will continue to favor increased biomass production. The recorded build-up of organic matter in the soil will reduce the likelihood of dunes becoming more mobile and increase the need for anthropogenic intervention to both sustain current levels of mobility and to possibly return to even greater levels of mobility on some key sites. The impact of sea level rise in the short term is unlikely to induce substantial mobility, while in the long term its impact on stability is dependent on a wide variety of factors, such as changes to long shore drift associated with sea defenses and changing wave exposure associated with bathymetric changes, making accurate predictions problematic.

Work to date has failed to quantify the relative contribution of different mechanisms driving stabilization, such as enhanced nutrient inputs or modified sediment transfer associated with climatic or sediment source issues. However, management intervention cannot wait for the development of a full understanding of the relative controls of dune mobility. Rather, a carefully monitored remobilization program needs to be pursued. In addition to safeguarding the species dependent on mobile dunes, this will no doubt provide additional understanding of factors affecting mobility. It seems that dunes have moved from being possibly unnaturally mobile in the 1950s to a state of stability today that has been driven at least in part by unnatural factors. If this process had not been influenced by humans and represented part of the natural evolution of dune systems, we would not be recommending intervention.

However, since this is clearly not the case, we would argue that there is an urgent requirement to partly reverse this trend or we will continue to witness the loss of suitable habitat for rare and uncommon species. Our strategic goal, therefore, is to increase levels of dune mobility on a range of targeted sites, but this will need to be supported by clear scientific rationale so that both regulators and stakeholders can appreciate why this is necessary. However, this will only be successful if resources are made available to compensate those affected by such measures.

References

Achermann B, Bobbink R (2003) Empirical critical loads for nitrogen. Proceedings of expert workshop held in Berne, Switzerland, 11–13 Nov 2002. Environmental Documentation No.164. Air., Swiss Agency for the Environment, Forests and Landscape SAEFL, Berne

Arens SM, Geelen LHWT (2006) Dune landscape rejuvenation by intended destabilization in the Amsterdam water supply dunes. J Coast Res 22:1094–1107

Arens SM, Slings QL, Geelen LHWT, Van der Hagen HGJM (2012) Restoration of dune mobility in The Netherlands. In: Martínez ML, Gallego-Fernandez JB, Hesp PA (eds) Coastal dune restoration. Ecological Series, Springer-Verlag (in press)

Bobbink R, Willems JH (1987) Increasing dominance of *Brachypodium pinnatum* (L.) Beaur. in chalk grasslands: a threat to a species-rich ecosystem. Biol Conserv 40:301–314

Bobbink R, Hornung M, Roelofs JGM (1998) The effects of air-borne nitrogen pollutants on species diversity in natural and semi-natural European vegetation—a review. J Ecol 86:717–738

Carter RWG (1991) Near-future sea level impacts on coastal dune landscapes. Landscape Ecol 6:29–39

Dargie TCD (1995) Sand dune vegetation survey of Great Britain. A national inventory. III. Wales, Joint Nature Conservation Committee, Peterborough

Davidson-Arnott RGD (2005) Conceptual model of the effects of sea-level rise on sandy coasts. J Coast Res 21:1166–1172

Erlandson JM, Rick TC, Peterson C (2005) A geoarchaeological chronology of Holocene dune building on San Miguel Island, California. Holocene 15:1227–1235

Everard M, Jones L, Watts B (2010) Have we neglected the societal importance of sand dunes?—an ecosystem services perspective. Aquat Conserv MarFreshwat Ecosyst 20:476–487

Fowler D, O'Donoghue M, Muller JBA, Smith RI, Dragosits U, Skiba U, Sutton MA, Brimblecombe P (2004) A chronology of nitrogen deposition in the UK between 1900 and 2000. Water Air Soil Pollut Focus 4:9–23

Hesp PA, Thom BG (1990) Geomorphology and evolution of transgressive dunefields. In: Nordstrom K, Psuty N, Carter W (eds) Coastal dunes: processes and morphology. Wiley, Chichester, pp 253–288

Hodgkin SE (1984) Scrub encroachment and its effects on soil fertility on Newborough Warren, Anglesey Wales. Biol Conserv 29:99–119

Howe MA, Knight GT, Clee C (2010) The importance of coastal sand dunes for terrestrial invertebrates in Wales and the UK with particular reference to aculeate Hymenoptera (bees, wasps and ants). J Coast Conservat 14(2):91–102

Jones MLM (2010) Impacts of an intensive poultry unit on a sand dune SAC. Centre for Ecology and Hydrology, Bangor. Project No: C02819

Jones MLM, Reynolds B, Stevens PA, Norris D, Emmett BA (2002a) Changing nutrient budget if sand dunes: consequences for nature conservation interest and dunes management. I: A review. Centre for Ecology and Hydrology, Bangor. CCW Contract Science Report no. 566a

Jones MLM, Hayes F, Brittain SA, Haria S, Williams PD, Ashenden TW, Norris DA, Reynolds, B (2002b) Changing nutrient budget if sand dunes: consequences for nature conservation interest and dunes management. II. Field Survey. Centre for Ecology and Hydrology, Bangor. CCW Contract Science Report no. 566b

Jones MLM, Wallace HL, Norris D, Haria SA, Jones RE, Rhind PM, Reynolds BR, Emmett BA (2004) Changes in vegetation and soil characteristics in coastal sand dunes along a gradient of atmospheric nitrogen deposition. Plant Biol 6:598–605

Jones MLM, Pilkington MG, Healey M, Norris DA, Brittain SA, Tang SY, Jones M, Reynolds B (2005) Determining a nitrogen budget for Merthyr Mawr sand dune system. Centre for Ecology and Hydrology. Countryside Council for Wales Review of Consents Report Number: 14

Jones MLM, Reynolds B, Brittain SA, Norris DA, Rhind PM, Jones RE (2006) Complex hydrological controls on wet dune slacks: the importance of local variability. Sci Total Environ 372:266–277

Jones MLM, Sowerby A, Wallace HA (2007) Better understanding of soil resources—dune stabilisation and rates of soil development on welsh dune systems. Final Report to Countryside Council for Wales. March 2007, Centre for Ecology and Hydrology, Bangor

Jones MLM, Sowerby A, Williams DL, Jones RE (2008) Factors controlling soil development in sand dunes: evidence from a coastal dune soil chronosequence. Plant Soil 307(1–2):219–234

Jones MLM, Sowerby A, Rhind PM (2010a) Factors affecting vegetation establishment and development in a sand dune chronosequence at Newborough Warren North Wales. J Coast Conserv 14(2):127–137

Jones MLM, Norman K, Rhind PM (2010b) Topsoil inversion as a restoration measure in sand dunes, early results from a UK field-trial. J Coast Conserv 14(2):139–151

Ketner-Oostra R (2006) Lichen-rich coastal and inland sand dunes (Corynephorion) in the Netherlands: vegetation dynamics and nature management. PhD Thesis, Wageningen University and Research Centre

Ketner-Oostra R, van der Peijl MJ, Sykora KV (2006) Restoration of lichen diversity in grass-dominated vegetation of coastal dunes after wildfire. J Veg Sci 17:147–156

Knight J, Burningham H (2011) Sand dune morphodynamics and prehistoric human occupation in NW Ireland. The Geological Society of America Special Paper 476

Kooijman AM (2004) Environmental problems and restoration measures in coastal dunes in the Netherlands. In: Martinez ML, Psuty NP (eds) Coastal dunes, ecology and conservation.Ecological Studies vol 171. Springer, Berlin, pp 243–258

Landlife (2008) Soil Inversion Works. Landlife, National Wildflower Centre, Court Hey Park, Liverpool

Plassmann K, Brown N, Jones MLM, Edwards-Jones G (2008) Can atmospheric input of nitrogen affect seed bank dynamics in habitats of conservation interest? The case of dune slacks. Appl Veg Sci 11:413–420

Plassmann K, Edwards-Jones G, Jones MLM (2009) The effects of low levels of nitrogen deposition and grazing on dune grassland. Sci Total Environ 407:1391–1404

Plassmann K, Jones MLM, Edwards-Jones G (2010) Effects of long-term grazing management on sand dune vegetation of high conservation interest. Appl Veg Sci 13:100–112

Provoost S, Jones MLM, Edmondson SE (2011) Changes in landscape and vegetation of coastal dunes in northwest Europe: a review. J Coast Conserv 15:207–226

Pye K (2001) Long-term geomorphological changes and how they may affect the dunes coasts of Europe. In: Houston JA, Edmondson SE, Rooney PJ (eds) Coastal dune management. Shared experience of European conservation practice. Proceedings of the European symposium coastal dunes of the Atlantic biogeographical region Southport, northwest England, Sept 1998, Liverpool University Press

Pye K, Saye S (2005) The geomorphological response of welsh sand dunes to sea level rise over the next 100 years and the management implications for SAC and SSSI Sites. Countryside Council of Wales Contract Science Report No 670

Radley GP (1994) Sand dune vegetation survey of Great Britain. A national inventory. Part 1: England Joint Nature Conservation Committee, Peterborough

Ranwell DS (1955) Slack vegetation, dune development and cyclical change at Newborough Warren, Anglesey. Ph.D Thesis, University of Wales, Bangor

Ranwell DS (1959) Newborough Warren, Anglesey I. The dune system and dune slack habitat. J Ecol 47:571–601

Ranwell DS (1960a) Newborough Warren, Anglesey II. Plant associes and succession cycles of the sand dune and dune slack vegetation. J Ecol 48:117–141

Ranwell DS (1960b) Newborough Warren, Anglesey III. Changes in the vegetation on parts of the dune system after the loss of rabbits by myxomatosis. J Ecol 48:385–395

Remke E, Brouwer E, Kooijman AM, Blindow I, Esselink H, Roelefs JGM (2009) Even low to medium nitrogen deposition impacts vegetation of dry, coastal dunes around the Baltic Sea. Environ Pollut 157:792–800

Rhind PM (1993) The spread of *Brachypodium pinnatum* in Wales. Bot Soci Brit Isles News 64:25–28

Rhind PM, Sandison W (1999) Burning the Warren—Peter Rhind and Wil Sandison explore deliberate burning as a management tool for dune grasslands. Enact 7:7–9

Rhind PM, Blackstock TH, Hardy HS, Jones RE, Sandison W (2001) The evolution of Newborough Warren dune system with particular reference to the past four decades. In: Houston JA, Edmondson, SE, Rooney PJ (eds) Coastal dune management. Shared experience of European conservation practice. Proceedings of the European symposium coastal dunes of the Atlantic Biogeographical Region Southport, northwest England, Sept 1998, Liverpool University Press

Ritchie W (2001) Coastal dunes: resultant dynamic position as a conservation managerial objective. In: Houston JA, Edmondson SE, Rooney PJ (eds) Coastal dune management. Shared experience of European conservation practice. Proceedings of the European symposium coastal dunes of the Atlantic biogeographical region Southport, northwest England, Liverpool University Press, Sept 1998

Rodwell JS (ed) (2000) British plant communities. Maritime communities and vegetation of open habitats. vol 5. Cambridge University Press, Cambridge

Talbot MR (1984) Late Pleistocene rainfall and dune building in the Sahel. Palaeoecol Afr 16:203–214

UKNEA (2011a) The UK National Ecosystem Assessment. Chapter 11: The coastal margins. March 2011. Defra

UKNEA (2011b) The UK National Ecosystem Assessment. Chapter 20: Wales synthesis. March 2011. Defra

UNECE (2010) Empirical critical loads and dose-response relationships. CCE report of working group on effects. Recent results and updating of scientific and technical knowledge. ECE/EB.AIR/WG.1/2010/14

van Boxel JH, Jungerius PD, Kieffer N, Hampele N (1997) Ecological effects of reactivation of artificially stabilised blowouts in coastal dunes. J Coast Conserv 3:57–62

van den Berg LJL, Tomassen HBM, Roelofs JGM, Bobbink R (2005) Effects of nitrogen enrichment on coastal dune grassland: a mesocosm study. Environ Pollut 138:77–85

van der Meulen F, Bakker TWM, Houston JA (2004) The costs of our coasts: examples of dynamic dune management from Western Europe. In: Martinez ML, Psuty P (eds) Coastal dunes. Ecology and conservation. Springer, Berlin (Ecol Stud 171:259–277) van Dijk HWJ (1989) Ecological impact of drinking-water production in Dutch coastal dunes. In: van der Meulen F, Jungerius PD, Visser J (eds) Perspectives in Coastal Dune management. Proceedings of the European Symposium, Leiden, The Netherlands. SPB Academic Publishing, The Hague

Wilson P, Orford JD, Knight J, Braley SM, Wintle AG (2001) Late-Holocene (post-4,000 years BP) coastal dune development in Northumberland, northeast England. Holocene 11:215–229

Chapter 9
Restoration of Andalusian Coastal Juniper Woodlands

J. C. Muñoz-Reinoso, C. Saavedra Azqueta and I. Redondo Morales

9.1 Introduction

Maritime juniper *Juniperus oxycedrus* L. subsp. *macrocarpa* (Sibth. & Sm.) Ball 1878 woodlands represent the late successional stage of the outer sandy dunes and cliffs of the Mediterranean coasts. They constitute a singular vegetation type in an environment that has been traditionally seen as a place for socio-economic development, mainly urbanization and tourism. Owing to the reduction of its distribution, its habitats were included in the 92/43 EU Habitat Directive (2250 Coastal dunes with *Juniperus* spp.) as prioritary (Anon 1992). Although some other *Juniperus* species are included in this type of habitat, maritime juniper may be the most characteristic of the Mediterranean sandy dunes and the most threatened.

This chapter reviews the ecology and conservation problems of the coastal juniper woodlands of the maritime juniper in Andalusia (southern Spain), shows the restoration efforts carried out by the Regional Government with regard to their conservation, and focuses on the research work done within one typical coastal juniper population showing the problems of those woodlands along the Andalusian coastal dunes.

J. C. Muñoz-Reinoso (✉)
Dpto. Biología Vegetal y Ecología, Universidad de Sevilla,
Apdo. 1095 41080 Sevilla, Spain
e-mail: reinoso@us.es

C. Saavedra Azqueta
Jardín Botánico Dunas del Odiel (RAJBEN, Consejería de Medio Ambiente),
Huelva, Spain
e-mail: jbotanico.dunasodiel.cma@juntaodiel.es

I. Redondo Morales
Consejera Técnica de la Dirección General de la Red de Espacios Naturales,
Avda. Manuel Siurot 50 41071 Sevilla, Spain
e-mail: isabelm.redondo@juntadeandalucia.es

M. L. Martínez et al. (eds.), *Restoration of Coastal Dunes*, Springer Series on Environmental Management, DOI: 10.1007/978-3-642-33445-0_9,

Fig. 9.1 Maritime juniper woodland in the dunes of Doñana (Almonte, Huelva)

9.2 The Maritime Juniper

From the Cupressaceae family, the maritime juniper is a dioecious, long-lived tree around 5–9 m tall, which usually grows in pyramidal or hemi-spherical shapes in the dunes, with branches developing from its base. Maritime juniper leaves are triverticillate, prickly, with a hard cuticle, and show two characteristic white lines on the face. It flowers from October to January, while fertilization takes place in May/June, and the cones are ripe in September/October of the second year (Arista et al. 2001). Their fleshy cones, called galbules, are relatively large (mean 15.5 mm), and give the name to that subspecies. The galbules, covered by pruina, have a strong scent that attracts small omnivorous species such as red foxes, badgers and wild boars that disperse lots of juniper seeds. Rabbits also eat the galbules and may disperse juniper seeds; those have also been found in the feces of some birds (blackbirds, thrushes).

Maritime juniper *Juniperus oxycedrus* subsp. *macrocarpa* has been considered a shrub, because in the dunes individuals reach a maximum height of only 3–5 m; however, in situations sheltered from the coastal environment these plants may grow up to 10 m high; thus, maritime junipers should be considered the upper stratum of the coastal woody community, and their formations as woodlands (Fig. 9.1).

Maritime juniper has a circum-Mediterranean distribution, from southwest Spain and North Africa to the Middle East, including large Mediterranean islands such as Mallorca, Sardinia, Corsica, Sicily, and Crete, and small ones such as Chrisi and Gavdos (Greece). In Spain, besides Andalusian populations, remnant populations survive in Mallorca and along the coast of the Valencia Region (Mayoral and Gómez 2003). This constrained distribution to coastal areas may be due to its high sensitivity to cool air temperatures (Rubio-Casal et al. 2010). In those areas, maritime juniper withstands the intense effect of coastal environmental stresses, such as sand mobility, salt spray deposition, relatively infertile soils, and low water availability (Crawford 1989; Brown and McLachlan 1990), and may be considered a stress-tolerant species (Grime 1979). Moreover, maritime junipers have shown vigorous growth as a response to burial in greenhouse experiments (Muñoz-Reinoso et al. 2000), which seems to demonstrate its adaptation to living on mobile dunes. Sprouting from roots has been observed in maritime junipers growing on sandy cliffs, but the vegetative multiplication by layering from decumbent branches as observed in *J. communis* (Clifton et al. 1997) have not been registered. Thus, coastal woodlands of maritime juniper represent the late successional stage in the harsh environment of Mediterranean sand dunes and cliffs.

9.3 Andalusian Coastal Juniper Woodlands

In Andalusia, maritime juniper woodlands (locally known as *enebrales*) are distributed along the Gulf of Cádiz, between El Rompido and Tarifa, in the provinces of Huelva and Cádiz respectively (Fig. 9.2). The climate of this coast is of a Mediterranean type with oceanic influences. The average temperature is 17 °C, with temperate winters (9 °C) and high temperatures (25.2 °C) in the summer. Mean precipitation is 600 mm, increasing from west to east (495.7 mm in Punta Umbría to 841.8 mm in Barbate). Highest precipitation occurs in November through January, with a second peak in spring (March/April). Summer drought is severe, with no rains during July and August, and scarce rainfall in June and September.

Three types of maritime juniper woodlands have been recognized along the western coast of Andalusia (Muñoz-Reinoso 2003):

1. Juniper woodlands on sandy dunes
2. Juniper woodlands on cliffs
3. Juniper woodlands as the understory of *Pinus pinea* plantations

Juniper woodlands on sandy dunes and cliffs may be considered the remnants of natural stands that should cover the sandy coast of Huelva and Cádiz. However, junipers appearing as the understory of pine plantations are the result of past land management practices carried out in order to stabilize the dunes since the second half of the nineteenth century (Granados and Ojeda 1994). In the present work, we will pay special attention to juniper woodlands on sandy dunes and under pine plantations.

Nowadays, the distribution of maritime juniper along the Atlantic coast of Andalusia is fragmented in heterogeneous populations of different sizes, coastal physiography, and management. The whole population has been estimated to comprise 25,000 individuals, who are concentrated mainly in the dunes of Doñana, the Chiclana-Conil coast, and La Breña Natural Park (Muñoz-Reinoso 2003). This relatively large number of individuals is not a guarantee of the conservation of the species if the habitats are not properly preserved and managed. Other populations are much smaller and more isolated, especially in Huelva province.

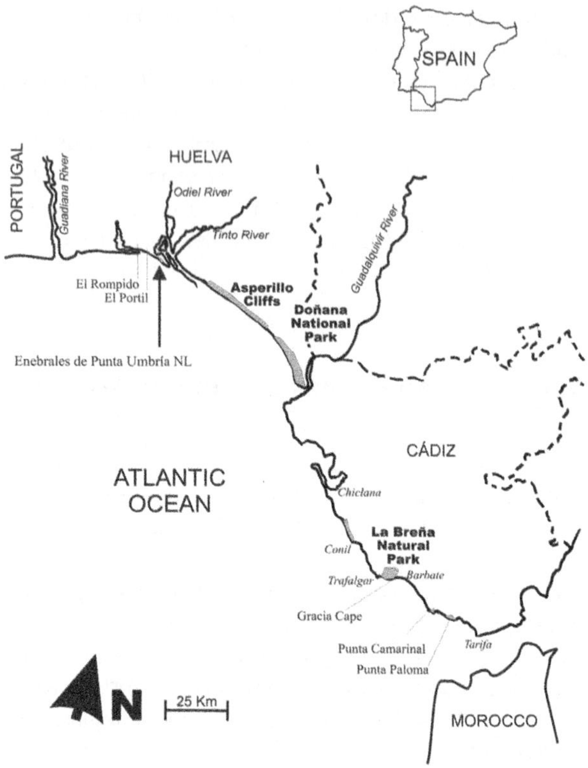

Fig. 9.2 Distribution of maritime juniper along the Gulf of Cádiz

Age–size distribution of the maritime juniper populations showed that those located on sandy dunes, with a certain degree of mobility, a substratum rich in calcium carbonate, and a high salt spray deposition, have high recruitment rates, with large proportions of young individuals, especially within protected, large or medium-sized populations (Muñoz-Reinoso 2003). Conversely, in juniper populations under a coastal pine tree plantation, especially in those from Huelva province, there is a lack of recruitment, showing only adult individuals. Several factors acting in a synergistic way may be responsible for that lack of natural regeneration. Among them, the deposition of pine needles, which prevent seedling establishment, competition with inland species (Muñoz-Reinoso 2003), and deficient pollination (Ortiz et al. 1998) have been singled out as potential inhibitors of juniper recruitment.

Most of the juniper populations along the Gulf of Cádiz show an unbiased sex ratio distribution. However, some populations close to the Gibraltar Strait show a male-biased ratio. This is hypothesized to be due to a stressful environment (Muñoz-Reinoso 2003), with strong eastern winds that may promote an increase in evapotranspiration and high water stress, which may favor male individuals owing to a lower cost of pollen production (Allen and Antos 1993).

As characteristic communities of coastal dunes (Crawford 1989; Goldstein et al. 1996), juniper woodlands are structurally open communities, with only a few woody plant species. Those juniper communities show variations in the composition of plant species over the Mediterranean Basin (Géhu et al. 1990). These variations are also apparent along the southwestern coast of Spain, where several factors seem to control the woody plant composition of juniper communities on different spatial scales (Muñoz-Reinoso 2003). On a large scale, climate and soil texture play an important role in controlling the soil water availability for plants, and in separating the xerophytic communities of Huelva, with *Halimium commutatum* and *Stauracanthus genistoides* among others, from the mesic communities of Cádiz, with *Chamaerops humilis*, *Phillyrea angustifolia*, *Olea europaea*, and *Quercus coccifera*.

On a smaller scale, sand mobility, pH values, and calcium carbonate content allow plant communities of the sand dunes and cliffs to be distinguished. Sandy dunes showed a moderate to high sand mobility, higher pH values, and high content in calcium carbonate, while cliffs showed a higher proportion of silt. Dunes were characterized by species such as *Ononis natrix*, *Sideritis arborescens*, and *Clematis flammula*, while *Cistus ladanifer*, *C. crispus* or *Ulex australis* characterized cliff communities.

Finally, the coastal pine plantations play an important role in modifying local environmental conditions, mainly decreasing sand mobility and salt spray deposition, which induces important changes in plant communities, such as the replacement of *Armeria pungens*, *Artemisia chritmifolia* or *Scrophularia frutescens* with *Halimium halimifolium*. Tekke and Salman (1995) have referred to the effects of pine plantations on the landscape, geomorphology, soils, and ecology of coastal ecosystems. Sand stabilization (the final purpose of pine plantations) results in the disappearance of the species adapted to a moving substrate, and the reduction in salt spray deposition under the pine tree canopy. Thus, these pine plantations allow colonization by species that are less tolerant to coastal stress, such as *Rhamnus lycioides*.

Differences in controlling environmental factors are not only shown by plant composition, but also by the richness and diversity of the species (Muñoz-Reinoso 2004). The harsher environment of coastal dunes produces a low diversity of woody species, and juniper communities are characterized by *Helichrysum picardii*, *Armeria pungens*, *Thymus carnosus*, *Sideritis arborescens*, and *Corema album*. On the cliffs, the elevation of the coast reduces the environmental stress promoted by the salt spray; then, the richness and diversity of the species increase and the composition of plant species is mainly controlled by lithology and climate (water availability).

Pinus pinea plantations attenuate the effect of the harsh coastal physical environment, causing the disappearance of endemic juniper woodland vegetation and promoting the establishment of landward species. Salt spray deposition falls off to less than 20 % of windward deposition (Fig. 9.3, Muñoz-Reinoso unpublished), and species of stabilized dunes such as *Halimium halimifolium* and *Rosmarinus officinalis* replace the species adapted to a moving substrate.

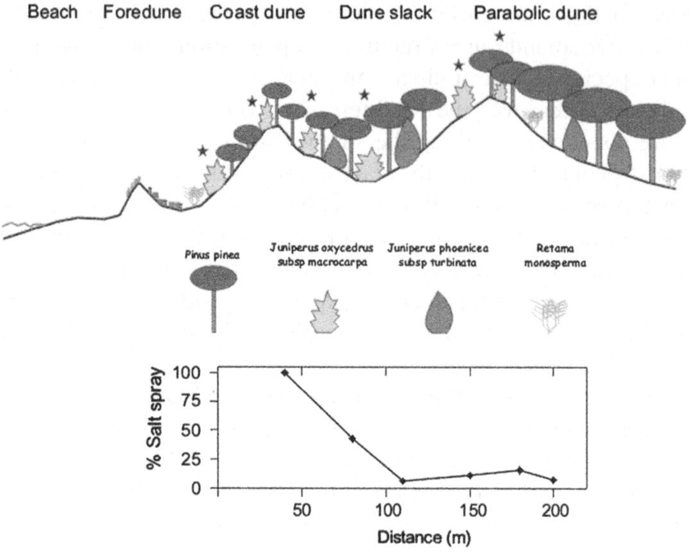

Fig. 9.3 Schematic representation of a transect in *Enebrales de Punta Umbría* (*Huelva*). *Graph* shows the decrease in salt spray deposition along the transect due to the pine tree plantation. *Asterisks* show the positions from which samples were taken for salt spray deposition

On the other hand, plant diversity of coastal pine plantations depends on the management of both the pine trees and the understory. Where pine trees are managed for wood and pine seed production or conservation proposals, and maritime juniper is favored over other woody species as understory, plant diversity is high, as in *La Breña* and *Enebrales de Punta Umbría*. However, removal of the understory or the abandonment of the exploitation often result in communities poor in plant species and of low diversity, as in the *El Portil* and *Punta Paloma* pine plantations (Muñoz-Reinoso 2004).

9.4 Threats to the Juniperus Woodlands

During the last century, coastal juniper woodlands have almost been destroyed or deeply disturbed by human activities. Nowadays, urban development and associated tourist pressures are still the main threat to coastal habitats where maritime junipers survive. Coastal urban development directly destroys the habitat, and produces fragmentation and isolation of juniper populations, interrupting ecological processes (Fig. 9.4). Thus, fragmentation, loss of connectivity (Muñoz-Reinoso et al. 2005; Puerto-Marchena et al. 2009), and intense anthropic use in small juniper stands have reduced the presence of mammals species that disperse juniper seeds.

Fig. 9.4 Maritime junipers in the garden of a detached house on the Chiclana-Conil coast (Cádiz), the oldest one has been pruned

Since the end of the nineteenth century, the plantation of pine trees (*Pinus pinea* mainly) and one-seed brooms (*Retama monosperma*) along the southwestern coast of Spain (Granados and Ojeda 1994) has been one of the main causes of juniper woodland decline. Pine plantation implied a radical change in the physical environment and in the plant composition of coastal communities, reducing seedling recruitment and increasing juniper mortality by pine needles deposition. Nowadays, most of those plantations are abandoned, without any exploitation, and juniper populations surviving as understory show a lack of natural regeneration.

The germination rate of maritime juniper seems to be low, as in other juniper species (Pack 1921; Young et al. 1988), although appropriated treatments may largely increase seed germination. Ortiz et al. (1998) showed that maritime junipers growing in a coastal pine plantation had a high proportion of empty seeds, which was explained as being due to xeric conditions and resource limitation, although the sex ratio and the spatial location of mates could also be important for pollination.

Other threats related to tourist pressure are the invasion by alien species, coastal erosion due to uncontrolled access by foot and motor vehicles, an increase in wild fires, the over-browsing by wild or domestic herbivorous mammals, and the eutrophication of sandy soils.

Given the endangered status of maritime juniper and its habitat, the recovery of juniperus woodlands from abandoned coastal pine plantations seems to be a necessary management strategy for the conservation of *Juniperus oxycedrus* subsp. *macrocarpa* woodlands in southwest Spain. If we consider the juniper woodlands of the dunes as natural reference situations (Tekke and Salman 1995), the results obtained by Muñoz-Reinoso (2003, 2004) suggest that the management of coastal pine plantations should focus on the restoration of the original environmental conditions. Pine tree clearing may allow wind flow, moderate sand mobility, and an increase in the effect of salt spray on the plant community favoring the establishment of stress-tolerant local endemic species. After tree felling, plant material (trunks and needles) must be removed as soon as possible to minimize the input of nutrients into sandy soils.

9.5 Conservation Program of Maritime Juniper Woodlands in Andalusia

Owing to its continuing decline, the Andalusian Catalogue of Endangered Wild Flora Species (Law 104/1994), and the Law 8/2003 on the Andalusian Wild Flora and Fauna included the maritime juniper as a species at risk of extinction. This implied the elaboration of a Recovery Plan for the species, which was materialized in the "Conservation Program of Maritime Juniper Woodlands (2002–2006)." The general purposes of the Plan were:

1. To guarantee the protection of extant populations and minimize the factors of threat,
2. To enable the growth of maritime juniper populations and the occupancy of both past and potential areas,
3. To obtain current information on the conservation status of the species,
4. To promote research lines that allow the redefinition of future conservation and recovery strategies,
5. To increase awareness of local groups of the problems, benefits, and necessity of maritime juniper conservation.

The Program included several actions on habitat restoration and complementary works of research, monitoring, dissemination of information, and environmental education. Habitat restoration activities comprise plant treatment (mainly pine plantations) in stands with junipers, juniper plantations, and the construction of new infrastructures for public use. Conservation measures affected 320 ha of protected and public lands belonging to 12 municipalities along the Gulf of Cádiz included in the past and potential distribution area of maritime juniper. Seventy thousand saplings (1–2 years old) were planted out from October to January in 2004/2005 and 2005/2006 in order to reduce the risk of genetic isolation and the progressive decline in juniper density of several populations. Saplings came from the Regional Government Nursery Net and were managed taking into account the locality of origin. A previous experience shows that planting depth is a critical factor and the saplings were buried to half of their height in order to be closer to the phreatic level. Tree guards of plastic mesh (60 cm in height, 21 cm in diameter) were used to reduce rabbit predation; the plastic mesh also avoided an overheating effect.

Survival monitoring of saplings showed a mean of 35 % with a high variation depending on local conditions. Adverse climatic conditions during the second hydrological year (less than 200 mm of annual precipitation) caused a high decrease in survival rate (46 % $\pm$ 10). On the other hand, the presence of a canopy of pine trees was identified as a favorable factor for sapling survival by reducing radiation and temperature in the initial phase of establishment. However, a long-term objective should be to substitute the pine plantation.

As noted above, coastal pine plantations change the physical environment, plant composition, and diversity of juniper communities, and affect mature juniper performance and natural regeneration. With the aim of increasing the natural regeneration of maritime juniper, several activities have been carried out in order to restore, at least partially, the physical environmental conditions of its habitat. In addition to these, alien species, such as *Carpobrotus edulis*, that occupy juniper areas and prevent seedling establishment were also removed. To reverse this situation, several treatments were carried out in pine plantations and in stands containing the neophyte *C. edulis*. These treatments affected 215 ha and consisted mainly in pine tree clearings, including pruning and felling of pine individuals selected according to their vicinity to junipers. Hand removal of *C. edulis* was also carried out, affecting 8.6 ha.

Actions to reduce the impact of human activities, such as trampling and use of motor vehicles, were carried out in the *Enebrales de Punta Umbría Natural Landscape*. Here, new wooden paths crossing the dune area between the beach and the road were built. Bike-lane protection, barriers for motor vehicles, and informative and interpretative signs were also installed. The purpose of these measures was to avoid motor vehicles passing through the juniper stand, and to channelize the walking paths of beach users to avoid sand erosion and direct damage to plants.

No doubt, an essential part of any conservation or restoration project, including habitat changes, is the involvement of the inhabitants of and visitors to the territory affected. Therefore, complementary activities to raise the awareness of local inhabitants within the distribution area of maritime juniper and environmental education were implemented. The purpose was to make several social groups aware of the necessity of preserving juniper trees and their habitat. Three different sub-programs were developed:

1. "*Enebrando dunas*," in which 20 schools and 810 children in elementary education participated.
2. "*Proyecto Enebro*," an environmental volunteering project.
3. "*Conoce tus enebrales*," guided interpretative visits to nearby maritime juniper woodlands.

The Program was developed between 2002 and 2006 with a total cost of 1,269,893 Euros. Various field work projects cost 860,250 Euros, and the technical works for the development of control and monitoring activities, research, dissemination of information, and environmental education 409,643 Euros in two phases (2002–2004: 146,000 Euros; 2004–2006: 263,643 Euros).

9.6 A Field Experiment in the *Enebrales de Punta Umbría* Natural Landscape

The *Enebrales de Punta Umbría Natural Landscape* (162 ha), located southward of Huelva city (southwest Spain) and protected since 1989, shows the problems of juniper woodlands under pine tree plantations (*Pinus pinea*). Here, along a sandy strip 2.5 km long and 200 m wide, approximately 300 mature maritime junipers without natural regeneration survive under a pine tree canopy. Anthropic pressure is intense, and the junipers are accompanied by Phoenicean juniper (*Juniperus phoenicea* subsp. *turbinata*) and a xerophytic shrub composed of *Cistus salvifolius*, *Halimium halimifolium*, *Rosmarinus officinalis*, *Rhamnus lycioides*, and *Pistacia lentiscus*, among others.

Within the framework of the "Conservation Program of Maritime Juniper Woodlands," a clearing experiment was carried out in this protected nature reserve. The general aim of the pine tree treatments was to reduce their negative effects on maritime junipers and to restore the environmental conditions to favor the regeneration of natural coastal plant communities. We analyzed the effects of three levels of pine tree clearing on the physical environment, the junipers' performance, and the coastal plant community.

Pine tree plantation clearings were carried out in January 2004, with three levels of pine removal:

1. Without pine clearfelling (dense canopy),
2. Removing the closest pine trees to the junipers (light clearing),
3. Removing all the pine trees and the largest Phoenicean junipers (clearings).

The original experimental design included three paired 50 × 50-m plots from the top of the first dune front, but it was not approved and needed to be reduced. Finally, treatments 1 and 3 were each applied to two adjacent 30 × 30-m plots located within the dune valley. The light clearing treatment was applied extensively in the rest of the area with junipers. The first exploratory analysis showed that plant cover and physical variables were similar in treatments 1 and 2.

The physical environment variables monitored were salt spray deposition, photosynthetic active radiation (PAR), and sand mobility. Juniper performance included cone and seed production in 2003 and 2006, growth of branches, and the density of pollen in the air. The establishment of sand dune plants in the clearings was also recorded.

Table 9.1 Mean ± standard error of environmental and "reproductive" variables in the three situations. Growth of branches estimated between May 2004 and February 2005

Situations	Clearings	Light clearing	Dense canopy
Salt spray (%)	22.9 ± 1.9	22.7 ± 1.8	20.4 ± 1.1
Sand mobility	No	No	No
PAR (μmol m^{-2} s^{-1})	1721.3 ± 4.2	747.1 ± 5.7	90.7 ± 9.9
Cones per tree 2003	1051.6 ± 261.3	530.6 ± 160.9	111.7 ± 30.7
Cones per tree 2006	947.4 ± 390.1	700.1 ± 136.5	193.7 ± 99.9
Seeds per cone 2003	3.1 ± 0.2	2.7 ± 0.1	2.5 ± 0.2
Seeds per cone 2006	3.2 ± 0.2	2.8 ± 0.0	2.8 ± 0.1
Air pollen density	1717.0 ± 159.0	–	378.7 ± 58.7
Branches growth (cm)	3.7 ± 0.9	–	1.2 ± 0.4

Salt spray deposition was estimated by washing in de-ionized water three twigs (10 cm long) per juniper out of the individuals located in different situations (six junipers per location) and measuring the electrical conductivity of the water. Data were presented as percentages in relation to the highest conductivity (Muñoz-Reinoso 2003). Because installed poles were stolen or pulled out, sand mobility was estimated subjectively through the presence of deposits of sand or evidence of erosion. Radiation was measured at noon with a LICOR LI-189 radiometer in the center of the plots in spring (18 recordings per situation).

At the end of the summer of 2003, prior to the clearing experiments, cone production was estimated by counting the fleshy cones present in juniper branches of individuals located in the three desired pine tree cover situations (dense canopy, light clearing, clearing). Samples were taken to estimate the number of seeds per cone (30–50 cones per juniper). In 2006, the sampling was repeated. Only 32 juniper trees that remained in the same situation during the studied period were considered (5 in clearings, 12 in light clearings, 15 in dense canopy).

The growth of branches and density of pollen in the air were estimated in the clearings and dense canopy plots. The growth of branches was estimated every 3 months in five individuals per treatment by measuring the increase in length of five twigs per individual between May 2004 and February 2005. Pollen traps consisted of Melinex tapes, fixed onto a wire mesh, and covered with transparent silicone. The traps were hung on the juniper branches (three traps per situation). They were hung during seven clear days in December 2004. Pollen counts were carried out with a microscope (×40) on a surface of 907.2 mm^2 per tape.

The results (Table 9.1) showed that the clearing treatments were inefficient at modifying the physical environment, such as salt spray deposition and sand mobility. This may have been caused by the location of the study plots (in the dune valley, protected by a high density of pine trees covering the first dunefront—coast dune—from the salt-laden winds, see Fig. 9.3), and their small size (30 × 30 m^2). In the dense canopy situation, incident radiation, growth of branches, and pollen density were significantly lower than in the clearings. Less radiation means fewer resources dedicated to growth and reproduction. As juniper flowers are formed on new branches, lower growth translates into a lower number of flowers, and thus, low cone production. On the other hand, the dense pine tree canopy seems to prevent the arrival of pollen, which is significantly lower in the dense canopy, decreasing fertilization and cone production (Ortiz et al. 1998).

In 2003, we observed a clear gradient in cone production, significantly varying from higher cone production in the cleared plots to lower production in the dense canopy plots. In 2006, the gradient still occurred, but there were no significant differences between the clearing and the light clearing situations. Differences in cone production within situations between years are relatively small and may be due to interannual variations in cone production. However, the increase observed in the dense situation may be due to higher pollen arrival because of the small size of the plots (edge effects).

The number of seeds per cone in 2003 showed a nonsignificant trend increasing the number of seeds from dense canopy to clearing plots. In 2006, the differences among the three situations were lower, which may be due to an increase in pollen arrival because of the small size of the plots (edge effects). The trend observed in the total number of monitored individuals (42) was toward a significant increase in cone production after clearing (from 409.9 $\pm$ 74.8 to 643.4 $\pm$ 171.9, Wilcoxon $Z = -2.105, p = 0.035$), but a nonsignificant increase in the number of seeds/cone (from 2.5 $\pm$ 0.1 to 2.8 $\pm$ 0.1, ANOVA $F = 3.126$, $p = 0.082$). The relatively scarce response in cone production may be due to the brief period of time in which junipers were exposed to the pine clearing treatments and also to an extremely dry year (2005: 200 mm in precipitation).

No successful new juniper establishment was detected in the Natural Landscape (only one juniper germination was observed). Species of the coastal community such as *Helichrysum picardii*, *Armeria pungens* or *Thymus carnosus* were not detected between 2003 and 2006 either, probably because of a shortage at the seed bank. However, species such as *Halimium halimifolium*, *Cistus salvifolius* or *Rosmarinus officinalis* colonized the clearings, which is consistent with the stability of the sands. It shows that clearing treatments had no effect on juniper germination and establishment. This may be due to a high proportion of empty seeds because of a deficient pollination, an intense seed predation, and a lack of seed-dispersing animals (Muñoz-Reinoso, unpublished).

9.7 Conclusions

The small size of the plots may be conditioning some results, such as salt spray deposition or cone production and seeds per cone in dense canopy plots. As stated earlier, clearing treatments do not initiate sand mobility or an increase in salt spray deposition in the dune valley of *Enebrales de Punta Umbría* Natural Landscape, where most of the junipers appear. It seems necessary that we remove, at least partially, the pine trees covering the coast dune (first dune front), and, thus, consider the dune system as a whole.

Generally, after the pine tree clearing, juniper performance was better, increasing their growth and cone and seed production, and it may continue to improve in the near future. However, the density of pollen in the air may still be very low. This, along with low pollen viability, may be responsible for a high proportion of empty or unviable seeds (approximately 90 %, Muñoz-Reinoso, unpublished).

Lack of establishment of species from the coastal dune plant community (*H. picardii*, *A. pungens*, *T. carnosus*) shows their absence in the seed bank, probably because of a long period of sand stabilization. On the other hand, under the pine canopy there is a thick layer of pine tree needles preventing the establishment of new seedlings (junipers or shrubs) that it seems necessary to remove. It is also necessary to remove the rest of the biomass from the pine trees to avoid soil eutrophication and stabilization.

Synergistic factors, such as pine trees shading, low pollen density, small cone and seed production, seem to prevent the natural regeneration of junipers inside *Enebrales de Punta Umbría* Natural Landscape. Nevertheless, there seems to be additional factors controlling the natural regeneration of junipers on a landscape scale. Fragmentation may be responsible for a high seed predation and a lack of seed-dispersing animals (Muñoz-Reinoso unpublished).

Successful restorers of maritime juniper woodlands need to consider the ecosystem (beach/dunes/inland) as a whole and understand the biology and the ecology of the species involved. Coastal pine plantations modify environmental conditions and affect the reproductive biology of junipers; therefore, they must be cleared properly to mimic the open communities that develop on Mediterranean dunes.

References

Allen GA, Antos JA (1993) Sex ratio variation in the dioecious shrub *Oemleria cerasiformis*. Am Nat 141:537–553

Anon (1992) Council Directive 92/43/EEC on the conservation of natural habitats and of wild fauna and flora. Off J Eur Community 35(L 206):7–50

Arista M, Ortiz PL, Talavera S (2001) Reproductive cycles of two allopatric subspecies of *Juniperus oxycedrus* (*Cupressaceae*). Flora 196:114–120

Brown AC, McLachlan A (1990) Ecology of sandy shores. Elsevier, Amsterdam

Crawford RMM (1989) Studies in plant survival. Ecological case histories of plant adaptation to adversity. Blackwell, Oxford

Clifton SJ, Ward LK, Ranner DS (1997) The status of juniper *Juniperus communis* L. in North-East England. Biol Conserv 79:67–77

Géhu JM, Costa M, Biondi E (1990) Les *Junipereta macrocarpae* sur sable. Acta Botánica Malacitana 15:303–309

Goldstein G, Drake DR, Alpha C, Melcher P, Heraux J, Azocar A (1996) Growth and photosynthetic responses of *Scaveola sericea*, a Hawaiian costal shrub, to substrate salinity and salt spray. Int J Plant Sci 157:171–179

Granados M, Ojeda JF (1994) Intervenciones públicas en el Litoral Atlántico Andaluz. Efectos territoriales. Agencia de Medio Ambiente, Junta de Andalucía, Sevilla

Grime JP (1979) Plant strategies and vegetation processes. Wiley, New York 222 p

Mayoral O, Gómez MA (2003) Nuevas poblaciones de *Juniperus oxycedrus* subsp *macrocarpa* (Sm.) Ball en la comunidad valenciana. Flora Montiberica 25:34–41
Muñoz-Reinoso JC (2003) *Juniperus oxycedrus* ssp *macrocarpa* in SW Spain: ecology and conservation problems. J Coast Conserv 9:113–122
Muñoz-Reinoso JC (2004) Diversity of maritime juniper woodlands. For Ecol Manag 192:267–276
Muñoz-Reinoso JC, Jiménez B, Ruíz A, Trillo M, Yanes A (2000) Respuesta del enebro marítimo al enterramiento. Simposio AEET, Granada, p 45
Muñoz-Reinoso JC, Asensio B, Rodríguez F (2005) Territorial integration of populations of threatened species and ecological connectivity in the Southwest coast of Spain. In: ICCM'05 proceedings, Tavira, Portugal, pp 341–345
Ortiz PL, Arista M, Talavera S (1998) Low reproductive success in two subspecies of *Juniperus oxycedrus* L. Int J Plant Sci 159:843–847
Pack DA (1921) After-ripening and germination of *Juniperus* seeds. Bot Gaz 71:32–60
Puerto-Marchena A, Asensio-Romero B, Rodríguez-Infante F, Muñoz-Reinoso JC (2009) Use of models to restore landscape connectivity. The case of the maritime juniper (*Juniperus macrocarpa*) in SW Spain. 52nd IAVS symposium abstracts, p 217
Rubio-Casal AE, Leira-Doce P, Figueroa ME, Castillo JM (2010) Contrasted tolerance to low and high temperatures of three taxa co-occurring on coastal dune forest under Mediterranean climate. J Arid Environ 74:429–439
Tekke RMH, Salman AHPM (1995) Coastal woodlands, forestry and conservation along the Atlantic and North Sea shores. In: Salman AHPM, Berends H, Bonanzoutas M (eds) Coastal management and habitat conservation. EUCC, Leiden, pp 396–409
Young JA, Evans RA, Budy JD, Palmquist DE (1988) Stratification of seeds of Western and Utah juniper. For Sci 34:1059–1066

Chapter 10
Dune Restoration Over Two Decades at the Lanphere and Ma-le'l Dunes in Northern California

Andrea J. Pickart

10.1 Introduction

Ecological restoration of coastal dunes on the west coast of North America had its beginnings in a relatively remote part of northern California. Although some earlier attempts had been made to revegetate disturbed dunes or stabilize moving dunes with native plants (Pickart 1988), the first documented efforts to restore dune processes as a component of restoration began at the Lanphere Dunes in the 1980s (Pickart and Sawyer 1998). In the ensuing decades, the region has become a model for dune restoration, and California has led the west coast in the development and implementation of restoration technologies. Most of this restoration has been carried out with the goal of restoring biodiversity through the removal of invasive plant species that have been shown to interrupt abiotic processes and reduce native diversity (Pickart and Barbour 2007). Many of these projects were sponsored by government agencies and NGOs with limited budgets, resulting in poor documentation and contributing to an under-appreciation of the progress made (Pickart and Barbour 2007). Even the "gray" literature vastly underreports restoration efforts because, unlike academics, restoration practitioners rely largely on a network of informal communication and information exchange (Baker 2005).

Historically, dune restoration success in California has been measured largely based on vegetation recovery as a proxy for ecosystem health. Much progress has been made in describing and classifying native vegetation types in California dunes, allowing managers to better evaluate restoration success in this context (Pickart and Barbour 2007; Sawyer et al. 2009). In other areas, such as New Zealand, monitoring has been driven more by physical processes, with the

A. J. Pickart (✉)
US Fish and Wildlife Service, 6800 Lanphere Rd,
Arcata, CA 95521, USA
e-mail: Andrea_Pickart@fws.gov

M. L. Martínez et al. (eds.), *Restoration of Coastal Dunes*, Springer Series on Environmental Management, DOI: 10.1007/978-3-642-33445-0_10,

assumption that ecosystem recovery will follow (Hesp and Hilton, this volume). There is a need to better integrate the physical and biological components of restoration monitoring, particularly in the light of global climate change (Psuty and Silveira 2010).

This chapter presents a case study of the restoration at the Lanphere and Ma-le'l Dunes in Humboldt Bay National Wildlife Refuge, California, USA. These areas are managed with the goals of conserving and restoring globally rare dune habitats, supporting the recovery of endangered dune plants, and promoting long-term viability of ecosystems (USFWS 2009). Restoration goals for these projects were articulated in the context of ecosystem processes, with measurable objectives and monitoring focused on biotic variables such as vegetation cover, species diversity, and endangered plant recovery. Dune morphological terminology in this chapter follows Hesp (2000). Plant nomenclature follows the Flora of North America (FNAEC 1993).

10.2 Dune system

The Lanphere and adjoining Ma-le'l Dunes units encompass 327 ha located at 40.88°N on the upper end of the North Spit of Humboldt Bay in northern California (Fig. 10.1). Littoral drift is dominated by northern, storm-driven transport in the winter, and by southern, wind/wave-driven transport in the summer (Winkelman et al. 1999). Net northern transport occurs during the El Niño events of the Pacific Decadal Oscillation (Moffat and Nichol 2011). The majority of sediments feeding the system originate from the Eel River, 28 km to the south, which drains 9,540 km^2 and is estimated to deliver approximately 1.8 million m^3 of sand annually (Patsch and Griggs 2007). The site is exposed to some of the most severe wave energy in the US, with predominant west–northwest swells from October through April (Costa and Glatzel 2002). The prevailing wind direction is north to northwest from spring through autumn, reversing to south to southeast in winter (Puffer 1998). Winds are strongest in April and May, reaching daily averages of 12.9 kph, and peak gusts up to 97 kph. An average of 96.7 cm of rainfall falls primarily between the months of October and May (Puffer 1998).

Surveyor charts from 1870 provide a record of the condition of the dune system prior to subsequent extensive modification of the area by EuroAmericans. The upper North Spit in 1870 consisted of two episodes of Holocene dune transgression, one older and stabilized by forest and a second overtaking the forest (Fig. 10.2). Reactivation of the forested dunes is hypothesized to be the result of the Cascadia subduction zone earthquake that occurred in 1700 (Atwater et al. 2005). Megaquakes along the subduction boundary occur at 100–1,000-year intervals and are tied to repeated dune activation at other dune sites in the Pacific Northwest (Wiedemann and Pickart 1996; Atwater et al. 1995). Relative sea level rise along the North Spit from 1977 to 2006 was 4.73 $\pm$ 1.58 mm per year, well above the global average (Moffat and Nichol 2011). Interseismic land level

Fig. 10.1 Location of the project site in California, USA

changes are a likely cause, but these changes are not consistent across Humboldt Bay (Moffat and Nichol 2011).

A record of topographic maps and aerial photographs has allowed documentation of the evolution of dune features in the more recent dune episode (Fig. 10.3). In 1939, the site consisted of a developing foredune–blowout–parabolic dune complex, with more developed features located upwind of discrete parabolic dunes, and more fragmented topography upwind of undifferentiated, transverse dune-dominated, transgressive dunefields. A decade later the less developed features had been eroded, and much of the southern portion of the site was unvegetated. The northern, more mature features supported vegetation now classified as the *Leymus mollis* herbaceous alliance on the foredune, and the *Abronia latifolia—Ambrosia chamissonis* herbaceous alliance on trailing ridges of parabolic dunes (Johnson 1963; Pickart and Barbour 2007; Sawyer et al. 2009). After 1970, *Ammophila arenaria* dispersed to the site from the south (Buell et al. 1995), resulting in the gradual development of an *Ammophila*-built foredune in the southern part of the site, and beginning the process of *Ammophila* invasion into the native vegetation to the north. By the 1980s, the Lanphere Dunes (consisting of the northern portion of the site) had been acquired and protected by The Nature Conservancy (later transferred to the US Fish and Wildlife Service), and this invasion was recognized as the primary threat to the preserve. *Ammophila* invasion results in steepening of the foredune, homogenization of vegetation, and suppression of the formation of new blowouts (Fig. 10.4). Although *Carpobrotus chilensis* had been present at the Lanphere Dunes at low densities since at least the 1960s (Johnson 1963), by the early 1990s the species was hybridizing with its invasive congener *Carpobrotus edulis.* These hybrids were observed to be

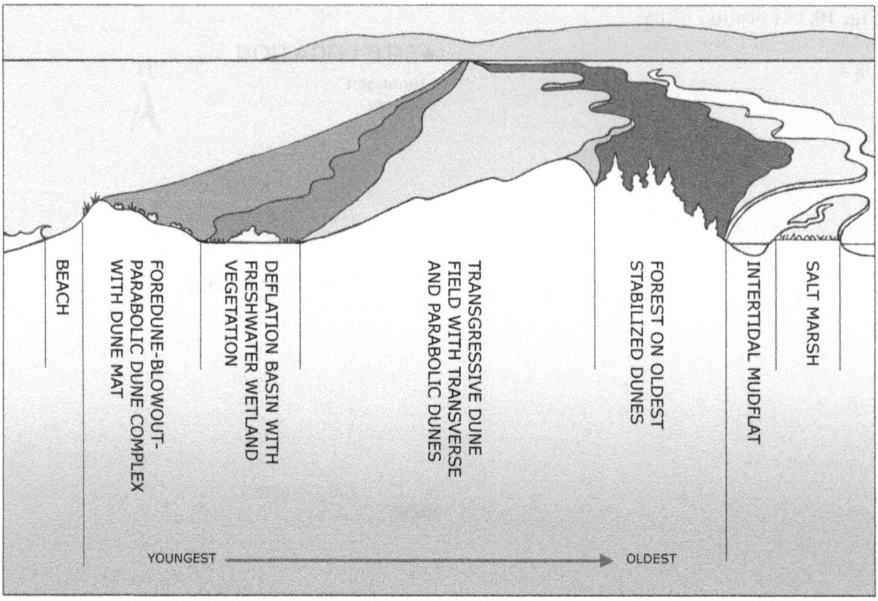

Fig. 10.2 A cross-section of the Lanphere and Ma-le'l Dunes system, showing the relationship between dune geomorphic forms and vegetation

spreading aggressively, especially on the erosional walls and depositional lobes of blowouts. Subsequent research confirmed that the population consisted of a hybrid swarm, with the potential for ecosystem-level impacts (Albert et al. 1998).

10.3 Methods

Successful methods to eradicate invasive plant species, and revegetate with native species, were developed during a series of research projects on the North Spit between 1982 and 1990 (Pickart and Sawyer 1998). Manual methods were chosen for the Lanphere and Ma-le'l Dunes based on their low impact to the extensive intact vegetation, including two federally listed endangered plants Humboldt Bay wallflower (*Erysimum menziesii* subsp. *eurekense*) and beach layia (*Layia carnosa*). Restoration included the removal of invasive species and, to a lesser extent, the re-introduction of native species. The two invasive taxa *Ammophila arenaria* and *C. edulis* × *C. chilensis* were the primary target for eradication. The Lanphere restoration was carried out between 1992 and 1998, over an area of 11 ha in the foredune–blowout–parabolic dune complex. The Ma-le'l Dunes were added to the refuge in 2005 and 14 ha of dunes were restored from 2005 to 2010 in the foredune–blowout–parabolic dunes as well as in the deflation basin. In both projects, stands of *Ammophila* and other invasives were mapped and georeferenced prior to the start of restoration.

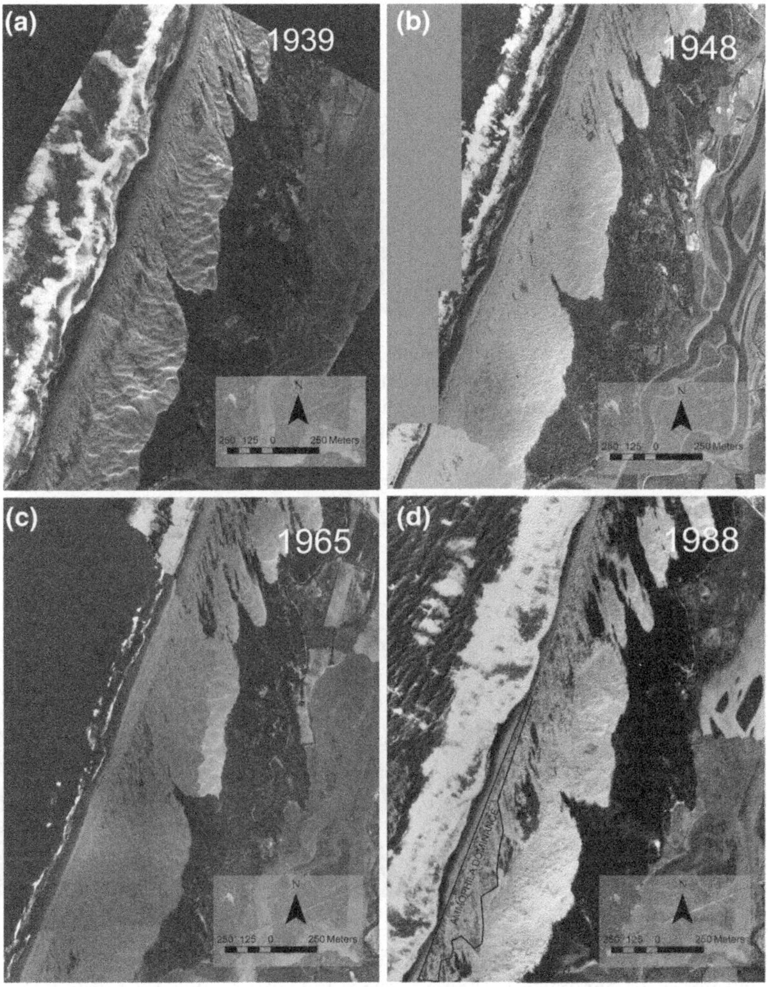

Fig. 10.3 Evolution of dune features at the Lanphere and Ma-le'l Dunes between 1939 and 1988. The *A. arenaria* invaded between 1965 and 1988, and was the dominant vegetation within the polygon in 1988

Ammophila was controlled through repeated digging/pulling, using a shovel to loosen or cut rhizomes below the surface (Fig. 10.5). Labor was provided by the California Conservation Corps, a state agency that provides a consistent labor source complete with crew supervisors and transportation. Digging commenced in February when carbohydrate reserves were low (plants had not yet broken dormancy), and continued through December. The greatest effort was required on the first dig, steeply diminishing thereafter. For the Lanphere project, treatments were applied over two growing seasons, with an average of eight treatments in the first

Fig. 10.4 *Ammophila arenaria*-built foredune at the Ma-le'l Dunes prior to restoration. The foredune to the left is transitioning from incipient to mature, and has few native species present, whereas the older, mature foredune to the right still retains some native plants. There are no blowouts along this stretch of invaded dunes; virtually all the sand is trapped in the foredune

Fig. 10.5 Members of the California Conservation Corps remove *A. arenaria* using a combination of pulling and digging at the Lanphere Dunes (1992)

year and seven in the second year. The Ma-le'l project employed fewer treatments per year over a 3-year period. Plants were piled up on site and burned after they had dried for a month or more.

In the earlier restoration at the Lanphere Dunes, a phased approach was employed to reduce the likelihood of reactivation of the dunes. *Ammophila* was removed in a patchwork fashion in three separate phases spaced 1–2 years apart, each receiving 2–3 years of treatment. Because the dead network of roots and rhizomes remained in the soil, the integrity of dune features was largely retained while native species recolonized from relict plants (Fig. 10.6). After initial deflation of several cm of sand, the dead stubble of the below-surface stems served to reduce wind speed and prevent further deflation of the surface. As a result, no revegetation was needed in the project.

At the Ma-le'l site, the restoration strategy for *Ammophila*-dominated dunes was altered somewhat. After observing the stabilizing effect of *Ammophila* stubble in the Lanphere restoration, this project was carried out without phasing it spatially

Fig. 10.6 The phased patchwork approach employed in the Lanphere restoration project prevented blowouts. Native species are already colonizing restored patches from the first phase, in the second year after initial removal (1994)

over time. However, after detecting less stability post-eradication compared with the Lanphere project, revegetation was added in the third year. Culms of the native dune grass *L. mollis* subsp. *mollis* with attached rhizomes or rhizome buds were harvested from a large nearby stand and planted during the winter months in 4-m^2 grids of 49 hills (two culms per hill) at spacings of 65 cm. The revegetated patches were distributed strategically along the foredune in areas of less stability and their position mapped with a GeoXT using ArcPad 7.1.

Carpobrotus control was also carried out using manual methods. The more aggressive hybrids form dense carpets that can be rolled up like a carpet. Shovels are used to sever rhizomes. When the plants are more scattered they are dug with a shovel and pulled, attempting to remove as much of the rhizome as possible. Several re-treatments are required to achieve complete mortality. As with *Ammophila*, the bulk of the labor is expended in the first treatment. *Carpobrotus* is difficult to burn owing to its high water content. Piles were dragged on tarps or in contractor bags to the beach and transported to a green-waste site by a four-wheel-drive truck.

Monitoring was carried out on both projects, including detailed labor tracking for the Lanphere project. Vegetation cover (total and by species) was sampled in 0.25-m^2 (Lanphere) or 1-m^2 (Ma-le'l) systematically placed plots. Sampling occurred at 1- to 2-year intervals through 2002 at Lanphere, and annually (still ongoing) at Ma-le'l, allowing documentation of progress toward the objectives of vegetation cover and species diversity. Surviving *Leymus* mounds were recorded in the summer following each planting. By the second summer, individual mounds could no longer be distinguished and the presence or absence of live *Leymus* in each grid was recorded.

In 2011 a new sampling design was implemented to estimate cover and species composition along a temporal gradient: the oldest vegetation that predated restoration and was never invaded, vegetation that originated as the result of restoration completed at the Lanphere Dunes in 1998, and the newly restored Ma-le'l Dunes

Fig. 10.7 *Leymus mollis* subsp. *mollis* planting areas at Ma-le'l Dunes, beginning to merge after 2 years

completed in 2010. Percentage cover of perennial species was estimated in 10-m^2 plots systematically distributed among the three strata. Only plots with greater than 5 % cover were included to avoid sampling in blowouts.

10.4 Results

In the Lanphere restoration, *Ammophila* was largely dead after 2 years of treatment, although there continued to be sparse resprouts requiring some treatment for a third year. After that, maintenance of the project was accomplished through annual "sweeps" for any remaining resprouts or new occurrences. Native species recovery was observed as early as the year after initial treatment, and the manual method allowed these plants to be avoided while treatments continued. *Ammophila* mortality took longer at Ma-le'l due to the less intensive treatment application. However, by year 5, *Ammophila* cover was less than 1 %. In both projects, *Carpobrotus* cover was reduced to less than 1 % after 5 years. More so than *Ammophila, Carpobrotus* has continued to reinvade from adjacent sites, requiring re-treatment at an interval of 3–5 years.

Leymus mound survival was 61 % after 1 year. In the second year (2011), when the original mounds could no longer be distinguished, there were live plants in 100 % of the planting areas, but high variation in the density and cover of surviving *Leymus* (Fig. 10.7). In January 2012, additional *Leymus* planting was carried out.

By 2011 there were no significant difference between mean total cover values and species richness of the Lanphere restoration compared with intact vegetation ($p = 0.59$). Only 1 year post-completion (but 6 years after the start of the restoration), the restored vegetation at Ma-le'l exhibited cover values not significantly different from intact vegetation ($p = 0.20$), but significantly and slightly lower than the Lanphere restoration ($p = 0.02$; Fig. 10.8). However, the Ma-le'l

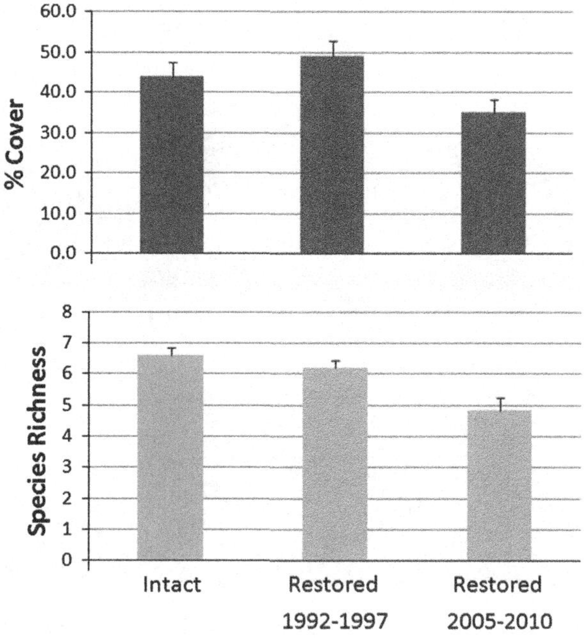

Fig. 10.8 Differences in the mean percentage total cover and species richness (of perennial species) in intact dune mat vegetation, compared with restoration carried out from 1992 to 1998 (Lanphere Dunes) and from 2005 to 2010 (Ma-le'l Dunes). Error bars represent the standard error

vegetation exhibited significantly lower species richness than both the intact vegetation and the older restoration ($p < 0.01$). Changes in vegetation over time are illustrated in Figs. 10.9 (Lanphere) and 10.10 (Ma-le'l).

10.5 Costs

The labor component of *Ammophila* eradication was carefully tracked for the Lanphere project. Most of the actual digging time (78 %) was expended in the first dig. Almost half of total labor time was expended on walking to and from the remote site. Removing this site-specific variable, eradication required a total of 3,600 ph (person-hours)/ha in year 1, a total of 843 ph/ha in year 2, and 188 ph/ha in year 3. In 1998 this translated to a combined cost of $54,590/ha (Pickart and Sawyer 1998). It is important to note that restoration managers use different approaches to quantifying costs. The approach employed at the Lanphere and Ma-le'l dunes based the costs not on the area of the total site, but rather on the area occupied by the invasive plants were they to occur in a contiguous fashion. Commonly, managers do not precisely map stands of invasive species, but simply use the area over which invasives are scattered. This is a valid approach in that the entire area benefits from the restored processes, but it makes costs difficult to compare among sites. When this second approach is used for the Lanphere and

Fig. 10.9 Photographs of an area in which dense *A. arenaria* was removed from the Lanphere foredune. The initial photograph shows piles of *Ammophila* waiting to be burned. An incipient foredune to the east was also treated. By 1997 native plants had volunteered at low densities, and the incipient foredune was eroding. Four years later, in 2001, native species, including the early successional *A. latifolia* and *Erigeron glaucus*, were present with abundant open sand between individual plants. By 2011, cover was much higher, and species composition had shifted in favor of *Poa macrantha* and *Lathyrus littoralis* in the foredune. A new incipient dune was forming, built by native *Leymus mollis* subsp. *mollis*

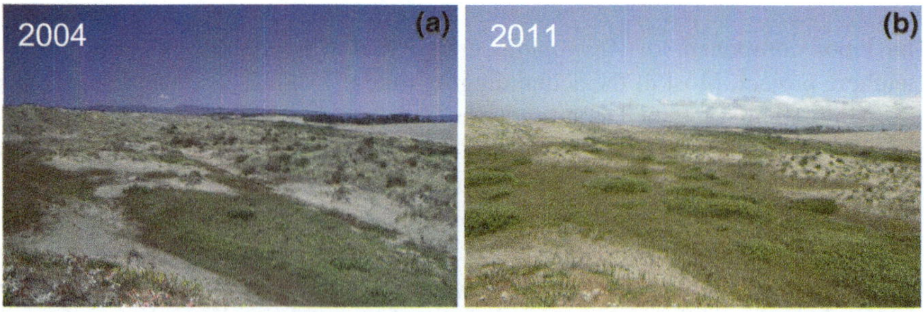

Fig. 10.10 Two photographs of the Ma-le'l Dunes, depicting the foredune (*left*), deflation basin (*center*), and parabolic trailing ridges (*right*). *Ammophila arenaria*-dominated vegetation has been replaced by native species. Behind the foredune, renewed sand movement allowed the deflation basin wetland to expand

Ma-le'l project, the cost of restoration over the entire affected area is reduced by almost two-thirds.

Using manual methods to remove *Ammophila* and other invasives is initially expensive, but carries many benefits that result in longer-term efficiency. Heavy equipment has been used successfully in northern California and elsewhere to eradicate *Ammophila*, but only where it achieves dominance over large areas. In an early-stage invasion into intact vegetation the impacts from heavy equipment on vegetation (including endangered species) would be too great, and this alternative would most likely not be permissible in the area under existing coastal laws and regulations. In addition, there is cause for concern at some sites that use of heavy equipment can reactivate dunes, slowing or preventing recovery of native vegetation. In contrast, manual methods can be targeted precisely, allowing the retention of interspersed native species. These surviving plants become a source of natural recolonization, reducing or even eliminating the need for revegetation. An effective herbicide treatment for *Ammophila* was not developed until the mid-1990s (Aptekar 2000), but more importantly, there was a strong community ethic opposing the use of herbicides. The restoration philosophy underpinning this effort embraced the support of the community, and this was perhaps one of the earliest community-based restoration projects (Jordan 1991; Pickart and Sawyer 1998). Over the past two decades, community support has crystallized in the form of an impressive volunteer labor force overseen by the NGO Friends of the Dunes, which supports restoration on lands throughout the Humboldt Bay region. Volunteer labor is ideal for maintaining restoration over the long term. *Ammophila* and other invasive species are able to reinvade at low rates from surrounding, unrestored areas. Given the extent of the regional *Ammophila* invasion as well as the constraints of private ownership, true eradication of *Ammophila* from the area is not currently feasible. However, continued management requires minimal resources if applied on a regular basis (in this case annually). The level of labor required to prevent reinvasion is suitable for volunteers who remain committed to this effort through a sense of investment. The support of the community adds to the likelihood of continued management even during periods of fiscal constraint for managing entities.

10.6 Conclusions

The manual methods employed in these two projects have resulted in the return of natural processes regulating the formation of foredunes, blowouts, and parabolic dunes, and the recovery of diverse native vegetation. Both projects were carried out over a 5- to 6-year period. The projects could have been accelerated given additional funding and resources, but the longer duration was ideal to allow a gradual transition from overstabilized dunes characterized by an almost monospecific stand of exotic vegetation to a more natural morphology without causing a level of instability precluding native species recovery.

Although both projects were relatively limited in area (together comprising a 2.5-km stretch of shoreline), the Lanphere project was the first of its kind in California, and gave rise to many other restoration projects in the region and beyond. As of 2011, a total of 12 km of coastline in the Humboldt and Del Norte counties have been cleared of *Ammophila* and other invasive species by a number of government agencies and NGOs. Methods have included both manual and mechanical (heavy equipment). Restoration has also been carried out, or is in progress, in a number of dune systems along the California coastline, including the Guadalupe-Nipomo Dunes, Morro Bay Dunes, San Francisco Bay Dunes, Point Reyes National Seashore, Bodega Bay, and MacKerricher State Park.

Acknowledgments The restoration projects described in this chapter are the result of much collaboration among many partners over a long period of time. I would like to acknowledge Linda Miller, Patricia Clifford, Trevor Goodman, and Kyle Wear, all of whom played critical roles in the planning, implementation, monitoring, and/or maintenance of these projects. Field work for new results presented in this chapter was carried out by Laurel Goldsmith and Kira Hawk. Funding for the projects was from a number of sources, including The Nature Conservancy, US Fish and Wildlife Service, National Fish and Wildlife Foundation, and the California Department of Corrections and Rehabilitation. Labor was provided by the California Conservation Corps., California Department of Forestry and Fire Protection, Sheriff's Work Alternative Program, Friends of the Dunes volunteers, and the staff of Humboldt Bay National Wildlife Refuge. Special thanks to Patrick Hesp for support, encouragement, and review of the manuscript.

References

Albert ME, D'Antonio CM, Schierenbeck KE (1998) Hybridization and introgression in *Carpobrotus* spp. (Aizoaceae) in California. Am J Bot 84:896–904

Aptekar R (2000) *Ammophila arenaria* (L.) Link. In: Bossard CC, Randall JM, Hoshovsky MC (eds) Invasive plants of California's wildlands. U.C. Press, Berkeley, pp 42–46

Atwater BF, Nelson AR, Clague JJ, Carver GA, Yamaguchi DK, Bobrowsky PT, Bourgeois J, Darienzo ME, Grant WC, Hempill-Haley E, Kelsey HM, Jacoby GC, Nishenko SP, Palmer SP, Peterson CD, Reinhart MA (1995) Summary of coastal geologic evidence for past great earthquakes at the Cascadia subduction zone. Earthq Spectra 11:1–18

Atwater BF, Musumi-Rokkaku S, Satake K, Tsuji Y, Ueda K, Yamaguchi DK (2005) The orphan tsunami of 1700—Japanese clues to a parent earthquake in North America. U.S. Geological Survey Professional Paper 1707. Reston, Virginia, USA

Baker MJ (2005) Socio-economic characteristics and contributions of the natural resources restoration system in Humboldt County, California. Ecol Restor 23:5–14

Buell AC, Pickart AJ, Stuart JD (1995) Introduction history and invasion patterns of *Ammophila arenaria* on the north coast of California. Conserv Biol 9:1587–1593

Costa SL, Glatzel KA (2002) Humboldt Bay, California, Entrance channel report 1: Data review, Coastal Inlets Research Program. U.S. Army Corps of Engineers, Engineer Research and Development Center. Washington, D.C., USA

Flora of North America Editorial Committee (ed) (1993) Flora of North America North of Mexico. 16 + vols. New York and Oxford

Hesp PA (2000) Coastal sand dunes: form and function. Coastal Dune vegetation network technical bulletin no. 4. Forest Research, Rotorua, NZ

Johnson JW (1963) An ecological study of the dune flora of the North Spit of Humboldt Bay. M.A. Thesis, Humboldt State University. Arcata, California, USA

Jordan WR (1991) A new paradigm. Restor Manage Notes 9:64–65

Moffat and Nichol. 2011. Coastal Regional Sediment Management Plan, Eureka littoral cell, California. U.S. Army Corps of Engineers M&N File No. 6731-06. Los Angeles, California

Patsch K, Griggs G (2007) Development of sand budgets for California's major littoral cells: Eureka, Santa Cruz, Southern Monterey Bay, Santa Barbara, Santa Monica (including Zuma), San Pedro, Laguna, Oceanside, Mission Bay, and Silver Strand littoral cells. California Coastal Sediment Management Workgroup, Sacramento, California, USA

Pickart AJ (1988) Dune restoration in California: a beginning. Restor Manage Notes 6:8–12

Pickart AJ, Barbour MG (2007) Beach and Dune. In: Barbour MG, Keeler-Wolf T, Schoenherr AA (eds) Terrestrial vegetation of California. U.C. Press, Berkeley, pp 155–179

Pickart AJ, Sawyer JO (1998) Ecology and restoration of northern California coastal dunes. California Native Plant Society Press, Sacramento

Psuty NP, Silveira TM (2010) Global climate change: an opportunity for coastal dunes? J Coast Conserv 14:153–160

Puffer AH (1998) Climate of Eureka, California. NOAA Technical Memorandum NWS WR-252. National Oceanic and Atmospheric Administration, U.S. Department of Commerce. Washington, D.C., USA

Sawyer JO, Keeler-Wolf T, Evans JM (2009) A manual of California vegetation, 2nd ed. California Native Plant Society Press, Sacramento

US Fish and Wildlife Service (2009) Humboldt Bay National wildlife refuge comprehensive conservation plan and final environmental assessment. Pacific Southwest Region, Sacramento

Wiedemann AM, Pickart AJ (1996) The *Ammophila* problem on the northwest coast of North America. Landsc Urban Plan 34:287–299

Winkelman JW, Schaaf D, Kendall TR (1999) Humboldt beach and dune monitoring. In: Ewing L, Magoon OT, Robertson S (eds) Proceedings sand rights '99 bringing back the beaches, 23–26 Sept 1999 at Ventura,. CA. American Society of Civil Engineers, Reston, pp 176–190

Chapter 11
Restoration of Coastal Sand Dunes for Conservation of Biodiversity: The Israeli Experience

Pua Bar (Kutiel)

11.1 Introduction

The Mediterranean coastal sand dunes in Israel stretch along 190 km, whereas their width at the south is about 7 km and narrows in the north to about 1 km (Fig. 11.1). The sediment originates from the Nile Delta, which is carried by the sea currents, deposited on the coast, and then blown inland by the wind. The common wind direction is south-west and the wind energy, as expressed by the drift potential index (DPI), which is proportional to the sand transport (Fryberger 1979), is 147. According to Fryberger's classification, 147 DPI is a very low wind energy environment, and is thus favorable for the stabilization of sand dunes. Therefore, the dunes at present are under a stabilization process. However, mobile sand dunes are still found alongside stabilized ones, whereas more than a half of the coastal sand dunes area is composed of semi-stabilized dunes and only about 20 % of mobile dunes. The process of stabilization, which can be considered a natural one, causes significant geomorphic and ecological changes that are expressed in changes of landscape features (fewer mobile dunes and more semi-stabilized and stabilized dunes with perennial vegetation coverage between 30 and 70 %) (Tsoar and Blumberg 2002), biodiversity (a shift of flora and fauna composition from xeric to mesic assemblages) (Kutiel et al. 2000a; Perry 2008; Ramot 2007), and reduction in sand-dwelling organisms (psammophiles) characteristic of mobile sand dunes, some of which are endemic either only to Israel or to the eastern Mediterranean coast—the Levant (Kutiel 2001; Kutiel et al. 2000a; Ramot 2007).

Two-thirds of the population of Israel live along the coast, as is true in much of the world and particularly along the Mediterranean Sea (van der Meulen and

P. Bar (Kutiel) (✉)
Department of Geography and Environmental Development, Ben-Gurion University of the Negev, P.O. Box 653 84105 Beer Sheva, Israel
e-mail: kutiel@exchange.bgu.ac.il

M. L. Martínez et al. (eds.), *Restoration of Coastal Dunes*, Springer Series on Environmental Management, DOI: 10.1007/978-3-642-33445-0_11,

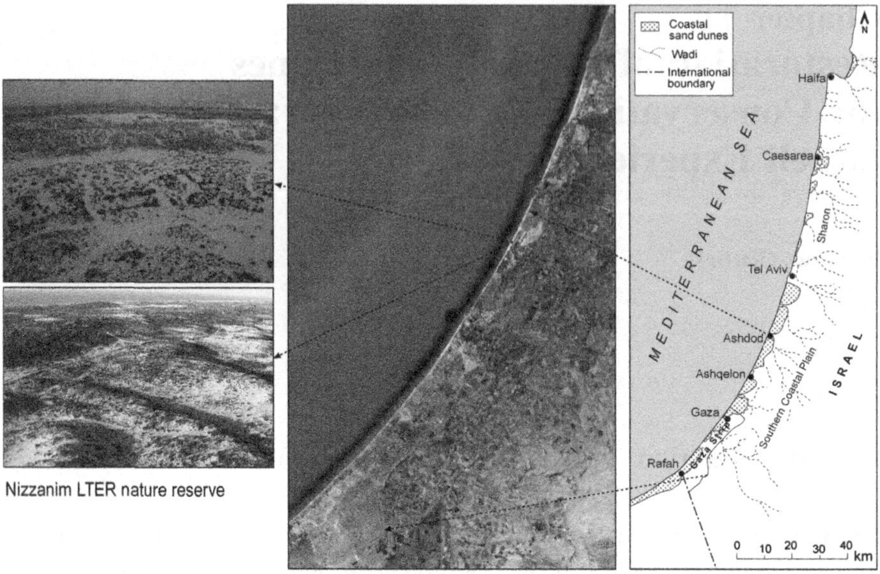

Fig. 11.1 The coastal sand dunes along the Mediterranean Sea in Israel. The attached photos are from Nizzanim LTER Nature Reserve

Salman 1996). Thus, most of the area is heavily developed and under man's impact, and only about 17 % of the Israeli coastal dunes are still ecologically intact, about half of which have protected status (Kutiel 2001). The largest protected area (2,000 ha) along the Mediterranean coast in Israel is the Nizzanim LTER (Long Term Ecosystem Research) Nature Reserve.

The Nature and Parks Authority in Israel aims to conserve the biodiversity (species, habitats, landform and ecosystem service richness) of the coastal sand dunes within the protected areas, which include the mobile sand dunes ecosystem. Therefore, the long-term challenge is to prevent the semi-stabilized dunes from becoming completely stabilized, and in the short term to restore some of the stabilized dunes to mobile dunes in order to expand and encourage the re-establishment of psammophilic plant and animals (sand-dwelling species). As ecologists we categorize the dune states (mobile, semi-stabilized, and stabilized) based either on the percentage of sand covered by vegetation, and also on species assemblages, or on bio-indicators, which indicate the affinity of species to the mobility rate of the dunes. Throughout 4 years of observation at Nizzanim LTER Nature Reserve we have collected data for four taxons: plants, arthropods, reptiles, and small mammals (mainly rodents). The results indicated that the plants and animals on almost bare sand (less than 15 % of perennial vegetation cover) and mobile dunes differ significantly from those on the other two dune states, especially from those on the stabilized dunes. The plants and animals on the mobile dunes are defined as psammophilic species. The psammophilic species known in our region (Sinai, Israel, and Lebanon) are either found on mobile dunes on the Northern African and Arab deserts or are endemic species to our region, but were derived from the psammophilic species known from these deserts. Therefore, in this chapter I refer to the different dune types (mobile, semi-stabilized, and stabilized) based on ecosystem features (plant and animal assemblages, psammophilic species, vegetation coverage, sand mobility, etc.) (Kutiel et al. 1980; Kutiel 1998; Ramot 2007; Perry 2008).

The object of this chapter is to contend with two main questions:

(1) Why restore and prevent the stabilization process if it is the natural trend?
(2) What does restoration intend to achieve, and is it feasible?

In order to answer these two questions, I will describe:

(1) The ecological and conservational importance of the coastal sand dunes in Israel, with an emphasis on the uniqueness of the mobile sand dune habitats.
(2) The evolution of these dunes and the role of human interference within this evolution.
(3) The initiative to restore the dunes and the trends observed 4 years later.

11.2 The Ecological Uniqueness of the Coastal Sand Dunes

The coastal sand dunes are located within the Mediterranean geographical region. However, because of the arid condition of the sand compared with adjacent Mediterranean clayish soil types and the fact that the dunes had been spatially geographically connected to the Sinai, the Arab, and the Sahara deserts, they have been practically isolated from their surroundings. The majority of the plant and animal species in the coastal sand dunes originated in the desert. The more remote they are from the source populations, the higher the probability of the evolution of sub- and new species (endemic to the region). Therefore, the Israeli coastal strip is considered to be a center of speciation for sand-dwelling small mammals, reptiles, and arthropods. However, the phylogenetic derivations of the coastal dune species have not been researched thoroughly, and knowledge about them is sparse.

Forty-two endemic plant species (most of them annuals) can be found in the sandy habitats of the coastal plain, including sand dunes. This constitutes 20 % of all endemic plant species in Israel and comprises the highest rate of endemism in any given habitat (Shmida 1982).

The only mammal species endemic to Israel, the Buxton's jird (*Meriones sacramenti*), and an endemic subspecies of Anderson's gerbil (*Gerbillus andersoni allenbyi*, previously referred to as *G. allenbyi*) are found in this region (Zahavi and Wahrman 1957; Yom-Tov and Mendelssohn 1988; Wasserberg 1997; Shalmon 2004). The sand-dwelling lizard, *Acanthodactylus shreiberi,* which is considered to be critically endangered according to IUCN definitions (Bouskila 2004), still exists in the sands of the coastal plain, but its habitat is fast disappearing. The sand-dwelling agamid, *Trapelus savignii*, whose status in Israel is defined as vulnerable, existed in the coastal dunes a few decades ago, but today it is completely absent from this area, and it survives in Israel only in the sands of the northwestern Negev Desert (Bouskila and Amitai 2001; Bouskila 2004).

Very little is known about arthropod assemblages in the Mediterranean Basin coastal sand dunes in relation to the state of dune stabilization. A detailed study at Nizzanim LTER Nature Reserve, aimed at defining such assemblages, found that among the arthropods (Ramot 2007), there are 131 species from more than 30 families, 10 orders, and 6 classes. The beetles (*Coleoptera*) are the dominant order with 24 species that belong to the *Tenebrionidae* family and 13 species from the *Carabidae* family. Two different assemblages can be observed that correlate significantly with sand mobility rate and plant perennial coverage: one on mobile sand dunes (10–15 % plant coverage) and the other on stabilized dunes (30–65 % plant coverage). The semi-stabilized dunes host species either characteristic to mobile dunes or stabilized ones, depending on the stabilization level of each particular dune. Likewise, a few indicator species were found whose abundance indicates the mobility rate of the dunes. For example, *Scarites striatus* (family *Carabidae*) and *Mecynotarsus bison* (family *Anthicidae*) are found only on mobile dunes, while *Carabus impressus hybridus* (family *Carabidae*) is found on stabilized dunes.

Twenty-two arthropod species are distributed in Israel only along the coastal sand dunes and all of them except one, *Ocypus ophthalmicus* (*Staphylinidae*), are characteristic to sand dunes. Thirteen species out of the 22 are considered to be endemic to the Levant (Syria, Lebanon, Jordan, Israel, Sinai, Cyprus, and Saudi Arabia), whereas two of them, *Arenivaga hebrea* (Polyphagidae) and *Chrysolina ruffoi benijaminica* (*Chrysomelidae*), are endemic only to Israel. Eight species out of the 13 favor sandy habitats with low vegetation coverage (mobile and semi-stabilized dunes) and the other five are found on mesic habitats with high vegetation coverage (stabilized dunes). Five of the nine non-endemic species that are also distributed in Europe and Asia are characteristic of mesic habitats. All nine species, except one, are characteristic of dune habitats.

Some stiletto flies and their relatives are also endemic to coastal dunes in Israel. One such example is *Stenomphrale teutandhameni,* a "window fly" (Diptera: *Scenopinidae*). Other species found include the stiletto flies *Neotherevella citrina* and *Rueppellii thoracica*, both confined to the coastal strip of North Africa and Israel. Several new insect species from Nizzanim Nature Reserve have been discovered, including leaf beetles and a fly (*Coenosia freidbergi*) (Lopatin and Chikatunov 1999; Pont and Grach 2008). These flying insects lay their eggs only in mobile dunes.

11.3 The Evolution of the Sand Dunes and the Role of Human Interference

Over the last 50 years, the shifting coastal dunes, which are considered ecologically valuable, have been increasingly subjected to shrub and dwarf-shrub encroachment. Aerial photographs taken in 1944 show that a large area of the present dunes was once made up of active barchans and transverse dunes, with

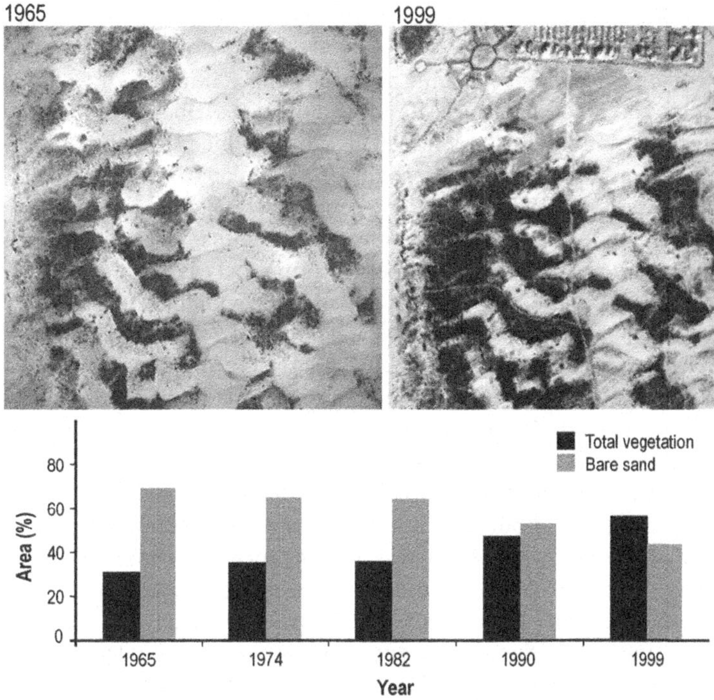

Fig. 11.2 Aerial photos of Nizzanim Nature Reserve in the years 1965 and 1999, and the reduction in the bare sand dune area compared with the vegetated areas

sparse vegetation coverage (Fig. 11.2). Over the last 50 years, woody vegetation has encroached and covered 80 % of the dune crests, resulting in gradual, but substantial dune stabilization and a change in dune crest shape from transverse to parabolic (Kutiel and Sharon 1996; Kutiel et al. 2004; Tsoar and Blumberg 2002). The question is: why were the dunes bare and mobile before 1944 and what has happened since then?

There are four major driving forces that might have shifted stabilized dunes into semi-stabilized and mobile dunes (Hesp and Thom 1990; Arens et al. 2007): climate (wind regime and precipitation amount), which affects vegetation establishment and growth, vegetation cover, sand supply, and human exploitation (cutting and grazing). As far as we know, our region has not experienced any climate change over the last 200 years, but still the dunes were bare and mobile, as described by Jakotin in 1818, and remained in this state until 1944, as can be seen from the air photos taken in that year. The sand supply along the coast has not changed during the past 8,000 years either (Zviely et al. 2007; Rohrlich and Goldsmith 1984). Most of the foredunes disappeared owing to intense construction close to the coast, and those that remain were either stabilized by the British, before the establishment of the State of Israel, or have been destroyed, as in the case of Nizzanim. Therefore, it is believed that the only force that has kept the dunes in a mobile state has been human exploitation, as is happening today along the border between Israel and Egypt.

There is conclusive evidence that the dunes have been impacted by humans for at least 200 years. The Jakotin map (1818), for example, describes the coastal dunes as "un-vegetated and mobile in character". This shifting dune situation was probably maintained for many years by the removal of significant amounts of vegetation. The vegetation was extensively utilized by people who inhabited the coastal region long ago, for example, nomadic Bedouins. The dominant native shrubs, desert broom (*Retama raetam*), and the monosperm wormwood (*Artemisia monosperma*), which is not palatable to sheep or goats owing to its high tannin content, was cut and used by the Bedouins for firewood, fences, and huts, as can be seen nowadays in the Sinai desert. Additionally, herds of sheep and goats trampled the physical and biogenic crust of the upper sand layer, helping to prevent sand stabilization (Kutiel 1998; Kutiel et al. 1999, 2000b, c; Levin and Ben-Dor 2004). Apart from their impact on sand mobility, sand crusts directly impact habitat suitability to animals (Zaady and Bouskila 2002). Removing the crusts led to a landscape dominated by shifting and semi-stabilized sand dunes because the vegetation responsible for dune stabilization was constantly being suppressed. The processes mentioned above acted to maintain largely active, mobile dunes almost devoid of vegetation until the second half of the twentieth century (when the State of Israel was already established), when grazing and cutting were dramatically curtailed all over the country, including the coastal plain (Levin and Ben-Dor 2004). Another impressive and very well-known example of human impact on the activation or stabilization of sand dunes is the Western Negev dunes. The Google Earth satellite image of this area exposes the significant difference between the two sides of the Israeli–Egyptian border, which highlights the border line and is thus known as the "line phenomenon". The dunes on the Israeli side are stabilized and covered by biogenic crusts because of the protected status and the almost complete absence of human interference, while on the Egyptian side; the dunes are bare and mobile owing to Bedouin cutting and grazing (Meir and Tsoar 1996; Qin et al. 2002).

The geomorphological and ecological significance of this stabilization process is expressed in changes in landscape features, particularly dune types (Kutiel 2001; Tsoar and Blumberg 2002), a decrease in landscape heterogeneity and biodiversity (Kutiel et al. 2004), and a reduction in the psammophilic component, in the flora and fauna composition of the ecosystem (Kutiel et al. 2000a, 2004; Kutiel 2001). Remarkably, however, less than 20 % of the dunes retain relatively little plant cover and considerable mobility, 60 % are semi-stabilized, and the other 20 % are stabilized (Fig. 11.3).

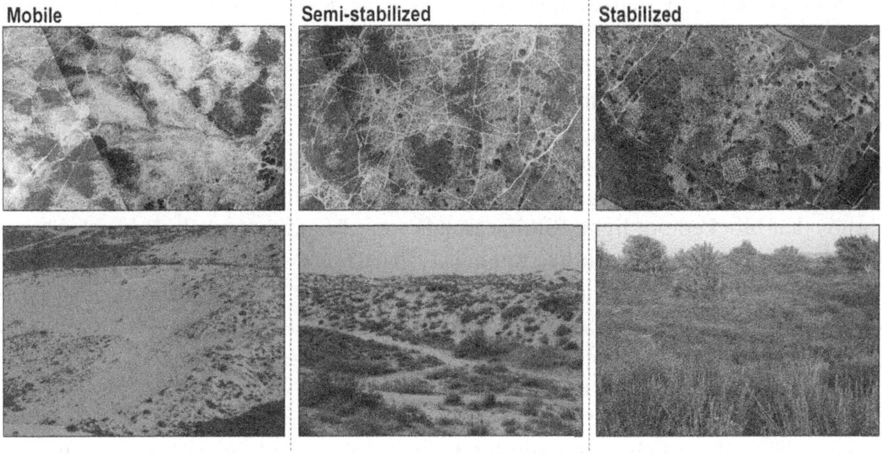

Fig. 11.3 Dune types: mobile, semi-stabilized, and stabilized

11.4 Restoration of Coastal Sand Dunes: Motivation, Initiatives, and Primary Results

The term restoration has many definitions (Jackson et al. 2006) and in some cases is confused with "recovery" and "rehabilitation". In our case, we refer to restoration as the attempt to conserve the high biodiversity of the coastal dunes that include various dune types in various states of stabilization. In order to apply this attempt there is a need to revive the stabilized dunes to the state they were in before the stabilization process. In a way it sounds paradoxical as we intend to prevent the natural process of stabilization and apply a "desertification" process on the dunes. However, this attempt imitates the "cutting and grazing policy" that existed within the coastal dunes for hundreds of years before the protection policy of the State of Israel, which stimulated the co-evolution of psammophilic species, particularly those that are considered endemic to the region, as described above. In other words, simple conservation of dune areas, which removes the cutting and grazing, will actually decrease their biodiversity values by turning all the dunes into stabilized ones. Thus, a more complex and dynamic type of management is required. The plan is to apply the re-activation management strategy to a certain percentage of the stabilized and semi-stabilized dunes in order to conserve the entire biodiversity of the coastal dunes, which includes the various types of dune stabilization levels.

The first study to examine the impact of the removal of woody plants on annual plants and rodents was conducted in 1995 at the Alexander Stream National Park located on the northern part of the coastal dunes strip (Kutiel et al. 2000a). The shrubs with 80 % coverage were completely removed manually (with the aid of saws and pruning hooks) in order to minimize damage to the soil. Three years of observation showed that annual plants such as *Rumex pictus*, *Daucus glaber*, *Crepis aculeata*, *Brassica tournefortii*, and *Cutandia philistaea*, which are considered characteristic of shifting dunes and semi-stabilized dunes, were most abundant in the cleared plots. In the first year after removal of the vegetation, *Senecio joppensis* (an annual endemic to the coastal dunes) was most abundant, while by the third year, *Rumex pictus* was the dominant species. The high abundance in the cleared plot was related to the opening of the area and the exclusion of shrubs that blocked out light. Similar trends were found for small mammals as was found for the psammophilic annuals. The non-psammophilic field mouse *(Mus musculus)* displayed a significant preference for the stabilized control plots while the psammophilic gerbil *(G. allenbyi)* showed a distinct preference for the treated open plots.

The experience gained at the Alexander Stream National Park was taken to Nizzanim LTER Nature Reserve to test it as a possible management tool for conserving the high and characteristic biodiversity of coastal dunes. The woody vegetation, with *Artemisia monosperma* as the dominant shrub species, was uprooted (partially from seven stabilized dunes and completely from another four stabilized dunes) by a bulldozer in 2005. Preliminary results indicated that during the first 2 years after treatment, the sand on the manipulated dunes shifted significantly more than it did on the non-manipulated dunes (Tsoar and Blumberg 2002). In addition, psammophile species had begun to establish themselves on the manipulated dunes, especially on those that were physically close to the non-manipulated, mobile and semi-mobile dunes. For example, the fringe-fingered lizard *(Acanthodactylus scutellatus)* was almost absent from stabilized and semi-stabilized dunes, but appeared in relatively large numbers on one of the dunes after the partial removal of the woody vegetation. This seems to be common only on manipulated dunes located near natural, mobile dunes. However, the data collected and analyzed for the next 3 years showed that there is a regression in the "psammophilic" trend and the original species of the stabilized dunes are re-establishing themselves on the manipulated dunes. The conclusion from this experiment is that while a slight shift in the community composition toward the shifting sand stage took place, it was insufficient (spatially and/or temporally) to allow recovery of the more specialized psammophiles chosen as indicators. Four years after the woody vegetation removal, the manipulated dunes are far from being similar to the mobile or semi-stabilized dunes in terms of plant and animal assemblages. These dunes are different, but still similar to the stabilized dunes (Figs. 11.4, 11.5).

There may be several reasons for the final results of this experiment:

(1) Scale and location—the vegetation was removed from a few dunes surrounded by stabilized ones with vegetation coverage between 40 and 60 %. Additionally, the dunes from which the vegetation was partly removed are located about 2 km from the coast and about 1 km from the mobile dunes. Therefore, the sand supply was low at the beginning and decreased shortly after to almost zero, enabling the establishment of vast numbers of annual plants.

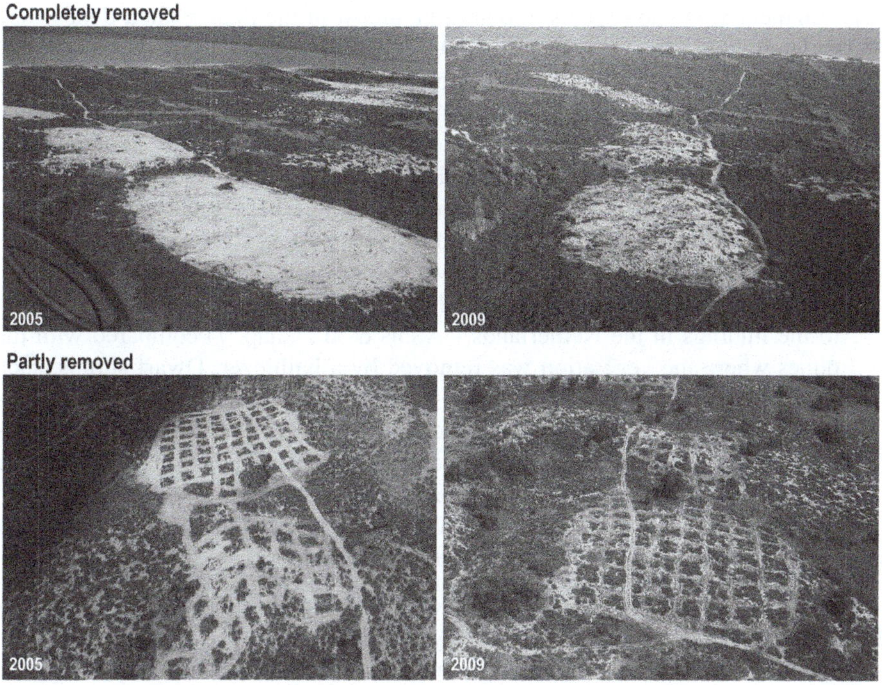

Fig. 11.4 Dune restoration at Nizzanim LTER Nature Reserve: above dunes from which the vegetation was completely removed; below dunes from which the vegetation was partly removed.

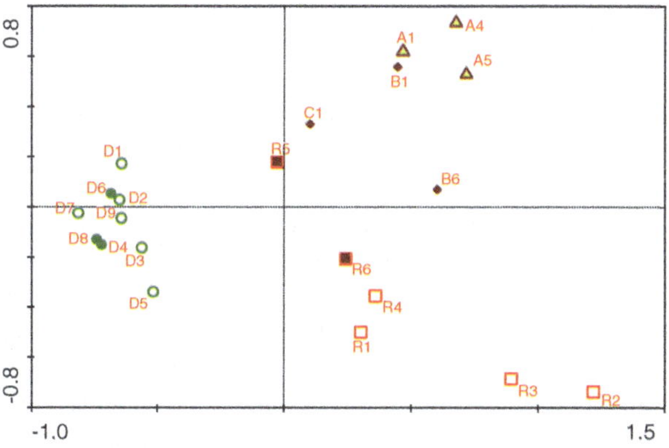

Fig. 11.5 CANOCO ordination of arthropods. There are few clusters of dunes: *yellow triangles* shifting sand dunes (reference dunes), *brown diamonds* and *red open squares* semi-stable dunes, *filled green circles* stabilized dunes, *open green circles* and *brown squares* manipulated dunes. The distance from each group type expresses the similarity between them. The closer the distance, the greater the similarity

(2) Fragmentation and organism migration—most of the manipulated dunes were isolated within the vegetated area. Only those dunes that were adjacent to the mobile or semi-stabilized dunes showed a slight establishment of psammophilic animals, mainly rodents. Our 4 years of observations demonstrated that rodents, which are marked, either do not show up at the next captured year (probably predated) or can be found at the same dune where they were previously captured or at the closest dune. Therefore, the spatial continuity of the mobile dunes is crucial for organism migration and establishment.
(3) Removal of perennial plant method—dunes in which the vegetation was clipped and uprooted showed a rapid regeneration of perennial plants (similar to the findings in the Netherlands—Arens et al., Chap. 7) compared with the dunes where the vegetation was removed by a bulldozer. Dwarf shrubs, such as *Artemisia monosperma*, which dominates the dunes, have long roots (at least 3 m) (Martínez et al. 1998; Zhenghu et al. 2004) with rejuvenate buds to cope with water limitations and sand burial or exposure. Hence, the remaining roots left under the surface after the plant removal allowed the plant re-growth throughout the covering layer of the sand. Therefore, the method of vegetation removal should ensure complete uprooting.

Similar to our results, Arens and Gleen (2006) conducted a large-scale destabilization of dunes in Netherlands by vegetation removal. The main measure was soil mobility. After 8 years of observation they did not succeed in remobilizing the dunes to the extent they expected owing to limitations of sand supply and rapid regeneration of the vegetation under humid conditions.

11.5 Conclusions

Coastal sand dunes are threatened worldwide because about two-thirds of the world's population live along the coast. Therefore, the conservation of this landscape is crucial and very challenging. In areas where the wind velocity is too low to cause significant movement of the dunes (as in Israel), and where the impact of humans has largely been removed, a stabilization process is taking place. However, some of these areas have maintained the mobile dunes' biodiversity as they co-evolved under traditional human interference, which ensured the existence of mobile dunes together with their characteristic ecosystems. Moreover, because these sand dunes have been fragmented, leaving "islands of mobile sand", psammophilic plants and animals have evolved independently from the original, spatially continuous population to form new subspecies and species that are considered endemic. In Israel, the majority of endemic plants and animals are found within the Mediterranean coastal dunes. Conservation policies in Israel that included non-active management in nature reserves, including in Nizzanim, were common until very recently.

The first initiative in re-activating stable dunes into mobile ones in order to restore this type of landscape and ecosystem was conducted in 1995 in one of the national parks in the northern part of the Israeli coastal strip. We hypothesized that by removing the woody vegetation, which is responsible for the stabilization process, we would encourage the re-establishment of the psammophile plant and animal species. Three consecutive years of observation indicated that such a trend can be achieved. Later, a larger program using a similar restoration strategy was tried at Nizzanim LTER Nature Reserve. The vegetation was partially removed from some stabilized dunes and completely removed from others all dominated by characteristic Mediterranean species alongside opportunist and generalist species.

There is no doubt that one-time vegetation removal is not enough to restore the mobile dune landscape and ecosystem (Arens et al., this book). There is a need for continuous management, as was done in the past by the nomadic tribes that lived in this area for decades and inhibited the dune stabilization and stimulated the psammophilic organism. The aim is to conserve the high biodiversity of the coastal dunes, which include various dune types in various states of stabilization. In order to prevent the transformation of the semi-stabilized dunes (covering most of the area of the nature reserve) into stabilized ones, an active policy must take place and grazing is one of the options. For the first time in Israel, there is a plan to use camels as browsers. The camels were selected as a disturbance agent because the dominant shrub species at Nizzanim are known to be non-palatable to goats and sheep owing to an abundance of tannins in the plants. According to Bedouin shepherds, camels readily eat those shrubs and consequently might significantly reduce their cover. We hope to succeed and restore one of the important landscapes and ecosystems of Israel.

Acknowledgments I want to sincerely thank the following organizations, colleagues, and students who funded and supported the restoration program and research at Nizzanim LTER Nature Reserve.

The Jewish National Fund (JNF) and the Nature and Parks Authority; my dear friends and colleagues, Dr. Amos Bouskila who was in charge of the small mammals and reptiles, Dr. Elli Groner who was in charge of the arthropods, and Dr. Amnon Friedberg who was in charge of the flying insects.

My great and ambitious students who were so keen to be part of this conservation and restoration program, Adi Ramot, Meirav Perry, Ittai Renan, Tarin Paz, Boaz Shacham, Constantin Grach, and all the undergraduate students who helped with the field work.

References

Arens SM, Geelen L (2006) Dune landscape rejuvenation by intended destabilization in the Amsterdam water supply dunes. J Coast Res 22:1094–1107

Arens SM, Slings QL, Gleen L, van der Hagen H (2007) Implications of environmental change for dune mobility in The Netherlands. paper presented at International Conference on Management and Restoration of Coastal Dunes, Minist. de Medio Ambiente, Santander, Spain, 3– 5 Oct 2007

Bouskila A (2004) Reptiles. In: Dolev A, Perevolotsky A (eds) Red data book of vertebrates in Israel. Yefeh Nof, Jerusalem (in English, revised edition)

Bouskila A, Amitai P (2001) A handbook of the reptiles and amphibians of Israel. Keter Publishing House, Jerusalem, p 345 (in Hebrew)

Fryberger SG (1979) Morphology and distribution of dunes in sand seas observed by remote sensing. In: McKee ED (eds) A study of global sand seas. US Geological Survey Professional Paper, Washington 1052, pp 137–169

Hesp PA, Thom BG (1990) Geomorphology and evolution of active transgressive dunefields. In: Nordstrom KF, Pstuy N, Carter RWG (eds) Coastal dunes: form and process. Wiley, New York, pp 253–288

Jackson LL, Lopoukhine N, Hillyard D (2006) Ecological restoration: a definition and comments. Restor Ecol 3:71–75

Kutiel P (1998) Annual vegetation of the coastal sand dunes, Israel. Isr J Plant Sci 46:287–298

Kutiel P (2001) Conservation and management of the Mediterranean coastal sand dunes in Israel. J Coast Conserv 7:183–192

Kutiel P, Sharon H (1996) Landscape changes in the last 50 years in the area of HaSharon Park, Israel. Ecol Environ 3:167–176 (in Hebrew, with an English abstract)

Kutiel P, Danin A, Orshan G (1980) Vegetation of the sandy soils near Caesarea, Israel. I. Plant communities' environment, and succession. Isr J Bot 28:20–35

Kutiel P, Zhevelev H, Harrison R (1999) The effect of recreational impacts on soil and vegetation of stabilized coastal dunes in the Sharon Park, Israel. Ocean Coast Manag 42:1041–1060

Kutiel P, Peled Y, Geffen E (2000a) The effect of removing shrub cover on annual plants and small mammals in a coastal sand dune ecosystem. Biol Conserv 94:235–242

Kutiel P, Zhevelev Y, Eden Z (2000b) Vegetation and soil response to experimental off-road motorcycle and pedestrian traffic in Mediterranean stabilized coastal dunes. Environ Conserv 27:14–23

Kutiel P, Zhevelev Y, Harrison R (2000c) Coastal dune ecosystems: management for conservation objective. III. Soil response to three vegetation types to recreational use. J Mediterr Ecol 1:171–179

Kutiel P, Cohen O, Shoshany M, Shub M (2004) Vegetation establishment on the southern Israeli coastal sand dunes between the years 1965–1999. Landsc Urban Plan 67:141–156

Levin N, Ben-Dor E (2004) Monitoring sand dune stabilization along the coastal dunes of Ashdod-Nizanim, Israel, 1945–1999. J Arid Environ 58:335–355

Lopatin IK, Chikatunov VI (1999) Two new species of Cryptocephalus from Israel (Coleoptera: Chrysomelidae). Zoosyst Ross 8:329–330

Martínez F, Merino O, Martín A, García Martín D, Merino J (1998) Belowground structure and production in a Mediterranean sand dune shrub community. Plant Soil 201:209–216

Meir A, Tsoar H (1996) International borders and range ecology: the case of Bedouin transborder grazing. Hum Ecol 24(1):39–36

Perry M (2008) Studding perennial plants impact on annual plants diversity in sand dunes in different spatial scales. Thesis submitted in partial fulfillment of the requirements for the degree of Master of Science, Ben-Gurion University

Pont AC, Grach C (2008) A new species of *Coenosia* Meigen from the Mediterranean coasts of Israel and Greece (Diptera: Muscidae). Isr J Entomol 38:115–124

Qin ZP, Berliner PR, Karnieli A (2002) Micometarological modeling to understand the thermal anomaly in the sand dunes across the Israel—Egypt border. J Arid Environ 51:281–318

Ramot A (2007) The effect of plant cover on arthropod community at Nizzanim coastal dunes. Thesis submitted in partial fulfillment of the requirements for the degree of Master of Arts, Ben-Gurion University

Rohrlich V, Goldsmith V (1984) Sediment transport along the southeast Mediterranean: a geological perspective. Geo Mar Lett 4:99–103

Shalmon B (2004) Mammals. In: Dolev A , Perevolotsky A (eds) Red data book of vertebrates in Israel. Yefeh Nof, Jerusalem (in English, revised edition)

Shmida A (1982) Endemic plants of Israel. Rotem 3:3–47 (in Hebrew, English abstract)

Tsoar H, Blumberg D (2002) Formation of parabolic dunes from Barchan and transverse dunes along Israel's Mediterranean coast. Earth Surf Proc Land 27:1147–1161

Van der Meulen F, Salman AHPM (1996) Management of Mediterranean coastal dunes. Ocean Coast Manag 30:177–195

Wasserberg G (1997) Centrifugal organization in a community of desert gerbils: revealing the mechanisms that underlies the pattern. M.Sc. thesis, Department of Life Sciences, Ben Gurion University

Yom-Tov Y, Mendelssohn H (1988) Changes in the distribution and abundance of vertebrates in Israel during the 20th century. In: Yom-Tov Y, Tchernov E (eds) The zoogeography of Israel. Dr W. Junks Publishers, Boston, pp 515–547

Zaady E, Bouskila A (2002) Lizard burrows association with successional stages of biological soil crusts in an arid sandy region. J Arid Environ 50:235–246

Zahavi A, Wahrman J (1957) The cytotaxonomy, ecology and evolution of the gerbils and jirds of Israel (Rodentia: Gerbillinae). Mammalia 21:342–380

Zhenghu D, Honglang X, Xinrong L, Zhibao D, Gang W (2004) Evolution of soil properties on stabilized sands in the Tengger desert, China. Geomorphology 59:237–246

Zviely D, Kit E, Klein M (2007) Longshore and transport estimates along the Mediterranean coast of Israel in the Holocene. Mar Geol 238:61–73

Chapter 12
Passive Recovery of Mediterranean Coastal Dunes Following Limitations to Human Trampling

Alicia Teresa Rosario Acosta, Tommaso Jucker, Irene Prisco and Riccardo Santoro

12.1 Mediterranean Coastal Dunes: A Wealth of Biodiversity and Ecosystem Services

All over the world sandy coastal dunes form transitional ecosystems that bridge the gap between land and sea. The strong environmental gradient that characterizes dune habitats has a strong influence in shaping the vegetation structure of these ecosystems and typically allows the coexistence of different plant communities in a relatively small area (Martínez and Psuty 2004). The complex and compressed zonation of plant communities along the sea-inland gradient is in fact one of the most interesting features of Mediterranean sandy shores (Acosta et al. 2003; Frederiksen et al. 2006) and gives rise to a unique biodiversity in terms of both the number of different habitats and species composition (Van der Maarel 2003). In addition to a diverse and ecologically unique flora, a suite of other ecological, environmental, and economic benefits are typically associated with well-preserved coastal dune systems. Dune habitats are an essential functional component of coastal ecosystems and their degradation is known to promote coastal erosion, since vegetation stabilizes soil against water and wind erosion (Martínez and Psuty 2004). Besides mitigating the loss of the sandy substrate and the abrasive effects of wind erosion, plant assemblages of dune complexes also offer other important ecosystem services, such as protection from sea storms, and provide refuge and habitat for several animal groups, such as arthropods, gastropods, reptiles, and birds (McLachlan and Brown 2006; Barbier et al. 2011). However, at the present time, coastal dune ecosystems are considered to be highly endangered worldwide

A. T. R. Acosta (✉) · T. Jucker · I. Prisco · R. Santoro
Department of Environmental Biology, Università di Roma Tre,
V.le Marconi 446 00146 Rome, Italy
e-mail: acosta@uniroma3.it

M. L. Martínez et al. (eds.), *Restoration of Coastal Dunes*, Springer Series on Environmental Management, DOI: 10.1007/978-3-642-33445-0_12,

Habitat sensu Directive 92/43/EEC	Code	Description
Quercus ilex and *Quercus rotundifolia* forests	Habitat 9340	Forests dominated by *Quercus ilex* usually with some deciduous species
Cisto-Lavanduletalia dune sclerophyllous scrubs	Habitat 2260	Evergreen macchia dominated by shrub species with high cover values and less exposed to the harsh coastal conditions
Humid dune slacks	Habitat 2190	Humid depressions of dunal systems
Coastal dunes with *Juniperus* spp.	Habitat 2250*	Scrub with sparse junipers on coastal sand dunes. Junipers are often relatively low and prostrate, sometimes erect
Malcolmietalia dune grasslands	Habitat 2230	Annual grass community, generally widespread where perennial communities have disappeared due to disturbance
Crucianellion maritimae fixed beach dunes	Habitat 2210	Perennial herb community, partially sheltered from winds and dominated by chamaephytic species
Shifting dunes along the shoreline with *Ammophila arenaria*	Habitat 2120	Perennial herb community growing on mobile dunes, dominated by the rhizomatous tussock grass, *Ammophila arenaria*
Embryonic shifting dunes	Habitat 2110	Pioneer, perennial and halophilous community, dominated by dune-forming plants with low vegetation cover and poor sandy substrate
Annual vegetation of drift lines	Habitat 1210	Annual, nitrophilous and ephemeral community, exposed to wind disturbance and flooding

Fig. 12.1 Theoretical scheme representing a well-preserved coastal dune vegetation zonation in central Italy. Habitats of conservation interest based on the European directive 92/43/EEC are shown

(Brown and McLachlan 2002), even if in Europe a few well-preserved dune systems still exist.

The European Directive 92/43/EEC, also known as the "Habitats Directive," provides a list of habitats and species of conservation interest in Europe, aiming to facilitate and improve conservation efforts and management strategies. At present, the Habitats Directive constitutes the most important legal instrument for biodiversity and nature conservation at the European level (Wätzold and Schwerdtner 2005). Habitats of conservation interest can be identified on the basis of the presence of characteristic and diagnostic plant species (European Commission 2007), a list of which has been standardized across Europe. About 20 coastal dune plant assemblages have been included among these habitats and most of them can be found in Italy (Biondi et al. 2009). An example of a well-preserved coastal dune habitat zonation in central Italy is shown in Fig. 12.1.

12.2 The Impact of Human Activities on Mediterranean Coastal Dunes and the Effects of Trampling

Several studies have highlighted the different stages of coastal dune deterioration throughout Europe (e.g., Acosta et al. 2000; Heslenfeld et al. 2004; Carboni et al. 2009). Various factors threaten coastal habitats and differ along the sea-inland gradient. Urbanization, cattle grazing, farming, and reforestation are among the leading factors that endanger landward side habitats, whereas beach tourism and coastal erosion threaten beach and foredune habitats (O'Shea and Kirkpatrick 2000; Brown and McLachlan 2002; Taveira Pinto 2004). In particular, human activities in coastal areas have intensified over the course of the twentieth century (Feola et al. 2011). At present, tourism is a growing activity with great economic potential, in terms of both employment and income. Many coastal areas depend largely on income derived from the recreational use of beaches, which especially during the warm and sunny summer months attract a great number of local visitors and international tourists (Curr et al. 2000; Houston et al. 2001). Although this recreational tourism is economically stimulating, it is also inevitably associated with a heavy impact on coastal areas and often makes a large contribution to the deterioration and loss of key dune habitats. On top of compromising the ecological functionality of these systems, intense tourist pressure may ultimately lead to a loss of their natural and recreational value as well as undermining other vital ecosystem services that they provide (Andersen 1995).

At present, many of the remnant coastal dune ecosystems of the Mediterranean have been converted into protected areas or incorporated into existing nature reserves, thereby limiting the most destructive activities that threaten these habitats such as sand extraction, or motorcycling (Guilcher and Hallégouët 1991). However, these areas still host a large number of visitors each year and in some cases human trampling of dune vegetation constitutes a serious threat to both the plant assemblages and the structural integrity of the dunes themselves (Bonte and Hoffman 2005; Hesp et al. 2010). Since the early studies of Liddle and Greig-Smith (1975), various authors have reported on the negative impacts that human trampling can have on dune vegetation communities (Bonte and Hoffman 2005). Dune plants have evolved various traits that allow them to thrive under stressful conditions and many of them can tolerant the harsh environmental conditions that characterize the upper beach and the foredunes (Wilson and Sykes 1999; Forey et al. 2008; Gallego-Fernández and Martínez 2011). However, these plants often appear to be particularly vulnerable to physical disturbances, such as those associated with trampling. Trampling on coastal dune vegetation directly affects the survival and reproductive output of individual plants though abrasion, thereby severely reducing the fitness of certain species (Gallet and Rozé 2001). In particular, previous studies have shown that human trampling on dune vegetation can seriously affect species diversity (Bonte and Hoffman 2005) and may also be associated with indirect negative effects owing to the sandy nature of the substrate, which makes it highly susceptible to erosion (Doody 1989).

12.3 Passive Recovery of Coastal Dune Vegetation Following Limitations to Human Trampling: A Case Study from Central Italy

Here, we present a study aiming to assess the effects of limitations to human trampling on dune vegetation in a protected coastal area of central Italy. Using a diachronic approach (studying community changes over time), we compared species cover and richness in a recently fenced sector (which strongly limits trampling) with that of an open sector that has been subjected to trampling over the course of a 4-year experiment. In other words, we have analyzed these two vegetation parameters over time (multitemporal analyses), searching for significant quantitative changes during the study period. As previously mentioned, the detrimental impact of trampling has already been widely documented (see the above references). Our main goal, therefore, was to assess whether species richness and cover in remnant foredune habitats could recover with the aid of a fence as a passive limitation to trampling and if so, try to estimate how long this recovery process is likely to take. We focused on foredune habitats both because they are likely to be exposed to the most trampling as visitors walk across the beach and the surrounding dunes, and because the vegetation that distinguishes them is likely to be the most susceptible to trampling. For the purposes of this study, we took passive recovery to mean recovery without human support, except for the initial act of erecting the fence.

12.3.1 Study Site and the Field Experiment

The experiment was conducted in the ''Torre Flavia wetland'' Natural Monument (Latium region, central Italy; 41°58′N; 12°03′E), a small protected coastal reserve (about 43 ha) located on the Tyrrhenian coast. Torre Flavia is a relic of a larger wetland recently drained (Battisti 2006). This coastal dune sector is longitudinally homogeneous and consists of a sandy beach with a row of degraded dunes and a wetland dominated by *Phragmites australis* on the landward side (Fig. 12.2). The degraded dunes are characterized by a narrow strip of strandline vegetation dominated by *Cakile maritima* and *Salsola kali* (Habitat 1210 of the "Habitats Directive"—annual vegetation of drift lines) intermingled with embryo dune vegetation dominated by *Elymus farctus* and *Anthemis maritima* (Habitat 2110—embryonic shifting dunes). Continuing inland along the degraded dune, the presence of several diagnostic species of a mobile dune habitat (Habitat 2120—shifting dunes along the shoreline) can be found (such as *Eryngium maritimun* and *Echinophora spinosa*). However, the most characteristic species of this habitat, *Ammophila arenaria*, is lacking. At the beginning of the experiment, even though remnants of the three aforementioned habitats were present, they were heavily

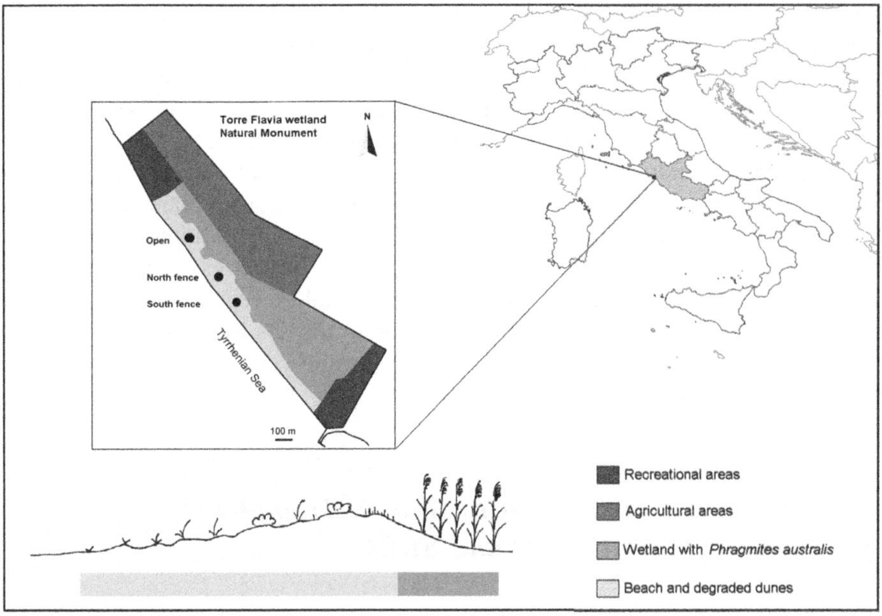

Fig. 12.2 Map indicating the location of the "Torre Flavia wetland" natural monument in relation to the tyrrhenian coast and showing the position of the three sectors within the study site. The vegetation profile at the beginning of the experiment is shown. This coastal dune sector is longitudinally homogeneous and consists of a sandy beach with a row of degraded dunes and a wetland with *Phragmites australis* on the landward side

fragmented and presented a patchy distribution interlaid into a matrix of bare sand. Therefore, sparse, degraded, and relatively homogeneous dune vegetation was observed in the field.

In the spring of 2007 the degraded dune belt was divided into three sectors of equal area (each sector measuring approximately 3,700 m^2) and each sector was assigned to a management scheme. To allow the nesting of two protected birds species (*Charadrius dubius* and *Charadrius alexandrinus*), and to reduce human trampling on dune vegetation, two of the three sectors were fenced off (hereafter, the north fence and the south fence). The third sector was instead kept open to the public (no restriction on trampling) and was used as a control area (hereafter, the open sector; Fig. 12.2). Fences were built using wooden poles (chestnut) with galvanized wire on top of them. To determine the effectiveness of the fences in reducing trampling intensity, the managers of the natural reserve recorded the number of visitors crossing each dune sector. The number of visitors present at the sea shore is highly variable throughout the year and is largely dependent on the season and the weather. To correctly gauge trampling intensity, observations were carried out at the beginning of the bathing season (April to June) during peak hours (10 to 12 a.m.). Following the erection of the fences, in the north and south fence the number of visitors crossing each dune sector was reduced from 5 to 10 people per hour (pre-fencing) to less than one per hour, while in the open sector the trampling intensity remained relatively constant during the study period.

In May 2007, immediately after the fencing systems had been put in place, random vegetation sampling was performed in the three dune sectors using georeferenced points located with a GPS (10 points in each sector, for a total of 30 points). At each randomly assigned point, all vascular plant species were sampled in a 2-m × 2-m (4-m^2) plot. In each of these plots we recorded plant species found along with the percentage cover of each species using a 10 %-interval rank scale. In 2007, just after the erection of the fences, the three sectors showed no significant difference in cover (linear mixed-effects model of vegetation cover in relation to the sector of each plot and distance from the sea as a correction factor). Sampling was repeated during the same period of the year and at the same points from 2007 to 2010.

12.3.2 Effects of Limitations to Human Trampling on Vegetation Cover and Richness

As of 2010, only 4 years after the enclosures were set up, the two fenced sectors showed a considerable increase in vegetation cover, which in both cases had become significantly higher than that of the open sector. Moreover, plant cover in both the north and south fence increased linearly during the study period (Fig. 12.3). For the most part this observed trend was due to an increase in cover of pre-existing species (species sampled in 2007 at the beginning of the study), and not to the colonization by new, incoming species.

Species richness was also strongly influenced by enclosure, but in a slightly different way. As was the case for plant cover, by 2010 species richness in both of the fenced areas had increased and had become significantly greater than that of the open sector. However, whereas cover increased linearly through time, the response in species richness showed a lag period and the increase in the number of species recorded per plot only occurred after the second year of enclosure (Fig. 12.4). In terms of the composition of the new species pool, results were particularly encouraging. Indeed, for the most part the increase in richness was attributable to colonization by species characteristic of coastal dune habitats. Among these were annuals such as *Lagurus ovatus*, *Silene canescens*, and *Vulpia fasciculata*, but also important sand dune geophytes like *Pancratium maritimum*. These species are widespread pioneers that produce a large number of seeds dispersed along the dunes by wind and waves (Watkinson 1978; Balestri and Cinelli 2004).

Another interesting result that emerged from the present study was the important role played by the environmental gradient that characterizes dune ecosystems. Factors of stress and disturbance, such as sea or wind erosion, salt spray, and sand burial, are typical of dune ecosystems and play a major role in shaping plant communities (Wilson and Sykes 1999; Forey et al. 2008; Maun 2009). The degree of

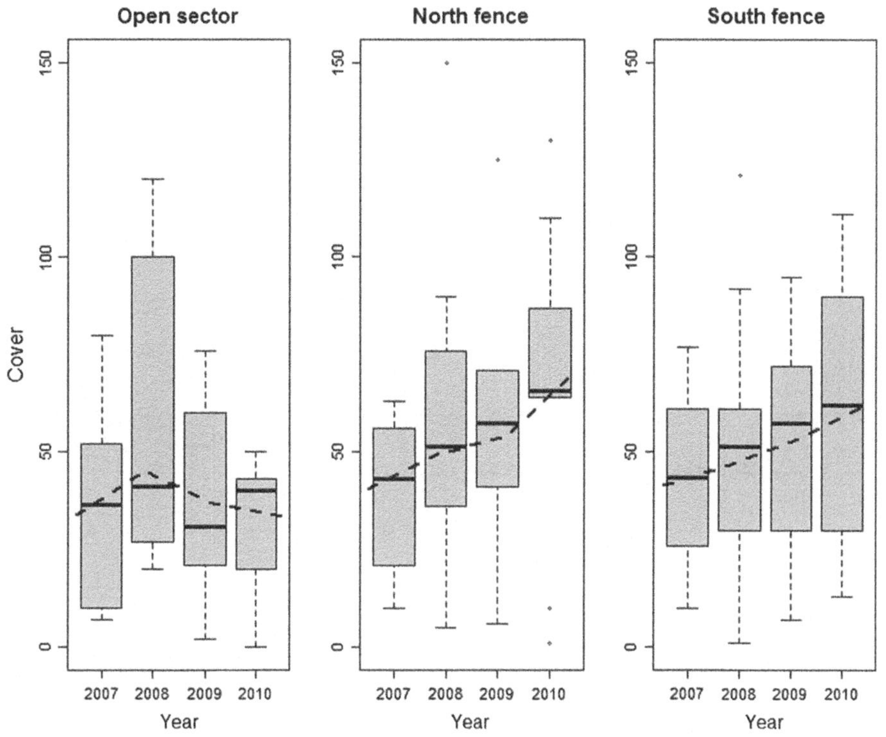

Fig. 12.3 Box-whisker plot of vegetation cover trends (2007–2010) in each of the three sectors. Linear mixed-effects models (LME) were fitted with the study year and distance to the sea as fixed effects and Year|Plot as a hierarchical random factor, allowing the model to fit a separate slope and intercept for each random plot. LME model showed that whereas in the open sector there was no significant difference in cover between 2007 and 2010 (*left*), in both of the fenced areas cover increased significantly ($p < 0.05$). Note that cover showed a rapid response to enclosure with a significant linear increase throughout the years ($p < 0.05$) in both of the fenced areas. A nonlinear smoother (LOWESS) was fitted to highlight trends (*dotted line*)

natural stress and disturbance typically decreases progressing further inland, thus giving rise to an environmental gradient that can be effectively summarized by a measure of distance from the sea (Lortie and Cushman 2007). When correlating the degree of change in species cover and richness to the distance from the sea of each plot, what we have seen is that in the two fenced areas cover and richness tended to increase progressively more over the study period the further away from the sea the plot fell. In the two fenced areas this trend was particularly strong for vegetation cover, whereas in the open sector no trend was observed for either (Fig. 12.5). In other words, it seems that by reducing the effect of trampling (south and north fence), the dune system started to recover some of the spatial structure along the sea–inland gradient that typifies coastal dune vegetation communities and which was absent in the degraded dune (open sector).

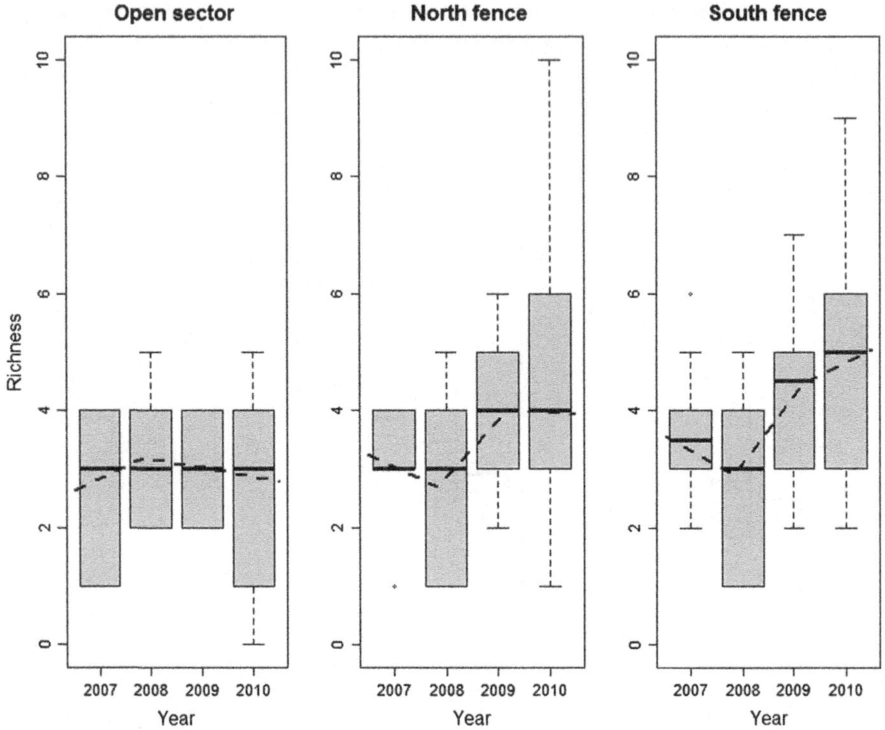

Fig. 12.4 Box-whisker plot of species richness trends (2007–2010) in each of the three sectors. LME of richness as a function of the study year and distance to the sea, and Year|Plot as a hierarchical random factor, were fitted. LME models showed that richness increased significantly ($p < 0.05$) in the two fenced sectors while remaining constant in the open sector (*left*). Note that colonization by new species mainly occurred during the second year of enclosure. A nonlinear smoother (LOWESS) was fitted to highlight trends (*dotted line*)

In summary, our results indicated a significant increase in species cover and richness in the fenced sectors, while both parameters remained relatively constant in the control. In other words, this 4-year experiment highlighted how fencing can be an effective method of promoting the passive recovery of coastal dune vegetation, in terms of both species cover and species richness, as well as their spatial organization (Fig. 12.6a, b).

12.3.3 Final Considerations

The findings of the present study highlighted the great potential of fencing as a passive recovery method for improving the conservation status of degraded and trampled Mediterranean foredune habitats, as well as helping to gain a clearer picture of how the recovery process takes place. By testing two different

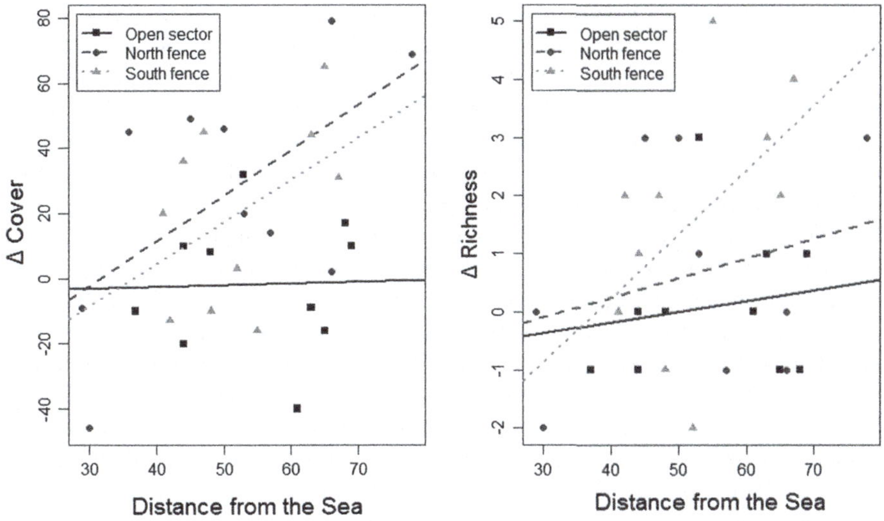

Fig. 12.5 Scatter plots relating the degree of change in cover and richness (Δ Cover and Δ Richness, measured as the difference in cover and richness between 2007 and 2010) to the distance from the sea for each plot. On the *left* is the plot for vegetation cover and on the *right* is the one for species richness. A line of best fit was applied for each sector separately to highlight trends. The *slashed line* refers to the plots in the north fence, the *dotted line* to those of the south fence and the *continuous line* to the open sector

Fig. 12.6 **a** The degraded coastal dune vegetation in 2007 in one of the fenced sectors. **b** The same fenced sector in 2010

parameters over time (species cover and richness) we obtained encouraging results, in terms of both effectiveness and rapidity of recovery. A net benefit to the biodiversity of the coastal dune plant assemblage was observed: during the study period, in those sectors protected by fencing, species richness and cover increased and the plant community regained a degree of spatial structuring along the sea–inland gradient.

Given that this recovery was the result of only 4 years of fencing, particularly striking was its speed. This capacity for rapid recovery is in fact a typical feature of many coastal dune plants, which are often well adapted to high levels of stress and disturbance (Wilson and Sykes 1999; Carboni et al. 2011), and are thus able to recover quickly once a source of disturbance has been limited. Furthermore, in addition to proving its efficacy as a passive recovery method, fencing is both time and cost effective. Indeed the fences are easy to put in place (a day's work with three employees), relatively inexpensive (in our case we calculate around US$700 for erecting the fence), and require little maintenance work, thus allowing a better allocation of resources (economic and employee time). Having said this, based on this 4-year study, it is difficult to estimate how long it would take to achieve cover and richness levels comparable to relatively undisturbed coastal dune habitats, or even whether or not a complete recovery could ever be possible. Also, even though fencing seems to have improved the spatial structuring of the plant assemblage, it is unclear whether the typical foredune habitats will be more easily identifiable in the future. Therefore, a long-term approach with repeat measurements over many different years is required to confirm our findings.

Because of the important functional role that plants play in dune environments, assessing their ability to recover following human disturbance is an important task for the restoration and conservation of coastal dune ecosystems (Emery and Rudgers 2010).

Careful management of coastal areas is needed to develop and implement strategies that reconcile human recreation demands with conservation and restoration priorities. In this context, the results of this study are encouraging for managers and stakeholders as they show how a cost-effective and passive practice, such as fencing, can help restore ecosystem functioning within a relatively short time-frame. Thus, even though our results should be corroborated by long-term studies, we believe passive recovery to be a useful tool for managers pursuing conservation goals in Mediterranean coastal dune systems, especially if implemented alongside other restoration practices.

To conclude, the present study demonstrates the suitability of the diachronic approach to monitoring threatened coastal habitats. We have demonstrated before that most of them are habitats of conservational interest at a European level (European Commission 2007). As human activities in coastal areas will probably intensify in the next few years, multitemporal monitoring should be strongly recommended and encouraged throughout European coastal dune areas.

References

Acosta A, Blasi C, Stanisci A (2000) Spatial connectivity and boundary patterns in coastal dune vegetation in the Circeo National Park, Central Italy. J Veg Sci 11:149–154

Acosta A, Stanisci A, Ercole S, Blasi C (2003) Sandy coastal landscape of the Lazio region (Central Italy). Phytocoenologia 33:715–726

Andersen UV (1995) Resistance of Danish coastal vegetation types to human trampling. Biol Conserv 71:223–230
Balestri E, Cinelli F (2004) Germination and early-seedling establishment capacity of Pancratium maritimum L. (Amaryllidaceae) on coastal dunes in the North-Western Mediterranean. J Coastal Res 20:761–770
Barbier EB, Hacker SD, Kennedy C, Koch EW, Stier AC, Silliman BR (2011) The value of estuarine and coastal ecosystem services. Ecol Monogr 81:169–193
Battisti C (2006) Biodiversità, gestione, conservazione di un'area umida del litorale tirrenico: la Palude di Torre Flavia. Gangemi Editore, Roma p 496
Biondi E, Blasi C, Burrascano S, Casavecchia S, Copiz R, Del Vico E, Galdenzi D, Gigante D, Lasen C, Spampinato G, Venanzoni R, Zivkovic L (2009) Italian Interpretation Manual of the 92/43/EEC Directive habitats. MATTM. Available from http://vnr.unipg.it/habitat/
Bonte D, Hoffmann M (2005) Are coastal dune management actions for biodiversity restoration and conservation underpinned by internationally published scientific research? In: Herrier JL, Mees J, Salman A, Seys J, Van Nieuwenhuyse H, Dobbelaere I (eds) Proceedings 'Dunes and Estuaries 2005'—international conference on nature restoration practices in European coastal habitats. Koksijde, Belgium, 685 pp
Brown AC, McLachlan A (2002) Sandy shore ecosystems and the threats facing them: some predictions for the year 2025. Environ Conserv 29:62–77
Carboni M, Carranza ML, Acosta A (2009) Assessing conservation status on coastal dunes: a multiscale approach. Landsc Urban Plann 91:17–25
Carboni M, Santoro R, Acosta A (2011) Dealing with scarce data to understand how environmental gradients and propagule pressure shape fine-scale alien distribution patterns on coastal dunes. J Veg Sci 22:751–765
Curr RHF, Koh A, Edwards E, Williams AT, Davies P (2000) Assessing anthropogenic impact on mediterranean sand dunes from aerial digital photography. J Coast Conserv 6:15–22
Doody P (1989) Conservation and development of the coastal dunes in Great Britain. In: Van der Meulen F, Jungerius PD, Visser JH (eds) Perspectives in coastal dune management. SPB Academic Publishing, The Hague, pp 53–67
Emery SM, Rudgers JA (2010) Ecological assessment of dune restorations in the great lakes region. Restor Ecol 18:184–194
European Commission (2007) Interpretation manual of European Union habitats, EUR 27. Available from http://ec.europa.eu/environment/nature/legislation/habitatsdirective
Feola S, Carranza ML, Schaminée JHJ, Janssen JAM, Acosta A (2011) EU habitats of interest: an insight into Atlantic and Mediterranean beach and foredunes. Biodivers Conserv 20:1457–1468
Forey E, Chapelet B, Vitasse Y, Tilquin M, Touzard B, Michalet R (2008) The relative importance of disturbance and environmental stress at local and regional scales in French coastal sand dunes. J Veg Sci 19:493–502
Frederiksen L, Kollmann J, Vestergaard P, Bruun HH (2006) A multivariate approach to plant community distribution in the coastal dune zonation of NW Denmark. Phytocoenologia 36:321–342
Gallego-Fernández JB, Martínez ML (2011) Environmental filtering and plant functional types on Mexican foredunes along the Gulf of Mexico. Ecoscience 18:52–62
Gallet S, Rozé F (2001) Resistance of Atlantic heathlands to trampling in Brittany (France): influence of vegetation type, season and weather conditions. Biol Conserv 97:189–198
Guilcher A, Hallégouët B (1991) Coastal dunes in Brittany and their management. J Coastal Res 7:517–533
Heslenfeld P, Jungerius PD, Klijn JA (2004) European coastal dunes: ecological values, threats, opportunities and policy development. In: Martinez ML, Psuty NP (eds) Coastal dunes ecology and conservation. Springer, Berlin
Hesp PA, Schmutz P, Martinez ML, Driskell L, Orgera R, Renken K (2010) The effect on coastal vegetation of trampling on a parabolic dune. Aeolian Res 2:105–111
Houston JA, Edmondson SE, Rooney PJ (2001) Coastal dune management. Shared experience of European Conservation Practice. Liverpool University Press, Liverpool, pp 316–325

Liddle MJ, Greig-Smith P (1975) A survey of tracks and paths in a sand dune ecosystem. J Appl Ecol 12:909–930
Lortie CJ, Cushman JH (2007) Effects of a directional abiotic gradient on plant community dynamics and invasion in a coastal dune system. J Ecol 95:468–481
Martínez ML, Psuty NP (2004) Coastal dunes. Ecology and conservation. Springer, Berlin 386 pp
Maun MA (2009) The biology of coastal sand dunes. Oxford University Press, New York, 288 pp
McLachlan A, Brown AC (2006) The ecology of sandy shores. Academic Press, Burlington 373 pp
O'Shea EM, Kirkpatrick JB (2000) The impact of suburbanization on remnant coastal vegetation in Hobart, Tasmania. Appl Veg Sci 3:243–252
Taveira Pinto F (2004) The practice of coastal zone management in Portugal. J Coastal Conserv 10:147–158
Van der Maarel E (2003) Some remarks on the functions of European coastal ecosystems. Phytocoenologia 33:187–202
Watkinson AR (1978) The demography of a sand dune annual: Vulpia fasciculata. II. The dynamics of seed populations. J Ecol 66:35–44
Wätzold F, Schwerdtner K (2005) Why be wasteful when preserving a valuable resource? A review article on cost-effectiveness of european biodiversity conservation policy. Conserv Biol 123:327–338
Wilson JB, Sykes MT (1999) Is zonation on coastal sand dunes determined primarily by sand burial or by salt spray? A test in New Zealand dunes. Ecol Lett 2:233–236

Chapter 13
Restoration of Dune Ecosystems Following Mining in Madagascar and Namibia: Contrasting Restoration Approaches Adopted in Regions of High and Low Human Population Density

Roy A. Lubke

13.1 Introduction

Two recent restoration studies are compared and contrasted. First, an assessment of a site in southwestern Madagascar that has been proposed for heavy mineral dune mining (CES 2006); and, second, an experimental study on desert dunes following mining for diamonds on the southwestern region of Namibia (van der Merwe 2005). The former study deals with pre-mining planning of future rehabilitation strategies in an area with a dense human population. The latter study, in an area with a very sparse human population dedicated to mining, assesses various rehabilitation techniques that had been attempted following mining.

Madagascar is very rich in flora and fauna and has become a center for studies on endemism and unusual biota (IUCN 1987). This has attracted the attention of many international conservation organisations, thus making it difficult for mining companies to become established in the area. The Namib Desert region of Namibia is equally as important as a conservation hotspot: the study area is within the Succulent Karoo, which extends from the southwestern corner of Namibia, along the Western Cape, and into the southern Cape (Fig. 13.1) (van Wyk and Smith 2001). The Orange River mouth region, on the border of Namibia and South Africa, has been mined for diamonds for many decades and mining commenced long before its conservation importance had been realized.

The Namibian and Madagascan study areas, although both in arid dune regions, are very different in terms of their physical, biological, and social environments. Different approaches are therefore required to planning restoration of the sites following mining. The aim of this chapter is to examine the following aspects:

R. A. Lubke (✉)
Department of Botany, Rhodes University, and Coastal & Environmental Services, Grahamstown, Eastern Cape, South Africa
e-mail: r.lubke@ru.ac.za

M. L. Martínez et al. (eds.), *Restoration of Coastal Dunes*, Springer Series on Environmental Management, DOI: 10.1007/978-3-642-33445-0_13,

Fig. 13.1 The west coast of southern Africa: the Sperrgebiet (*bounded by the square*) in relation to the Succulent Karoo biome (*gray*)

1. Variation in terms of physical features, such as temperature, rainfall, and sand movement, and how this affects restoration processes.
2. Ecosystems in the two regions; determining the formulation of restoration objectives.
3. The role of local communities and other interested and affected parties (IAPs) (or lack thereof) in terms of influencing restoration processes.
4. Contrasting the types of restoration projects suitable for the two sites.

13.1.1 The Two Study Sites

The two study sites are in the West Malagasy (Madagascar) and the Sperrgebiet in the Namib Desert, southwestern Namibia (Pallett 1995). The latter site lies within the coastal zone (Fig. 13.1), a 2-km wide strip between the sea and the 300-m contour (Williamson 1997). Both sites are situated in dry regions with annual rainfall of less than 400 mm, and both are characterized by desert-adapted vegetation. Nevertheless, there are major differences between these sites, in terms of vegetation and the character of environmental impacts.

Madagascar is a mountainous island of the western Indian Ocean that rises to 1800 m above sea level (Fig. 13.2). It is separated from the African mainland by the Mozambique Channel through which the warm Agulhas Current flows

Fig. 13.2 Madagascar landforms showing elevation and large river deltas on coast. The label T shows the approximate position of the Toliara Sands mine site

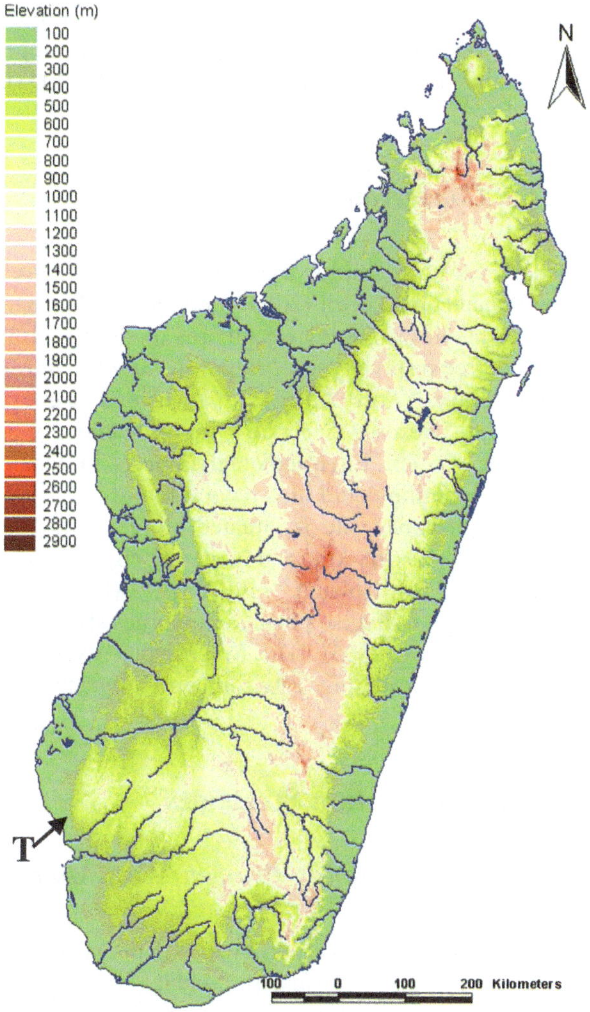

southward. In contrast, Namibia is situated on the Atlantic southwestern coast of Africa, along which the cold Benguela Current flows northward (Britannica Online 2010; Readers Digest Atlas of the World 1991).

In Madagascar the annual rainfall varies from <500 mm in the southwest to levels exceeding 2,500 mm at high elevations (Elouard and Gibon 2003). Air temperatures in the dry southwest vary from 20 to 27 °C (Jury 2003). The Namibian climate is influenced by the cold Bengula Current (Seely, unpublished report). Humidity at the coast is high and average temperatures are moderate (17.3 °C: South African Weather Bureau 1990), although temperatures above 30 °C can occur during easterly "*bergwind*" conditions (Seely unpublished report). The average annual rainfall in the Sperrgebiet (study area) is less than 100 mm (Pallet 1995). Dense coastal fog is an important source of moisture and fog precipitation often exceeds rainfall (CSIR 2004).

13.1.2 Conservation

Madagascar ranks third in the world in terms of terrestrial endemism per unit area. Randrianandianina et al. (2003) reported that Madagascar had 46 legally protected areas at 44 sites, covering a total area of 1,698,639 ha (2.9 % of total area). These include three reserve types: "*Reserve Naturelle Intrégrale*", with a "strict" degree of protection; "*Parc National*," for preserving exceptional natural or cultural areas; and "*Réserve Spéciale*" for protection of specific ecosystems, sites or particular species. Environmental impacts in Madagascar are largely anthropogenic, but mining could also have a future impact.

The Sperrgebiet site is situated close to the Namib-Naukluft National Park, which covers 49,768 km^2 and is mainly characterized by the burnt orange-colored sand dunes. High isolated "*inselbergs*," which, in the Central Namib region are known to harbor diverse plant communities (Burke 2002), also occur in this region (van der Merwe 2005). The Spierrgebiet exclusion area is protected from all activities except mining. Environmental impacts are largely due to mining.

13.2 Rehabilitation Strategies

13.2.1 Toliara Sands Project, West Malagasy, Madagascar

Exxaro Sands (previously known as Kumba Resources) established the Toliara Sands Project in southwestern Madagascar (Fig. 13.2) for the purposes of the exploration of heavy mineral deposits at Ranobe, north of the town of Toliara; Coastal and Environmental Services carried out an Environmental Impact Assessment of the proposed mining (CES 2006).

The climate is hot and dry and the vegetation predominantly a dense, deciduous thicket or deciduous forest (Fig. 13.3), generally referred to as "spiny thicket" (CES 2006). The landscape is characterized by a series of ancient sand dunes lying between the coast and a limestone plateau to the east. The mine deposit lies to the west of a limestone cliff and extends 1–2 km westward in a narrow north–south valley across the sandy plain toward the coast. The study area falls within the driest zone of the sub-arid region, where the annual rainfall averages 350 mm, increasing gradually toward the north. Most rainfall is experienced from December to February, although it is erratic from season to season. Other than a few pools formed around springs, the area is devoid of surface waters.

Fig. 13.3 Deciduous spiny thicket vegetation of the southwestern regions of Madagascar, north of Toliara

Indigenous vegetation cover of much of southwestern Madagascar comprises short dense forest or deciduous thicket that is degraded in many areas owing to expansion of the local people, from the coast eastward. The study area is characterized by Xeric Thicket, wooded grasslands and established fallow lands with scattered Baobab trees. As a result of continued human activities (primarily grazing by domestic livestock, burning, and removal of wood for fuel and charcoal production) plant communities are usually maintained in a sub-climax condition. Extensive areas, believed to have once been covered by thicket and forest communities, now support grassland, wooded grassland, and scrub thicket or forest. Wild plants and animals are important resources for the daily survival of villagers: besides the use of woody plants, as described above, plants are also utilized as food, medicine, and for spiritual purposes (some examples in Table 13.1), and wild animals (such as the ring-tailed lemur and mice and other rodents) are hunted and consumed.

13.2.1.1 Rehabilitation Strategy

The first stage in planning for restoration following mining was to formulate a sound strategic framework (Richards et al. 1993) that could be used to:

1. Establish management vision and objectives for sustainable land-use after mine closure
2. Establish the type of rehabilitation that would be required or desired by IAPs
3. Formulate a management vision that guides:
   (a) Mining methods to minimize damage to adjacent areas
   (b) Rehabilitation work in active post-mining areas
   (c) Rehabilitation work in rehabilitation areas

Table 13.1 Proposed pioneer species suitable for ecosystem rehabilitation in the Toliara Sands area, Madagascar

Species	Uses
Combretum grandidieri	Edible fruit
Fernandoa madagascariensis	Nectar plant for birds
Euclinia suavissima	Edible fruit
Entada chrysostachys	Nodulating roots
Gyrocarpus americanus	Honey plant
Jatropha mahafaliensis	Medicinal plant, potential bio-fuel
Flacourtia lamoutchi	Edible fruit

4. Formulate a rehabilitation plan that would minimize damage to adjacent areas during mining and guide rehabilitation in active post-mining areas
5. Assess the rehabilitation potential of new deposits to assist decision making
6. Guide the development of environmental management programmes relating to new deposits
7. Guide discussions with local communities, authorities, nongovernmental organizations, and other relevant stakeholders during the various stages of the project

13.2.1.2 Rehabilitation Implementation Plan

Once the strategic plan for rehabilitation had been drafted and discussed with the IAPs, a working plan for its implementation was formulated. Following any disturbance, such as mining, there is an opportunity to redefine the use of the land, based on carefully considered social and environmental needs. Using an integrated planning approach, we proposed a management plan that benefitted development within the society, while providing meaningful ecological restoration (Box 1). Meetings were held with a variety of stakeholders within Madagascar to determine which land use option, or range of options, would be suitable for the Toliara Sands project. The following options were explored: woodlots for charcoal and wood production; orchards for local communities, and indigenous vegetation for conservation and ecotourism.

Indigenous Vegetation–Ecosystem Restoration

The general aim of a rehabilitation programme is to recreate a natural ecosystem. This encompasses the following progressive steps: stabilizing land; establishing a cover crop of pioneer species, and establishing a self-sustaining ecosystem with a diversity of plants and animals. Species selection for initial rehabilitation focused the following characteristics:

Fig. 13.4 Disturbed area in the Toliara Xeric Thicket with an almost mono-specific stand of the herbaceous legume, *Tephrosia purpurea*, which could be used as a cover crop species

1. Species (mainly woody types) with clear anthropological uses
2. Robust plants with available seed and high germination success
3. Fodder species with animal-dispersed seed.

Lists were compiled and seeds collected from species on Red Lists, Species of Concern, and CITES-protected species.

Cover crop species were selected from earlier studies. Nonforested zones are often covered by the dense growth of *Tephrosia purpurea,* an annual legume that invades open habits that have been cleared of trees (Fig. 13.4). It produces seed freely, has nitrogen-fixing nodules on the roots, is unpalatable to cattle, and appears suitable as a cover crop. Freshly collected seeds germinate evenly and rapidly, with the mean germination time being 8 days.

Indigenous plant species selection focused on "pioneer" species that promote rapid natural succession, and those considered as "framework" (Goosem and Tucker 1995) or "structural" species. Pioneer species are those plants that colonize a disturbed habitat in the initial stages of restoration and are typically abundant and widespread species. Framework species maybe herbs, trees or lianas that occur in established vegetation and are important components of the structure of the plant community. A list of suitable species is provided in Table 13.1.

Alternative Rehabilitation Options

As an alternative to recreating a natural ecosystem, the "replacement option" aims to establish an alternative to indigenous vegetation. The following options were considered in experimental trials:

1. Creation of woodlots using indigenous species (such as in Table 13.1) and exotic cultivars (such as eucalypts)
2. Establishment of orchards of fruit trees, including citrus and non-citrus trees
3. Cultivation of agricultural crops and grazing pastures

Plot Size and Sample Plot Location

A mobile modular pilot plant was designed and built to undertake pilot scale continuous process evaluations for the proposed mine. In order to minimize disturbance to the natural environment, the processing plant and bulk sample sites were located in degraded vegetation, which had been cleared for agriculture or used for selective harvesting of tree species. Thus, a major limitation for rehabilitation experimentation is that the ideal sites were not particularly ideal, especially in that they were not selected for experimentation in rehabilitation. However, this pilot plant and bulk sample program offered the opportunity to test the potential rehabilitation strategies throughout the prospecting area.

These bulk sampling sites simulated mined conditions, where topsoil had been removed, stacked in shallow mounds, and the sand mined to a variable depth between 1 and 5 m. A total of seven pits were dug. Owing to size constraints at the site, an optimal plot size of 3 × 3 m was chosen in order to set up sufficient plots in the available space.

Planting Treatments

The constant factors or non-variables in each plot were as follows: all plots to be covered with top soil and mulched (Fig. 13.5); all remnants of foreign debris to be removed from the site; compacted soil to be ripped to a depth of greater than 250 mm and the final prepared surface to be furrowed to follow the natural contours of the land.

Fig. 13.5 Young plants laid out in plots in a bulk sample pit at the Toliara Sands using a seaweed and sugar cane waste mulch

Table 13.2 Arrangement of Plots, Plot Numbers and other details in the Toliara Sands Rehabilitation Trials on the Pilot Plant sites

The plots arranged as follows: Pit location	Plot Numbers	Pit dimensions E/W, N/S	Number of plots
Pilot Plant Pit 1	Plot 1 – Plot 25	16 m x 17 m	5 x 5 = 25
Pilot Plant Pit 2	Plot 26 – 74	21m x 22m	7 x 7 = 49
Pilot Plant Pit 3	Plot 75 – 86	9m x 14.6m	3 x 4 = 12
Pilot Plant Pit 4	Plot 87 – 104	16m x 20m	3 x 6 = 18
Pilot Plant Pit 5	Plot 105 – 176	28m x 26m	9 x 8 = 72
Western/Ranobe Site Pit	Plot 177 – 182	5.7m x 19m	1 x 6 = 6
Western/Ranobe Site *Control*	Plot 183 - 200	Approx.10m x 19m	3 x 6 = 18
Northern/Tsifanoka Site Pit	Plot 201 – 212	22.5m x 12m	4 x 3 = 12
Total mined plots			212
Total control plots			212

Plants that had been established in nurseries having undergone a period of "hardening-off" during which they had been exposed to full, direct sunlight under a reduced watering regime, were then ready to be planted out in the experimental plots (Fig. 13.5). Seedlings or cuttings were planted at a density of five plants per square meter. The allocation of pioneer, framework, woodlot, and exotic species was done on a random basis, the plots established according to the details in Table 13.2.

13.2.1.3 Discussion

Although this project is still in progress, and the results of the experimental work are not yet available, the importance of the experimental strategy is presented, whereby:

1. The details of the aims of the rehabilitation program first need to be established
2. The feasibility of the proposed aims needs to be assessed
3. The interests of the local people and other IAPs needs to be tested
4. Experimental trials need to then be established to see whether a large scale program of this type can be established

Detailed planning, assessment, and trials and monitoring are too rarely carried out when mining is proposed. The process should become the norm rather than the exception.

Fig. 13.6 Vegetation in the NAMDEB Pocket Beach area. **a** General view of the dwarf shrubland with *Othonna furcata* and wind dispersed seeds collecting on the lee side of the shrub. **b** The dune grass, *Cladoraphis cyperoides*, on mobile sand dunes

13.2.2 Rehabilitation Following Diamond Mining in the Pocket Beach Areas, Sperrgebiet, Namibia

13.2.2.1 Vegetation

Indigenous vegetation in the Sperrgebiet region of the Succulent Karoo can be summarized as follows: 1,038 flowering plants, 13 ferns and 22 moss and liverwort species, representing a total of 21.5 % of the flowering plant diversity in the entire Succulent Karoo Biome (Burke and Mannheimer 2004); 184 endemic plants have been recorded, 17.5 % of which are endemic to Namibia (Burke 2004). Only 12 species, however, were recorded in the study area by van der Merwe (2005). Vegetation in this region—categorized as "Outer Namib Fog Desert"—is sparse, consisting of herbs and grasses and scattered cushions of succulent shrubs such as *Salsola nollothensis*. Perennials survive in microhabitats within rocky outcrops that are generally covered with lichens. Stem succulents, e.g., *Othonna furcata*, and leaf succulents such as *Lithops* subspp., are common (van Damme 1991; van der Merwe 2005) and patches of dune grasses, *Cladoraphis cyperoides* on mobile sand dunes (Fig. 13.6).

13.2.2.2 Physical Characteristics and Development History

In contrast to the intense planning exercise currently being undertaken at the Toliara Sands site, no impact assessments were undertaken prior to mining at the Sperrgebiet site, which started more than 70 years ago. Restricted access to much of this area, which commenced in 1908 to protect the interests of the diamond mining industry, has, however, inadvertently led to the protection of the biodiversity of a unique wilderness area. In new mining areas rigorous restoration techniques are favored (Burke and Cloete 2004) and van der Merwe (2005) demonstrated how specific techniques could be implemented to re-establish indigenous plant species after the cessation of mining.

Namdeb Diamond Corporation (Pty) Ltd. identified a stretch of coastline for present and future mining activities that consists of sandy beach deposits, referred to as the "Pocket Beach" areas, i.e., sediment-filled embayments situated between headlands (Burke et al. 2002).

The southern part of the Sperrgebiet has an average annual rainfall of 10–89 mm (Pallet 1995). Rain, falling mainly in the winter months, is essential for the germination of plants, and moisture from fog can maintain growth of perennial plants for many years (Seely, unpublished report). Fog, with a mean occurrence of 89 days per year, is important as a source of moisture, an effect that is enhanced by the prevailing low temperatures in the coastal region. Wind has a major influence on coastal vegetation (Seely, unpublished report) with velocities being highest in summer, and wind speeds ranging between 30 and 80 km h^{-1} (Williamson 1997).

In the Sperrgebiet, mining has had a significant impact on coastal dune vegetation—characterized by *Salsola nollothensis* and *Cladoraphis cyperoides* hummocks (Fig. 13.7)—and on sand plain vegetation: "dwarf shrubland" dominated by *Amphibolia rupis-arcuatae* and *Othonna furcata* (Fig. 13.6) (Burke 1997). Van der Merwe (2005) investigated the efficacy of transplanting *Othonna fuscata* as a means of re-establishing the natural dwarf shrubland.

13.2.2.3 Methods

Plot Designs and Experimental Conditions

At Pocket Beach Site 2 (Fig. 13.1) 48 plots measuring 25 × 15 m were laid in 16 sets of 3, in a rectangular shape with the longest axis of each plot being perpendicular to the prevailing wind from the south (Fig. 13.7). Plots were constructed on a mined-out area using a split-plot design to investigate different methods of rehabilitation. In the first split-plot, the presence or absence of nets and their effectiveness as windbreaks was tested. In other tests three different substrate types ("landscaped," "flat," and "pebbled") were incorporated into the design, which is summarized in Fig. 13.7.

In addition to those involving the above variables, further tests were carried out to assess the impact of applying a hydrogel to the roots of *Othonna furcata*. Ninety-six *O. furcata* specimens, salvaged from three different locations, were transplanted into the plots. The experiment commenced in August 2004, coinciding with the dormancy period—from late spring into summer (Anon 2004)—so that the transplant shock would be reduced. The growth of transplanted specimens, changes in sand accumulation, and counts of invading germinating plants in the plots were recorded over a period of 12 months until August 2005.

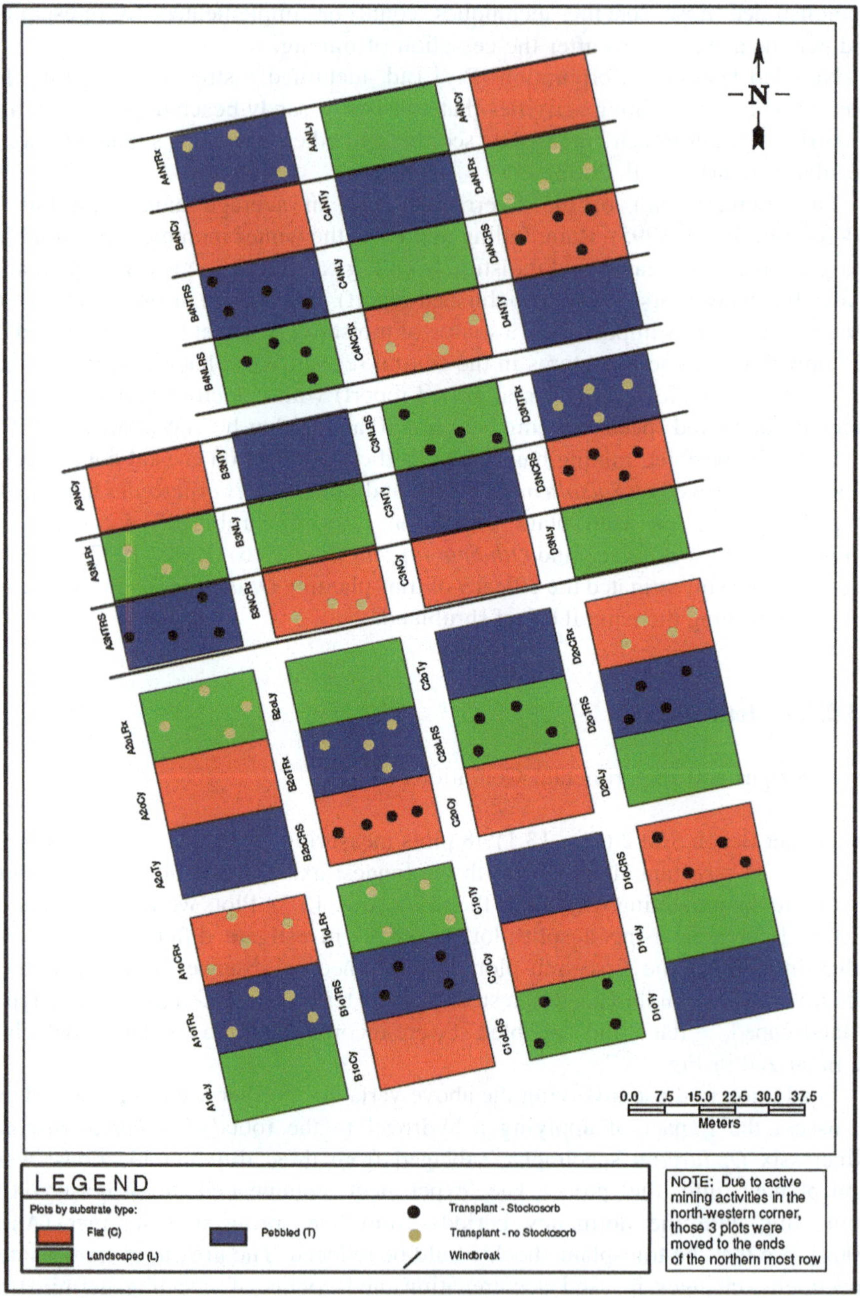

Fig. 13.7 Diagrammatic representation of the experimental layout of the treatments in the rehabilitation site, NAMDEB Pocket Beach Area

Fig. 13.8 Rehabilitation Plot A3NTRS viewed from west to east. **a** August 2004; **b** August 2005

Survival and Growth Measurements of Plants in Experimental Plots

Growth performance of *O. furcata* transplants in experimental plots, were assessed in terms of the following parameters:

1. Presence and survival: plants were categorized as "alive" (with green leaves), "dead/dormant," or "missing."
2. Growth and canopy cover: estimated in terms of canopy area (mm^2) from canopy width measurements.

Invasion of plots by indigenous species from adjacent areas was assessed by means of identifying and counting germinated seedlings in experimental plots.

13.2.2.4 Results

Sand Accumulation

Sand accumulation was significantly higher in areas where nets had been employed as a windbreak (Fig. 13.8). Plot A3NTRS was photographed in August 2004 and 2005. The plot is situated in the netted area, with pebbles as a substrate and transplants added (see Fig. 13.7 for the locality). Sand deposition is considerable in the extreme southwestern corner of the netted area. Measurements of various growth parameters of *O. furcata* transplants indicated no significant difference in plots with different substrate types. No significant interaction effects were observed between the presence or absence of nets and the substrate type (Fig. 13.9).

Survival data for the transplanted *O. furcata* indicated that 4 % of transplants disappeared in the netted area, whereas all plants were present in the un-netted area. In the netted areas, missing plants were buried beneath accumulating sand (Fig. 13.9). This presented a risk, whereas in the un-netted area there was a risk that plants would be blown out due to wind erosion of sand around stems and

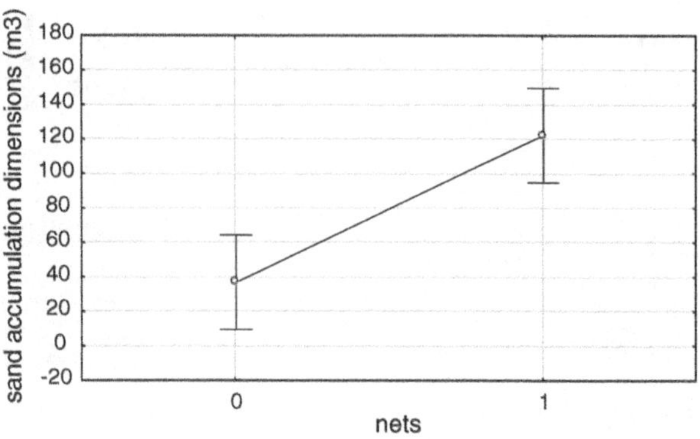

Fig. 13.9 Mean sand accumulation (m^3) in rehabilitation plots without or with nets; LS Means Current effect: $F(1,42)=19.628$, $p=.00007$ Effective hypothesis decomposition Vertical bars denote 0.95 confidence intervals

roots. Most plants in the netted area had green leaves in August 2005 and were considered to be alive.

Growth of *Othonna furcata*

Canopy cover showed a significant increase inside netted plots. This was probably owing to two inter-related factors. First, that the nets provided effective protection against the continuous abrasive effects of the prevailing wind. This not only resulted in a majority of plants surviving, but also in a more stable environment with fewer disturbances. Thus, the transplants had the opportunity to overcome water stress, adapt to new conditions, and grow. Second, the increased sand deposition could have enhanced the growth of *O. furcata,* which is adapted to desert conditions.

Spontaneous Restoration by Seed Germination and Seedling Growth

It was not possible to sample the study site prior to the commencement of mining, but 12 species were recorded in the undisturbed vegetation south of the study area. By the end of the study 9 of these had recolonized the experimental area, indicating that 75 % of species from surrounding plant communities had naturally recolonized the site. Species richness was significantly higher in the netted area, second to the undisturbed area (i.e., the region that had not been mined). This high percentage of returning species indicates a reasonable amount of spontaneous success of the trials, which was over and above what was expected.

Costs

Operational costs for machinery maintenance, fuel, and manpower per hour were calculated using the average time for trips between the plant and the study site for

Table 13.3 Arrangement of Plots, Plot Numbers and other details in the Toliara Sands Rehabilitation Trials on the Pilot Plant sites

		cost (N$)	area(m^2)	N$cost/$m^2$
netted	*nets, pebbles, transplants+stockosorb*	5620	750	7.49
	nets, pebbles, transplant+no stockosorb	5608	750	7.48
	nets, landscaped, transplants+stockosorb	5400	750	7.20
	nets, landscaped, transplant+no stockosorb	5388	750	7.18
	nets, flat, transplants+stockosorb	4860	750	6.48
	nets, flat, transplant+no stockosorb	4848	750	6.46
	nets, pebbles, no transplants	11140	1500	7.43
	nets, landscaped, no transplants	10700	1500	7.13
	nets, flat, no transplants	9620	1500	6.41
un-netted	No nets, pebbles, transplants+stockosorb	2736	750	3.65
	No nets, pebbles, transplant+no stockosorb	2724	750	3.63
	No nets, landscaped, transplants+stockosorb	2516	750	3.35
	No nets, landscaped, transplant+no stockosorb	2504	750	3.34
	No nets, flat, transplants+stockosorb	1976	750	2.63
	No nets, flat, transplant+no stockosorb	1964	750	2.62
	nets, pebbles, no transplants	5372	1500	3.58
	nets, landscaped, no transplants	4932	1500	3.29
	nets, flat, no transplants	3852	1500	2.57

the hauling of pebbles, as well as for the flattening, smoothing or furrowing of each plot. Additional costs were the use of Stockabsorb (calculated using the amount used per plant), windbreaks, which included material and labor, the cost of employing a consultant and technician at an average daily rate for 40 days. Costs were worked out per combination of treatment, which was then multiplied by the number of plots of that treatment and then reworked to the area covered by those plots, which yielded a cost for each treatment in (N$)/$m^2$ (Table 13.3). Note that the erection of nets as windbreaks caused the largest difference in costs.

13.2.2.5 Conclusions

Results suggest that transplanting dwarf shrubs and leaving the area to recover naturally by spontaneous restoration (succession of wind-blown seeds of plants into the mined areas) can result in a community composition similar to that of the pre-mining state, provided that a viable and appropriately positioned seed source is kept intact. Factors limiting establishment and survival of seedlings in coastal dune systems include nutrient deficiency, lack of moisture, predation, and salt spray (Maun 1994). Rapid seedling recruitment is usually triggered by increased moisture events. Seedlings are able to survive partial burial and exhibit increased growth in conjunction with sand accumulation. Thus, rehabilitation sites should be established in the "wet" winter period, as was the case in these trials.

13.3 General Discussion and Conclusions

Very different approaches are required for post-mining rehabilitation in Madagascar and Namibia. Because of the interests and concerns of IAPs residing near the Madagascan study site as well as the high human population density in the area, much planning and a variety of options were necessary for conservation and for establishing plants for the use of local communities. In contrast, because of the harsher climate in the Namib and the extremely low human population density in the area, conservation was the only land-use goal to consider, and innovative methods were developed to achieve these goals.

Acknowledgments Thanks to Rhodes University for research grants to carry out coastal research, Exxaro Sands (Kumba Resources) and NAMDEB for their cooperation and allowing me to report on the studies, and colleagues Ronel van der Merwe, Peter Phillipson, Blaise Cooke, Peter Wood, Carla Strydom and those at CES for their input. Finally, most sincere thanks to Dr Juan Gallego-Fernández and the committee for the invitation to attend the ICCD 2007 conference, where this work was presented in a plenary session.

References

Anon (2004) www.africansucculents.com. Accessed 10 May 2004

Britannica Online: Madagascar and Mozambique Channel. www.britannica.com/ EBchecked/ topic/355562/Madagascar. Accessed April 2010

Burke A (1997) Environmental baseline study and implications for mining of pocket beaches in the Sperrgebiet, Namibia. Unpublished report

Burke A (2002) Plant communities of a central Namib inselberg landscape. J Veg Sci 13:483–492

Burke A (2004) A preliminary account of patterns of endemism in Namibia's Sperrgebiet—the succulent Karoo. J Biogeogr 31:1613–1622

Burke A, Cloete J (2004) Restoring tire tracks: lessons from the Southern Namib Desert. Ecol Restor 22:269–274

Burke A, Mannheimer C (2004) Plant species of the Sperrgebiet (Diamond Area 1). Dinteria 29:79–09

Burke A, Raimondo J, Wiesel I, Clark B, Pulfrich A, Noli D, Griffen M, Griffen E, Simmons R (2002) Pocket beach areas study—environmental impact assessment and environmental management programme. Unpublished report, Windhoek

CES (2006) Rehabilitation strategy for the Toliara sands project. Part 1: Strategic approach and Part 2: implementation plan. Coastal & Environmental Services, Grahamstown

CSIR: Envirotek (2004) Environmental impact assessment. Proposed Kudu CCGT Power plant at Oranjemund. Namibia Power Corporation (Pty) Ltd, Windhoek

Elouard J-M, Gibon F-M (2003) Ecology of aquatic insects. In: Goodman SM, Benstead JP (eds) The natural history of Madagascar, vol 1709. University of Chicago Press, Chicago, pp 511–157

Goosem S, Tucker NIJ (1995) Repairing the rainforest. Cassowary Publications, Cairns

IUCN (1987) Madagascar: an environmental profile. Gresham Press, Surrey

Jury MR (2003) Chapter 3: climate. In: Goodman SM, Benstead JP (eds) The Natural History of Madagascar, vol 1709. University of Chicago Press, Chicago, pp 75–87

Maun MA (1994) Adaptations enhancing survival and establishment of seedlings on coastal dune systems. Vegetatio 11:59–70

Pallett J (1995) The Sperrgebiet—Namibia's least known wilderness. Namdeb Diamond Corporation (Pty) Ltd., Oranjemund

Randrianandianina, BN, Andriamahaly, LR, Harisoa, FM, Nicoll ME (2003) In: Goodman, SM, Benstead JP (eds) pp 1423–1437. The Natural History of Madagascar. University of Chicago Press, Chicago, 1709 pp

Readers Digest Atlas of the World (1991) Pope JA (ed) The Reader's Digest Association Limited, London, 260 pp

Richards IG, Palmer JP, Barratt PA (1993) The reclamation of former coal mines and steelworks. Elsevier, Amsterdam

Van Damme P (1991) Plant ecology of the Namib desert. Afrika Focus 7(4):355–400

Van der Merwe R (2005) Rehabilitation following diamond mining in the pocket beach areas, Sperrgebiet, Namibia. BSc (Hons.) project. Rhodes University, Grahamstown

Van Wyk AE, Smith GF (2001) Regions of floristic endemism in Southern Africa—a review with emphasis on Succulents. UMDAUS Press, Hatfield

Williamson G (1997) Preliminary account of the floristic zones of the Sperrgebiet (Protected diamond area) in Southwest Namibia. Dinteria 25:1–68

Chapter 14
The Impacts on Natural Vegetation Following the Establishment of Exotic *Casuarina* Plantations

Patricia Moreno-Casasola, M. Luisa Martínez, Gonzalo Castillo-Campos and Adolfo Campos

14.1 Introduction

Casuarina equisetifolia Forst. (Casuarinaceae) (hereafter referred to as *Casuarina*) also called Australian-pine or beach she-oak, is a nitrogen-fixing, medium to large evergreen tree, 15–30 m or more high and with a diameter at breast height (DBM) of up to 50 cm. It is native to the tropical and subtropical coastlines of Australia, Southeast Asia, Melanesia, Polynesia, and New Caledonia. However, it has become pantropical as its native range has expanded through its introduction and later naturalization, because of its ability to reproduce in dense stands from abundant self-seeding and its high tolerance to extreme environmental conditions, such as drought and low nutrient availability. It is also tolerant to burial by sand. In Mexico, the natural expansion of *Casuarina* on coastal dunes seems to be limited, but nevertheless in the past decades this species was widely planted, and nowadays it is considered as an invasive in the national invasive species inventory (March-Mifsut and Martínez-Jiménez 2007).

Casuarina plantations have frequently been used outside its distribution range in order to stabilize coastal dunes in Senegal (Mailly and Margolis 1992) and Mexico; for coastal protection (Mascarenhas and Jayakumar 2008); and in coastal dune restoration projects (meaning revegetation) in India (Homji 1995), Cuba (Izquierdo et al. 2005), Puerto Rico (Parrotta 1995), and Mexico (Espejel and Ojeda 1995).

P. Moreno-Casasola · M. L. Martínez (✉) · A. Campos
Red de Ecología Funcional, Instituto de Ecología, A.C, Antigua Carretera a Coatepec no. 351 91070 El Haya, Xalapa, Ver, Mexico
e-mail: marisa.martinez@inecol.edu.mx

G. Castillo-Campos
Red de Biodiversidad, Instituto de Ecología, A.C, Antigua Carretera a Coatepec no. 351 91070 El Haya, Xalapa, Ver, Mexico

M. L. Martínez et al. (eds.), *Restoration of Coastal Dunes*, Springer Series on Environmental Management, DOI: 10.1007/978-3-642-33445-0_14,

These introductions have had a strong negative impact on local soil attributes, as well as on the population, community, and ecosystem levels (Gordon 1998).

14.1.1 Impact of Casuarina on Soil Attributes

Casuarina is an actinorhizal species that is able to grow in nutrient-poor soils, mainly because of its ability to fix N_2 in the root nodules through a mutualistic symbiosis with the actinomycete *Frankia*, and by its capacity to develop mycorrhizae (Gordon 1998). Because of this, *Casuarina* plantations may largely modify soil characteristics, especially in coastal dunes.

Sand dune soils have high leaching rates, with a resulting low level of nitrogen. When nitrogen fixers form large stands, local nitrogen cycling may be altered significantly (Vitousek 1986), resulting in higher phosphorous and nitrogen concentrations in leaves as well as enriched soil-nutrient levels (Versfeld and van Wilgen 1986). This enriched soil has been observed in *Casuarina* stands too. Leaf and branch litter beneath *Casuarina* trees can be 5–10 cm thick (Fernald and Barnett 1991), with organic matter accumulating at rates that are similar to those recorded in boreal and subalpine forests (Mailly and Margolis 1992; Izquierdo et al. 2005).

14.1.2 Impact of Casuarina on Vegetation

The high rates of litterfall accumulation and nutrient accumulation beneath *Casuarina* stands can have contrasting effects at the community level. On the one hand, it has been observed that a higher nutrient content in the soil (organic matter and nitrogen) can actually facilitate natural regeneration and thus catalyze succession in deforested and degraded sites (Parrotta 1995). However, contrasting patterns have also been observed by Bond (1993); Duever et al. (1986); Gordon (1998); and Abdel Wahab (1980). These authors found that *Casuarina* plantations can also suppress recruitment of other species, although the mechanisms causing inhibition or shifts in recruitment patterns have not been investigated in detail. Evidence suggests that because *Casuarina* trees reduce light availability, and have higher evapotranspiration rates than the native vegetation (Gordon 1998), they are likely to modify competitive interactions and can out-compete native species (Abdel Wahab 1980; Gordon 1998), by suppressing seedling growth and establishment. Because very few plants grow beneath the canopy of *Casuarina*, Bond (1993) has named this tree a "keystone weed".

14.1.3 Impact of Casuarina on the Fauna

Introduced *Casuarina* plantations also affect the native fauna. For example, when these trees are planted on foredunes they fall over easily during strong winds, making nesting habitat inaccessible to sea turtles (*Caretta caretta*) (Schmid et al. 2008). In addition, the removal of *Casuarina* plantations was favorable to the skink (*Eutropis bibronii*) in India, increasing its population, probably because of the availability of more suitable space for basking activity. It was also attributable to the regeneration of *Spinifex littoralis* at sites previously occupied by *Casuarina*, since skinks were found to be associated with this plant (Subramanean and Vikram Reddy 2010).

In their native distribution range *Casuarina* stands may have both negative and positive impacts on the bird community. Arnold (1988) found that the total number of birds was lower in *Casuarina* forests than elsewhere (*Eucalyptus wandoo* open forest and *E. accedens* open forest with *Dryandra sessilis*). In addition, because *Casuarina* provides little food for most nectar-feeders, their densities are usually lower in *Casuarina* forests than in other communities (Arnold et al. 1987). On the other hand, Frith (1979) studied the ecology of land birds in Aldabra Atoll and found that many endemic birds were closely associated with *Casuarina*. In southeastern Queensland Bentley and Catterall (1987) found that *Casuarina* corridors and linear remnants of forests were used by birds to move between forest fragments.

In brief, *Casuarina* is usually considered an invasive species outside its native distribution range, with a negative impact on the population, community, and ecosystem functioning, generally resulting in arrested natural succession. Because it is oftentimes used in Mexico to stabilize coastal dunes and increase vegetation cover.The goal of this chapter was to analyze the impact of *Casuarina* stands on the natural regeneration process occurring beneath the canopy of this tree. We also wanted to explore whether local conditions modify the impact of this species in terms of passive restoration and natural regeneration on tropical dunes in Mexico. We asked: is it possible to recover the natural coastal dune vegetation after reforestation with these exotic–invasive *Casuarina* trees?

14.2 Study Site

The study site is located to the north of the port of Veracruz, in the outskirts of the city (Fig. 14.1). The area is being actively transformed because of the growing infrastructure for commercial activities in the Port. This area is an extensive sand plain where the river Río Medio drains. Nowadays the site is mostly urbanized (Fig. 14.1).

The coastline is oriented in an almost east-west direction, and faces north-west. Sand is transported both by ocean currents and by wind, and extensive dunefields

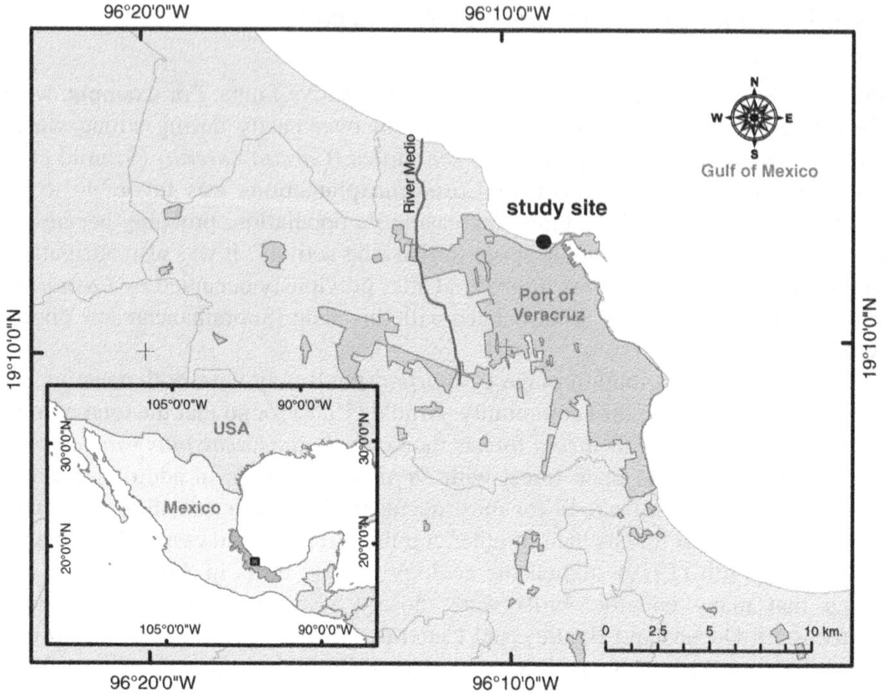

Fig. 14.1 Location of the study site

with transgressive and parabolic dunes can be found in the region (López-Portillo et al. 2011). Long-term (40 years) means that annual precipitation is 1,500 mm, and mean monthly temperatures range from 17 °C in January to 33 °C in June. Eighty percent of the rain falls between May and October and the dry season occurs from November to April.

In 1885, Ignacio Ochoa Villagómez, an agronomist, constructed an artificial dune at the study site with the basic idea that it would protect urban infrastructure by stopping the sand blowing from nearby mobile dunes, until the trees and grasses that were introduced stabilized the sand. Australian sea-pines (*C. equisetifolia*) were planted, since they tolerate considerable sand accretion and the low nutrient and moisture conditions of sand dunes. This plantation largely prevented the sand from blowing into the neighboring city and then became protected and remained a cornerstone for the local inhabitants (Siemens et al. 2006). It was known as "*La Pinera*" (the pine grove), until the present century when the port began to expand: *La Pinera* was partially cut down and its surface contracted to less than 50 % of its original area.

14.3 Methods

14.3.1 Vegetation

The *Casuarina* artificial forest was subdivided into three stands with different mean trunk width. Diameter at breast height (DBH; approximately 1.50 m) was measured in five 10-m × 10-m plots in each of these three forest stands, allowing us to define them as young (smallest DBH values), mature, and old *Casuarina* plots (largest DBH values). These *Casuarina* stands were planted on different dates and in rows parallel to the coast. In each of these different-aged stands two transects of five 10 × 10 m adjacent quadrats were marked, parallel to the shoreline, so that we could observe changes in the community composition in each of these stands. In each quadrat, species were collected for identification, and plant cover per species was estimated with the Westhoff and van der Maarel (1978) scale. For comparison, the same type of sampling took place in a coastal tropical dry forest growing on sand dunes further inland, on a slope of the same dune system, but where *Casuarina* trees had not been planted. Finally, another set of ten 10 × 10 m quadrats was sampled on a foredune area stabilized with *Casuarinas*, 5 km north of the study site, with a DBH structure similar to the mature stand (referred to here as *Casuarina*). For each stand Shannon's diversity index was calculated. In all cases, once species were identified, the type of fruit was determined, either from herbarium species or from the literature and then a dispersal syndrome was assigned to each species. A matrix was built with the species cover values and analyzed with a principle component analysis (PCA) ordination using the program PCOrd (McCune et al. 2002).

14.3.2 Soil Attributes

Soil profiles 1 m deep were excavated at two sites in each stand, close to the quadrats where floristic composition was sampled. In each horizon soil samples were collected and transported to the laboratory. Samples were air dried and sieved (up to 2 mm) before chemical analyses were performed. The following analyses were carried out following standard procedures as described by Sparks (1996): pH in water was measured using a 1:2 relation, total organic carbon (Walkley and Black 1934), total nitrogen using the Kjeldahl method, exchangeable cations (Ca, Mg, Na, K) through soil lixiviation with ammonium acetate 1 N pH 7, $CaCO_3$ percentage was calculated using the neutralization method of HCl with a known concentration and a posteriori titulation of the excess acid with NaOH. In a saturation extract the following chemical properties were determined: electrical conductivity (dS m^{-1}), calcium (mg L^{-1}), magnesium (mg L^{-1}), potassium (mg L^{-1}), sodium (mg L^{-1}), chloride (mg L^{-1}), bicarbonates (mg L^{-1}), and sulfates (mg L^{-1}). The saturation extract is an important aqueous solution because many soil properties, such as the composition of soluble salts and electrical conductivity, are related to the plant's response to salinity.

14.4 Results

14.4.1 Vegetation

The three *Casuarina* stands (young, mature, and old) were dominated by *C. equisetifolia* trees, but these showed very different DBH sizes (Fig. 14.2). Although they could not be correlated with planting date or age class (because we could not find that information in the local archives), they showed a different size structure, and thus it could be assumed that they were planted at different times.

A total of 100 species were collected in the area, and eight species were not identified. Species richness and diversity varied among stands (Fig. 14.3). Species richness and diversity were very low in the *Casuarina* plots growing on the dry sand of foredunes. These values increased considerably on the young *Casuarina* plots in humid areas, where several herbaceous, shrub, and tree species begin to appear. With increasing age, species richness, and diversity beneath *Casuarina* continued increasing (Fig. 14.3). Highest diversity and mean species richness were recorded in the tropical dry forest and in the old *Casuarina* plots. The highest belong to the tropical dry forest, which represents the original vegetation, or at least the less disturbed vegetation at the site.

The plant community structure was also different between the study sites (Fig. 14.4). For comparison, trees, shrubs, and woody lianas were grouped into a single class, and herbs and creepers into another. In the *Casuarina* plots located in the drier area, the only tree species that we found was the Australian pine, with only three herbaceous species (Fig. 14.4a). Young and mature plots showed an

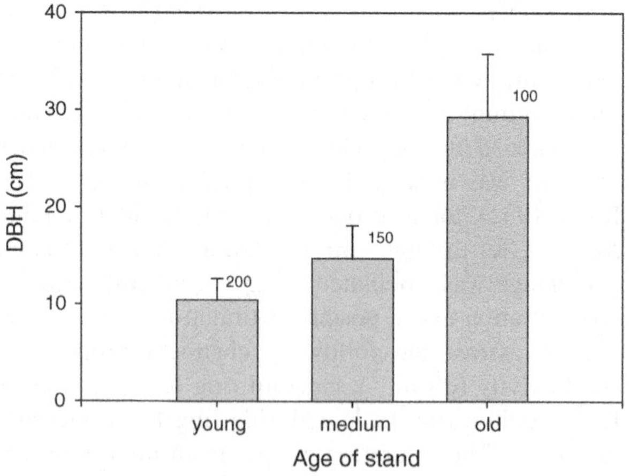

Fig. 14.2 Mean diameter at breast height (DBH) and standard deviation of *Casuarina* trees, in a *Casuarina* plantation growing on coastal sand dunes. Numbers above bars indicate total number of individuals measured in ten 10-m × 10-m plots. The first diameter class was interpreted as a young stand, the medium-sized class as a mature stand, and the larger-stemmed trees as the oldest plantation

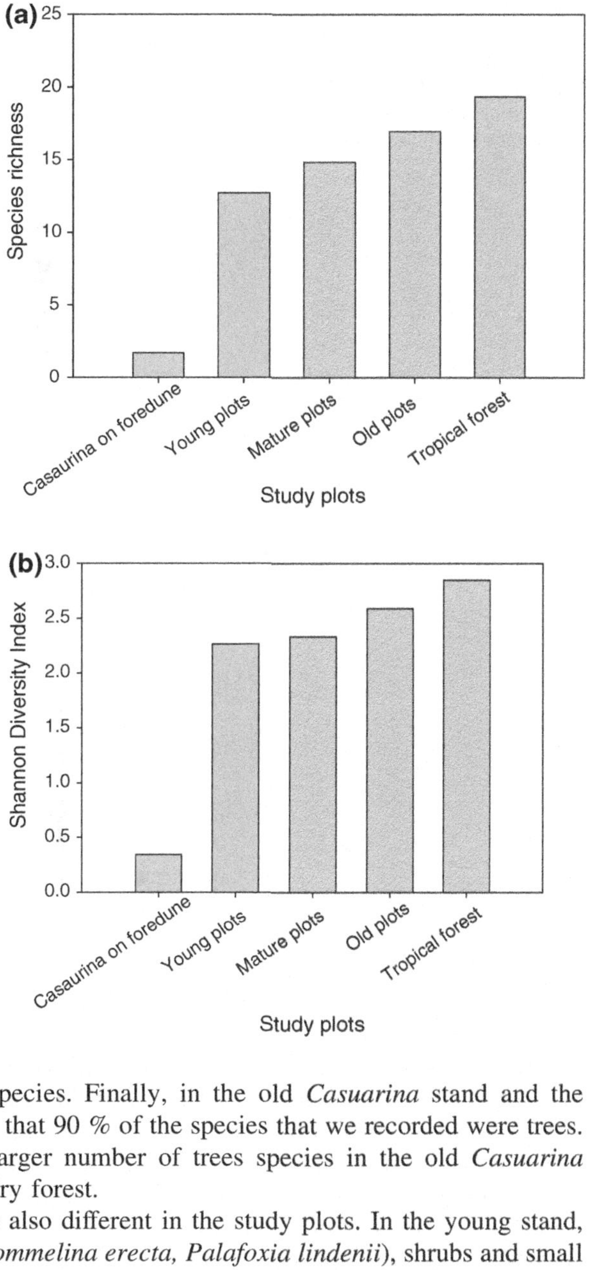

Fig. 14.3 **a** Species richness and **b** diversity values (Shannon diversity index) for the five stands. Highest diversity and species number were found in the tropical dry forest and in the older *Casuarina* plots

increased number of tree species. Finally, in the old *Casuarina* stand and the tropical dry forest we found that 90 % of the species that we recorded were trees. Interestingly, we found a larger number of trees species in the old *Casuarina* stands than in the tropical dry forest.

Species composition was also different in the study plots. In the young stand, herbs (*Urochloa maxima, Commelina erecta, Palafoxia lindenii*), shrubs and small trees (*Citharexylum ellipticum, Solanum diphyllum, Trixis inula, Lycianthes lenta, Malvaviscus arboreus, Bursera simaruba*), and woody climbers (*Gonolobus barbatus* and Passilfora spp.) were found. Mature stands have a similar number of herbaceous and woody species as the young stand, but species richness and

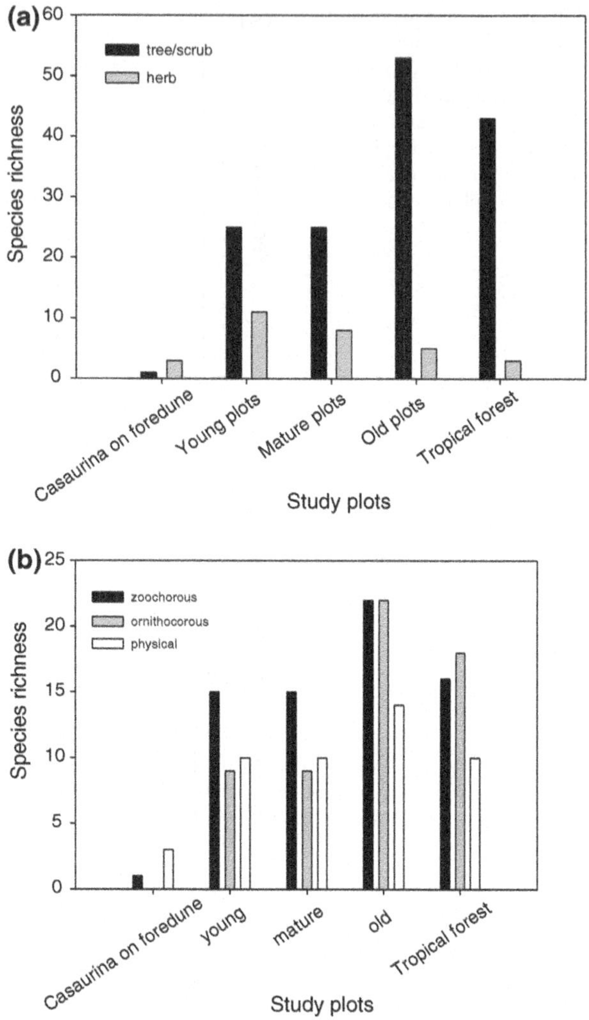

Fig. 14.4 **a** Species number of the different growth forms recorded in each stand. **b** Species number in relation to different seed dispersal syndromes in different stands

diversity was slightly higher (Fig. 14.4a). In the older *Casuarina* plots, the number of woody species increased considerably (*B. simaruba, Sideroxylon celastrinum, Cestrum dumetorum, Chiococca coriacea, Jacquinia macrocarpa, Piper amalago*, and *Passiflora serratifolia)* and were more abundant. In this case, species richness and diversity values were higher and closer to those found for the tropical dry forest. In contrast with the above, the *Casuarina* plantation on a nearby foredune showed a very low species richness, with only one tree species (and besides *Casuarina*) and three typical dune herbaceous pioneers (*Ipomoea pes caprae, P. lindenii, Bidens pilosa*; Fig. 14.4a). In brief, community richness and composition has changed drastically over time (Fig. 14.5).

Fig. 14.5 The vegetation in the understory of a planted *Casuarina* forest has changed over time. **a** Panoramic view of the forest growing on previously mobile dunes. **b** Understory species growing beneath *Casuarina* stands. Many species are from nearby tropical rain forest. **c** Understory plants growing beneath *Casuarina* and **d** increasing vegetation cover, height, and diversity

14.4.2 Dispersal Syndrome

The analysis of the dispersal syndrome of the fruits from all the species that we found in the sampled plots shows differences among the five stands. Species whose seeds are dispersed by physical vectors (wind, water, explosion or gravity) were predominant in the *Casuarina* plots located on the foredunes (75 %) (Fig. 14.4b). In the old plots and tropical dry forest, the percentage of species with seeds dispersed by abiotic vectors decreased to 24 and 23 % respectively. Zoochorous species increased in all four *Casuarina* stands in comparison to the *Casuarina* forest on foredunes. In particular, ornithochorous species were most abundant in the oldest stands, as well as in the tropical dry forest (38 and 41 % respectively). It is evident that the most frequent species have an ornithochorous or zoochorous dispersal syndrome, with the exception of *C. equisetifolia*.

The species that we observed in more than ten quadrats, except for *Casuarina*, are commonly found in tropical dry forest and in old secondary forests. Most of them are trees or shrubs, except *G. barbatus* and *P. serratifolia*, two woody lianas that were also found in the tropical dry forest. Most of these species are dispersed by animals (Fig. 14.6; Table 14.1). Woody species were predominantly

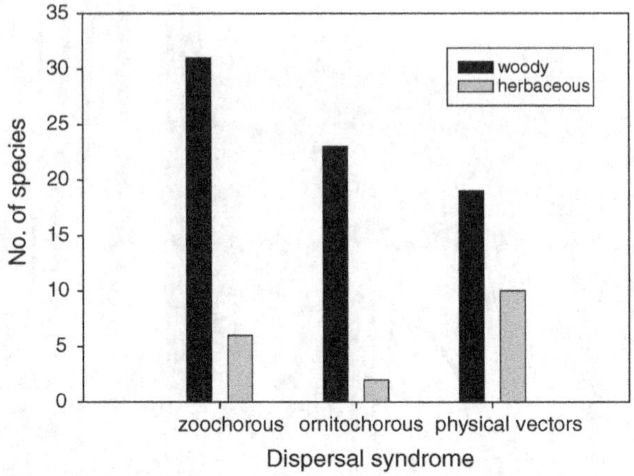

Fig. 14.6 Relationship between seed dispersal syndromes and species growth forms (woody and herbaceous)

Table 14.1 List of the most frequent species and their seed dispersal syndrome

Zoochorous	Ornithochorous	Physical vectors
Acacia cornigera (L.) Willd.	*Bursera simaruba* (L.) Sarg.	*Casuarina equisetifolia* L.
Chiococca coriacea M. Martens & Galeotti	*Casearia corymbosa* Kunth	*Gonolobus barbatus* Kunth
Hamelia patens Jacq.	*Cestrum dumetorum* Schltdl.	*Matayba clavelligera* Radlk.
Lycianthes lenta (Cav.) Bitter	*Citharexylum ellipticum* D. Don	
Malvaviscus arboreus Cav.	*Coccoloba barbadensis* Jacq.	
Passiflora serratifolia L.	*Jacquinia macrocarpa* Cav.	
Piper amalago L.	*Rhacoma uragoga* (Jacq.) Baill.	
Solanum diphyllum L.	*Rivina humilis* L.	
Xylosma panamensis Turcz.	*Sideroxylon celastrinum* (Kunth) T.D. Penn.	
	Stemmadenia donnell-smithii (Rose) Woodson	

zoochorous and ornithochorous, while herbaceous species were mostly dispersed by physical vectors and a few were zoochorous (Fig. 14.6).

The PCA analyses of the 49 plots located in the five study sites explained 85.20 % of the accumulated variance (Axis 1 and 2, Fig. 14.2). Axis 1 shows a gradient from samples that combine *Casuarina* with other species (in the old *Casuarina* stands) and tropical dry forest (located on the left and in the center of the ordination space). In this case, the occurrence and dominance of *Casuarina* in the old stands separates them from those located in the tropical dry forest. In turn, plots from young and mature *Casuarina* stands with very few species were located in the center of the ordination space along Axis 1, and clearly separated from the

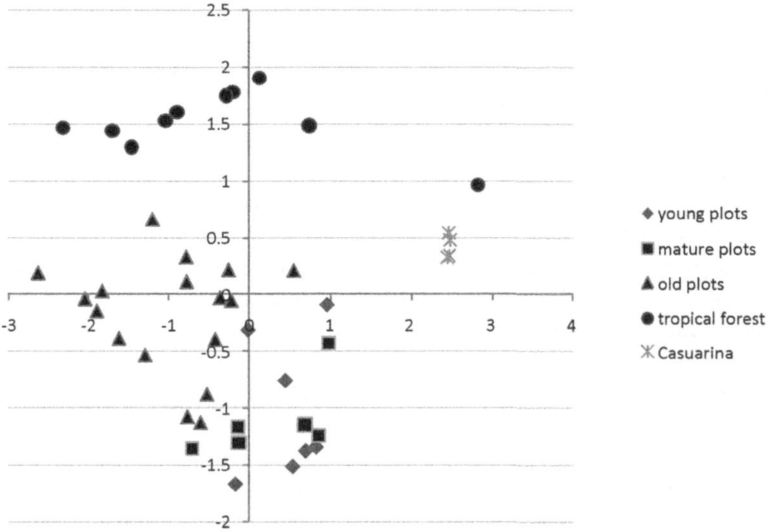

Fig. 14.7 Principle component analysis (PCA) variance–covariance explaining 85.20 % of the accumulated variance. *Axis 1* shows a gradient from samples of tropical dry forest and *Casuarina* plots with other species (old, mature, young) and plots from *Casuarina* stand with hardly any accompanying species. *Axis 2* shows a gradient of species richness, from tropical dry forest plots, rich in species, to species-poor young and mature plots

old stands and the tropical dry forest. *Casuarina* stands in nearby foredunes showed very few accompanying species and were located at the extreme positive end of Axis 1. That is, foredune plots with almost a monospecific *Casuarina* stand were different from the rest of the quadrats that we sampled (Fig. 14.7). Axis 2 shows a gradient from tropical dry forest plots (at the top), rich in species and with no *Casuarina* trees, old plots (also rich in species, along the center) and mature and young plots (species poor) at the negative end of Axis 2.

14.4.3 Soil Analysis

Dune soils where *Casuarina* stands were planted are considered Endogleyic Arenosols, in which horizons are formed mainly by fine, loose sand, representing 75 % of the mineral fraction. A thin homogeneous layer of fresh organic remains, from the litter of *Casuarina* trees, was found to be in direct contact with the mineral sandy soil. In the young *Casuarina* stand, we observed an AC horizon (0–8 cm) in the soil, with a very dark brown color (10YR2/2) dominant in 80 % of the horizon and a more pale brown color (10YR6/3) in 20 % of the soil. We observed very few fine roots (<1.0 mm). The color of horizon Cg (8–56 cm) was a dominant dark brownish grey (10YR4/2) mixed with yellowish brown diffuse mottles (10YR5/8), which is evidence of a slight hydromorphism.

In the plots with mature *Casuarina* trees, horizon A1 (0–5 cm), showed a dark brown color (10YR2/2), with organic remains slightly decomposed and numerous very fine roots, less than 1 mm in diameter. The lower horizon, Cg (5–60 cm) is predominantly pale brown (60 %–10YR6/3) and brown (40 %; 10YR4/3), indicating a slight hydromorphism. This last horizon had very few roots.

In the old *Casuarina* plots, some of the trees have already fallen down and the wood is dry. In this area, the phreatic level is only 30 cm deep. Horizon A1 (0–9 cm) in this site is very dark brown (10YR2/2) and medium-sized roots (with mean diameter ranging from 2 to 5 mm) are common. The next horizon, Cg (9–35 cm) was dark brownish grey (10YR4/2) with dark yellowish brown veins (10YR4/6) that follow the root distribution. This soil shows a greater hydromorphism than that found at the other two sites.

The chemical characteristics of the soil showed differences between the *Casuarina* stands. In general, pH values varied strongly from alkaline to extremely alkaline in the old *Casuarina* plots. The carbon concentration, in the first horizon, increased in the old plots. In this site organic matter is incorporated into the soil, thus more root activity was observed. The concentrations of exchangeable cations (Ca, Mg, K, Na) were, in all sites, higher in the upper horizons than in the lower, emphasizing the high to very high calcium and sodium values. The calcium carbonate concentration showed no significant variation between sites. The chemical composition of the saturation extract showed significant differences among sites (Table 14.2). For example, the ion concentration measured with electrical conductivity shows that there was a tendency to decrease from the young *Casuarina* stands to the old ones, a situation that can be attributable to the increment in organic matter content (FAO 1998). In this case, the electrical conductivity values indicate that soils are saline to moderately saline (Hillel 2000), so it is likely that the established plant communities are semi-tolerant to salinity. Soluble salts in the saturation extract are virtually restricted to sodium and calcium among the cations, and chloride and sulfate between the anions. This is consistent with the marine influence on the dune soils. The relative dominance of sodium is associated with the high levels of chlorine, while the abundance of calcium is related to high levels of sulfate (Richards 1954). Bicarbonate concentration tends to increase from the young plots to the old plots, a situation explained by the greater root and microbial activity in the older stands with a high CO_2 production, which generates bicarbonate.

14.5 Discussion

As expected, *Casuarina* was the dominant tree in all our study plots (except for the tropical dry forest, of course), creating an almost uniform canopy cover. Nevertheless, we found that natural regeneration is occurring in the hydromorphic soils, especially in the old *Casuarina* stands. Here, diversity and species composition were similar to the tropical dry forest growing on coastal dunes (tropical dry forest

Table 14.2 Chemical composition of the sand from *Casuarina* stands of different ages in two soil horizons (AC and Cg)

Site	pH	C	N	$CaCO_3$	Ca^{2+}	Mg^{2+}	K^+	Na^+	EC	Cl^-	HCO_3	SO_4	Ca2$^+$	Mg^{2+}	K^+	Na^+
		(%)			($cmol_c\ kg^{-1}$)				(mS/cm)	(meq/L)						
					Exchangeable cations					Soluble anions and cations						
Young *Casuarina*																
AC (0–8 cm)	8.3	1.8	0.10	10.5	11.3	2.5	0.2	2.5	5.4	42.2	9.3	5.9	18.1	10.6	2.3	69.1
Cg (8–56 cm)	9.7	1.2	0.02	12.4	4.9	0.9	0.1	0.4	1.0	6.4	3.2	0.96	3.8	2.3	0.2	8.3
Mature *Casuarina*																
A1 (0–5 cm)	7.6	2.8	0.18	9.0	20.4	3.1	0.2	1.0	4.2	25.7	13.8	2.7	36.8	14.5	0.9	17.4
Cg (5–60 cm)	8.6	0.12	0.02	12.9	5.1	0.8	0.1	0.4	2.4	19.3	2.7	1.5	16.6	6.8	0.4	12.8
Old *Casuarina*																
A1 (0–5 cm)	9.6	7.1	0.37	10.3	12.4	2.6	0.3	5.4	3.7	20.3	25.7	4.3	9.5	20.6	0.9	51.7
Cg (9–35 cm)	9.1	0.18	0.03	12.8	5.3	1.4	0.1	0.5	1.2	8.8	4.6	1.1	4.8	3.2	0.2	10.0

stand): many of the tropical dry forest species were found in the old stands as well as tropical dry forests from close-by dunes (Castillo-Campos and Medina-Abreo 2005; Castillo-Campos and Travieso-Bello 2006). However, *Casuarina* stands on foredunes were species-poor, although the sizes of these trees were similar to the mature and old *Casuarina* stands on hydromorphic soils. Foredunes were between 2.5 and 3.5 m tall and soils were not hydromorphic.

Our results show that only under conditions of high moisture in the soil such as those occurring in the hydromorphic sand of the Port of Veracruz, can native species colonize the understory of *Casuarina* stands. These species are also found in nearby tropical dry forests growing on sand dunes. Thus, under these circumstances, removal of the exotic trees would probably accelerate the natural recovery of native species since they were already growing beneath the trees. It is also important to bear in mind that although recovery of natural vegetation is possible in the circumstances mentioned above, native species should preferably be promoted as the species for plantations because the risk of an aggressive exotic is always latent and potentially very damaging to biodiversity and the natural functioning of ecosystems. Native species can probably only grow beneath *Casuarina* trees under specific circumstances (hydromorphic soils).

Our studies concur with those of Parrotta (1995), who confirmed that in a degraded tropical coastal dune in Puerto Rico, up to 19 native forest species were established beneath a canopy of three exotic species: *C. equisetifolia, Eucalyptus robusta*, and *Leucaena leucocephala*. In our case, however, it was interesting that in the *Casuarina* stands growing on relatively drier foredunes and nonhydromorphic soils, we did not find any tropical dry forest species growing beneath the canopy of *Casuarina*. In fact, we observed significant differences between foredune and hydromorphic stands, both in growth forms and species richness and diversity.

We also found species that are common to coastal wetlands (*Hydrocotyle bonariensis, S. diphyllum, Syngonium podophyllum*). This provides further evidence of a humid substrate owing to the closeness of the water table, which is also shown by the hydromorphism of the soils analyzed. In the older *Casuarina* stands, the improved organic matter conditions of the substrate were enhanced by the hydromorphic attributes of the soil that tend to reduce mineralization rates of the organic residues.

Seed dispersal syndromes also changed in the different stands. In all of them we found species dispersed by wind and animals, although the number of species in each one differed. The relative importance of zoochorous species during colonization by native species in *Casuarina* plantations in degraded coastal dune vegetation was also found by Parrotta (1995) in Costa Rica. Here, and in coincidence with our results, the plantations that were introduced are probably acting as bird perching sites and seeds are constantly being introduced into the plots.

In summary, we observed that tropical dry forest species may colonize the understory beneath *Casuarina* stands under particular circumstances:

(1) There are neighboring tropical dry forests or remnants with abundant seeds to be dispersed into the abandoned stands.
(2) There are important bird populations.
(3) There are perching sites for birds.
(4) Soils are hydromorphic and water is not too limiting.

Certainly, in Mexico there is a great need to restore tropical dry forests growing on dunes, because most of them have disappeared in the region, and because these woody species are useful and people have cut down the trees without planting more to replace them (Moreno-Casasola and Paradowska 2009). Also, the introduction of cattle has resulted in vegetation loss. Frequently, and because the species grows fast and tolerates the extreme coastal dune environment, *Casuarina* trees have been used in reforestation programs in the tropics, ignoring the fact that it is an exotic and invasive species. However, in the humid swales and low-lying areas where litter can be decomposed at a relatively fast rate, restoration actions could be implemented to enhance the growth and establishment of tropical dry forest vegetation already growing in the understory. The gradual removal of the exotic tree would lead to an advanced state of natural regeneration.

It is noteworthy to highlight that the establishment of the tropical dry forest may take a long time. Locally, it is known that *C. equisetifolia* trees live 40–45 years. We observed that some of the trees in the old plots were already dying, which means that the natural regeneration and colonization process in the old stands has been occurring for about 40–45 years and still there were significant differences with the tropical dry forest stands. This implies that restoration by natural regeneration is a long process that lasts more than 4 decades.

Finally, although in high moisture conditions there is a potential for restoration beneath *Casuarina* stands, it is important to bear in mind that native species should preferably be promoted as the species for plantations since the risk of an aggressive exotic is always latent and potentially very damaging to biodiversity and the natural functioning of ecosystems.

Acknowledgments We are very grateful to P.A. Hesp and J.B. Gallego-Fernández for their thorough revision and very useful comments and recommendations made in earlier versions of this chapter. R. Monroy elaborated Fig. 14.1. Thanks are also due to D. Infante for her help with the field work.

References

Abdel Wahab AM (1980) Nitrogen-fixing nonlegumes in Egypt. I. Nodulation and N_2(C2H2) fixation by *Casuarina equisetifolia*. Zeitschrift fur Allgemeine Mikrobiologie 20:3–12

Arnold GW (1988) The effects of habitat structure and floristics on the densities of bird species in wandoo woodland. Aust Wildl Res 15(5):499–510

Arnold GW, Malle RA, Litchfield R (1987) Comparison of bird populations in remnants of wandoo woodland and in adjacent farmland. Aust Wildl Res 14(3):331–341

Bentley JM, Catterall CP (1987) The use of bushland, corridors, and linear remnants by birds in Southeastern Queensland. Aust Conserv Biol 11(5):1173–1189

Bond WJ (1993) Keystone species. In: Schulze ED, Mooney HA (eds) Biodiversity and ecosystem function. Ecological studies, vol 99. Springer Verlag, New York, pp 237–253

Castillo-Campos G, Medina-Abreo MA (2005) Arboles y arbustos de la Reserva Natural La Mancha, Veracruz. Manual para la identificacion de las especies. Instituto de Ecología A.C, Xalapa, p 144

Castillo-Campos G, Travieso-Bello AC (2006) La Flora. In: Moreno-Casasola P (ed) Entornos Veracruzanos: la costa de La Mancha. Instituto de Ecología A.C, Xalapa, pp 171–204

Duever MJ, Carlson JE, Meeder JF et al (1986) The big cypress national preserve. Research report number 8. National Audubon Society, New York, p 455

Espejel I, Ojeda L (1995) Native plants for recreation and conservation in Mexico. Ecol Restor 13:84–89

FAO (1998) Salt-affected soils and their management. Soil Bull 39. FAO, Rome

Fernald RT, Barnett BS (1991) Establishment of native hammock vegetation on spoil islands dominated by Australian pine (*Casuarina equisetifolia*) and Brazilian pepper (*Schinus terebinthifolius*). In: Center TD, Doren RF, Hofstetter RL, Myers RL, Whiteaker LD (eds) Proceedings of the symposium on exotic pest plants. Technical report NPS/NREVER/NRTR-91/06. US Department of the Interior National Park Service, Washington, DC, pp 131–150

Frith CB (1979) Feeding ecology of land birds on West Island, Aldabra Atoll, Indian Ocean: a preliminary survey. Philos Trans R Soc London B Biol Sci 286(1011):195–210

Gordon DA (1998) Effects of invasive, non-indigenous plant species on ecosystem processes: lessons from Florida. Ecol Appl 8(4):975–989

Hillel D (2000) Salinity management for sustainable irrigation: integrating science, environment, and economics. The World Bank, Washington, DC

Homji VMM (1995) Curbing coastal erosion—example of Udvada (South Gujarat). Nat Acad Sci Lett India 9–10:193–198

Izquierdo I, Caravaca F, Alguacil MM, Hernandez G, Roldan A (2005) Use of microbiological indicators for evaluating success in soil restoration after revegetation of a mining area under subtropical conditions. Appl Soil Ecol 30:3–10

López-Portillo J, Martínez ML, Hesp PA et al (2011) Atlas de las costas de Veracruz: manglares y dunas. Secretaría de Educación y Cultura del estado de Veracruz

Mailly D, Margolis HA (1992) Forest floor and mineral soil development in *Casuarina equisetifolia* plantations on the coastal sand dunes of Senegal. For Ecol Manag 55(1):259–278

March-Mifsut IJ, Martínez-Jiménez M (eds) (2007) Especies invasoras de alto impacto a la biodiversidad. Prioridades en México. IMTA-Conabio-GECI-AridAmerica. The Nature Conservancy, p 73

Mascarenhas A, Jayakumar S (2008) An environmental perspective of the post-tsunami scenario along the coast of Tamil Nadu, India: role of sand dunes and forests. J Environ Manag 1:24–34

McCune B, Grace JB, Urban DL (2002) Analysis of ecological communities. MJM Software Design, Gleneden Beach

Moreno-Casasola P, Paradowska K (2009) El uso de los árboles de las dunas costeras del centro de Veracruz. Madera y Bosques 16(1):21–44

Parrotta JA (1995) Influence of overstory composition on understory colonization by native species in plantations on a degraded tropical site. J Veg Sci 6:627–636

Richards LA (ed) (1954) Diagnosis and improvement of saline and alkaline soils. USDA Handbook No. 60, Washington, DC

Schmid JL, David S Addison DS, Donnelly MA, Shirley MA, Wibbels T (2008) The effect of Australian Pine (*Casuarina equisetifolia*) removal on loggerhead sea turtle (*Caretta caretta*) incubation temperatures on Keewaydin Island, Florida. J Coastal Res 55:214–220

Siemens A, Moreno-Casasola P, Sarabia C (2006) The metabolization of wetlands by the city of Veracruz, Mexico. J Lat Am Geogr 5(1):7–29

Sparks DL (ed) (1996) Methods of soil analysis. III. Chemical methods. SSSA and ASA. No 5, Madison, p 1390

Subramanean J, Vikram Reddy M (2010) Effect of casuarina (*Casuarina equisetifolia*) plantation on the sand skink (*Eutropis bibronii* Gray 1839) population. Curr Sci 98(5):604–605

Versfeld DB, van Wilgen BW (1986) Impact of woody aliens on ecosystem properties. In: Macdonald IAW, Kruger FJ, Ferrar AA (eds) The ecology and management of biological invasions in southern Africa. Oxford University Press, Cape Town, pp 239–246

Vitousek PM (1986) Biological invasions and ecosystem properties: can species make a difference? In: Mooney HA, Drake JA (eds) Ecology of biological invasions of North America and Hawaii. Ecological studies 58. Springer-Verlag, New York, pp 163–176

Walkley A, Black TA (1934) An examination of the Degtjarett method for determining soil organic matter and a proposed modification of the chromic acid titration method. Soil Sci 37:29–38

Westhoff V, van der Maarel E (1978) The Braun-Blanquet approach. In: Whittaker R (ed) Classification of plant communities. Dr. W. Junk Publishers, Dordrecht, pp 287–399

Chapter 15
Restoration of Dune Vegetation in the Netherlands

Ab P. Grootjans, Bikila S. Dullo, Annemieke M. Kooijman, Renée M. Bekker and Camiel Aggenbach

15.1 Introduction

15.1.1 Historic Dune Landscapes

Despite the fact that a large proportion of the Dutch population lives very close to the coastal dune areas, these areas have been to date among the best-preserved nature reserves in the country, mainly because modern agriculture was never practiced in the dune areas. Fertile soils are found behind the dunes, where the topography is plain and soils consist of clay or peat. However, dune areas have

A. P. Grootjans (✉) · B. S. Dullo
Center for Energy and Environmental Sciences, Energy and Sustainability Research Institute, University of Groningen, Nijenborgh 4 9747 AG Groningen, The Netherlands
e-mail: a.p.grootjans@rug.nl

A. P. Grootjans
Department of Aquatic Ecology & Environmental Biology, Institute for Water and Wetland Research, Radboud University Nijmegen, Heyendaalseweg 135 6525 AJ Nijmegen, The Netherlands

A. M. Kooijman
Institute for Biodiversity and Ecosystem Dynamics, University of Amsterdam, 94062 1090 GB Amsterdam, The Netherlands

R. M. Bekker
National Authority for Data concerning Nature, ATLAS building 104, Droevendaalsesteeg 4, Wageningen, The Netherlands

R. M. Bekker
Community and Conservation Ecology group, University of Groningen, Nijenborgh 7 9747 AG Groningen, The Netherlands

C. Aggenbach
Kwr Water Cycle Research Institute, 1072 3430 BB Nieuwegein, The Netherlands

M. L. Martínez et al. (eds.), *Restoration of Coastal Dunes*, Springer Series on Environmental Management, DOI: 10.1007/978-3-642-33445-0_15,

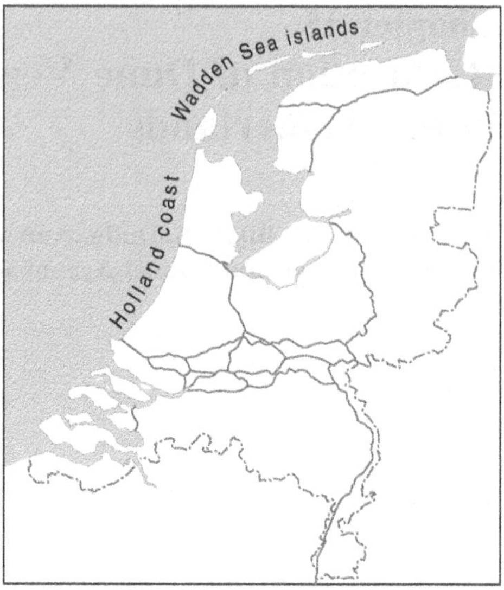

Fig. 15.1 Position of the Holland coast and the Dutch Wadden Sea islands

been used for centuries, often intensively, for purposes such as grazing, which often initiated intensive sand blowing. Dune areas in the Netherlands were sparsely vegetated for centuries with a mix of blowing dunes, low productive herbaceous vegetation and many freshwater wetlands. These habitats were sheltered from the direct influence of the sea by large dune ridges. Patches of forests grew mainly in areas further away from the shores where the influence of the sea was minimal (Provoost et al. 2009).

Along the Holland coast (Fig. 15.1), dry dunes have been preserved relatively well because they serve as a reservoir of drinking water for the surrounding large urban areas.

However, the enormous abstraction of groundwater desiccated almost all interdunal wetlands (swales and dune slacks), which prompted development of a new technique in order to increase the amount of water in the dunes by channeling surface water from the Rhine and Meuse Rivers into sand dunes. The technique involved pumping of surface water from the rivers, which is polluted, into artificial infiltration channels. The water passes through the sandy soil and is then more or less purified and collected again in so-called seepage slacks to be further processed for drinking water. This production process caused large fluctuations in water tables in the whole dune area. In addition, most of the natural dune slacks were severely polluted owing to the increased availability of nutrients, in particular phosphate. One advantage of this method was that it had helped to rewet the dried dune slacks in the area, but at the same time large parts of the dune areas had been eutrophicated by the polluted river water (Van Dijk and Grootjans 1993).

Public opinion about the degradation of the dune area forced the water companies to invest in more environmentally friendly techniques for producing drinking water. In some areas the "deep infiltration technique," was introduced. This requires pre-purification of surface water, which is then pumped into the deep subsoil, below large clay layers (see Box 15.1). Drinking water could then be collected from this new water reservoir when needed. The use of such subsoil reservoirs led to much smaller water level fluctuations in the local, more superficial hydrological systems, and enabled successful restoration of affected dune slacks.

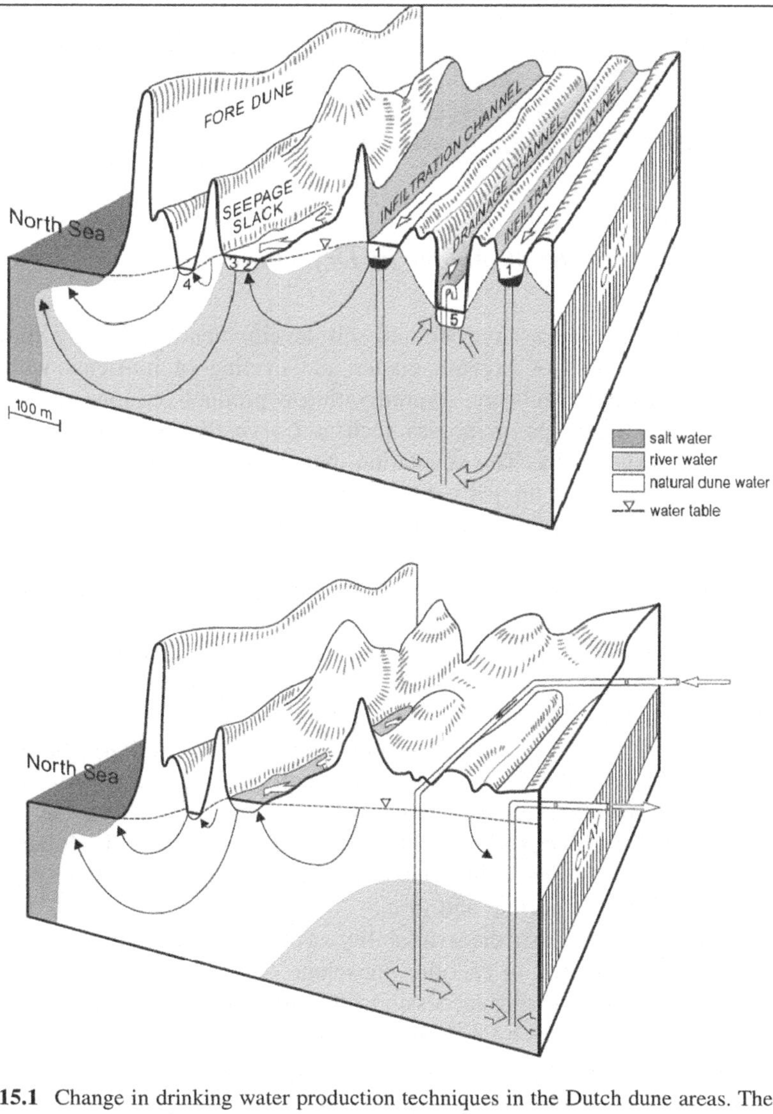

Box 15.1 Change in drinking water production techniques in the Dutch dune areas. The first image shows the traditional method of drinking water production (changed after Stuyfzand 1994). The second shows the improved "deep infiltration" method (see text). The clay layer, indicated on the right, lies beneath the whole system

In the past decades, biodiversity of dune areas in the Netherlands decreased further because of large-scale pine plantation during the first half of the last century. The planting of pine forest was mainly done to stop sand blowing, but it prevented the formation of new dune habitats. A surge in mass recreation and high atmospheric nitrogen deposition from agricultural areas, and even industrial areas in neighboring countries contributed to the further decline in biodiversity in the coastal areas (Van Dijk and Grootjans 1993). These changes have led to a severe reduction in the natural regeneration potential owing to the lack of aeolian dynamics. As a result, open dune habitats have become very rare; the pH of the topsoil has decreased owing to increased decalcification, which has generally led to loss of characteristic plant and animal species (Van der Maarel et al. 1985). Similar changes have been recorded in other countries in Europe (Provoost et al. 2009).

15.1.2 Ecosystem Development in Dry Dunes

In dry dune systems, succession rate and soil development are largely controlled by decomposition rates of organic matter and cycling of nutrients within the ecosystem. Young dune soils are covered with low productive pioneer vegetation, consisting of small grasses or mosses such as *Corynephorus canescens* or *Syntrichia ruralis* respectively. The vegetation is usually N-limited, because organic matter content is low and the pH is high, especially in areas with calcareous sand (Kooijman and Besse 2002). At high pH levels, microbial communities in the soil are dominated by bacteria, which have higher N demand than the fungi predominating in acid soils. A high microbial N demand may lead to relatively low N availability to the vegetation, even when decomposition and gross N mineralization are high. During succession, soil organic matter content increases and pH decreases owing to the dissolution of lime and leaching of base cations from the soil exchange complex. The availability of N increases, partly because of greater nitrogen pools in the soil's organic matter (Gerlach et al. 1994), but more so because of the higher litter input from the vegetation (Kooijman and Besse 2002). In later successional stages the vegetation becomes more productive, but there are differences in soil formation between the calcareous and iron-rich Holland coast and the rather acidic soils of the Dutch Wadden Sea islands, where the initial beach sand is very low in lime and iron.

Along the Holland coast, the availability of P is very low in pioneer stages, since available P is stored in calcium phosphate, which is insoluble at high pH (Kooijman et al. 1998). During succession, soils are leached owing to a net surplus of precipitation over evapotranspiration. Dissolution of calcium and phosphate occurs and the soil gradually becomes more acidic. However, along the Holland coast, this acidification process takes a long time, because the initial lime content is high (5–10 %). Calcareous dunes typically contain 60 mg m^{-2} P in calcium phosphate per millimeter of soil (Kooijman et al. 1998), which becomes available during the decalcification process. Partly decalcified soils thus have high availability of both P and N (Kooijman and Besse 2002). At even lower pH, iron concentrations in the soil solution increase, and P may precipitate in the form of iron phosphates. However, this process only occurs in soils with low organic matter content, where iron is mainly present in mineral form (Kooijman and Hedenäs 2009). When soil organic matter content is high, iron is incorporated in iron–organic matter complexes, which can still adsorb P, but in a much weaker way. In acidic, iron-rich soils with high organic matter content, productivity will still be high. Consequently, nitrogen and phosphorus availability increases during succession along the Holland coast.

On the Wadden Sea islands, lime content is usually too low for substantial calcium phosphate formation; thus, even in pioneer stages with low amounts of organic matter the productivity of the dune ecosystems is limited only by nitrogen (Lammerts et al. 1999). N availability may increase during succession, owing to soil acidification and shift in microbial communities from bacteria to fungi with lower N demand. However, before the 1980s, when high atmospheric N deposition became a serious problem, N availability may have remained relatively low. Low microbial N demand may lead to high N availability to the vegetation especially under high N deposition, when N input has increased, but microbial uptake remains low. Before the 1980s, decomposition rates were low, and N input was still mainly dependent on the relatively low litter input, which further reduced gross N mineralization. However, in the last few decades, atmospheric N deposition and N input strongly increased in northwestern Europe, in particular in The Netherlands, with their high numbers of cows and pig stables (Bobbink et al. 1998).

15.1.3 Effects of Increased Nitrogen Deposition

The dry dunes in the Netherlands have been very negatively affected by atmospheric deposition of nitrogen due to its stimulating effect on biomass production. Stuyfzand (1993) showed that the atmospheric nitrogen deposition in the Holland dune areas had increased from 5 to 36–55 kg N ha^{-1} $year^{-1}$ between 1939 and 1981. In one area with a large amount of industrial activity even 80 kg N ha^{-1} $year^{-1}$ was measured. Values on the Dutch Wadden Sea islands were slightly lower; 20–35 kg (Sival and Strijkstra-Kalk 1999). Grass encroachment has reduced open, species-rich dune grasslands to tall grass vegetation with low biodiversity. Grass encroachment further increased with declining grazing activity, including rabbit grazing. Rabbit populations have strongly declined because of viral diseases, such as myxomatosis and viral hemorrhagic disease (VHD). Less than 50 years ago dry dunes had a very high biodiversity in terms of both flora and vegetation, and harbored 70 % of the total flora of the Netherlands. Nowadays, almost 100 of them are endangered and have been placed on the Red List.

As indicated above, calcareous and acid dunes may differ in sensitivity to high N deposition. In the Netherlands, N deposition is higher along the Holland coast than on the Wadden islands, but grass encroachment generally showed opposite patterns

(Kooijman et al. 2009). Calcareous dune grasslands can supposedly stand critical loads of 20 kg N ha^{-1} $year^{-1}$, but acid dune grasslands only 10 kg N ha^{-1} $year^{-1}$. For acid dune grasslands, even this may be too high, because in the Baltic Sea area grass encroachment in acidic soils already seemed to occur at 5–8 kg N ha^{-1} $year^{-1}$ (Remke et al. 2009). In calcareous dunes, high microbial N demand may lead to relatively high storage of N in the soil, but in acid dunes, low microbial N demand may lead to relatively high allocation of N to the vegetation and thus to grass encroachment (Kooijman and Besse 2002; Kooijman et al. 2009). The Wadden sea area is therefore more sensitive to atmospheric nitrogen deposition.

15.1.4 Crucial Role of the Hydrology in Dune Slacks

In most dune slacks high water tables prevail during winter and spring, while summer water tables may drop 50–100 cm below the surface, depending on the weather conditions. Dune slacks are usually influenced by groundwater, which is usually calcareous, and by precipitation, which is mostly acidic.

The groundwater enters the dune slacks from different hydrological systems (Fig. 15.2) (Munoz-Reinoso 2001; Grootjans et al. 2002). Dune slacks situated at the periphery of the main dune system are often fed by calcareous groundwater, which comes from the main hydrological system higher up in the landscape. Such dune slacks are considered to be "flow-through lakes" because groundwater discharge takes place in one part of the slack, and infiltration of surface water in another (Stuyfzand 1993). When several slacks occur side by side, slight differences in water levels between slacks may initiate groundwater flow from one slack to another (Grootjans et al. 2002). Therefore, in order to understand the dynamics of dune slack vegetation, knowledge about the hydrological dynamics is fundamental (Grootjans et al. 2004; Jones et al. 2006).

Likewise, dune slack vegetation shows diverse adaptations in order to survive the extreme fluctuations in water table. Long-term water logging in winter or early spring creates anoxic conditions around the roots, exposing them to toxic substances like reduced sulphide, iron, and manganese. Summer drought may improve

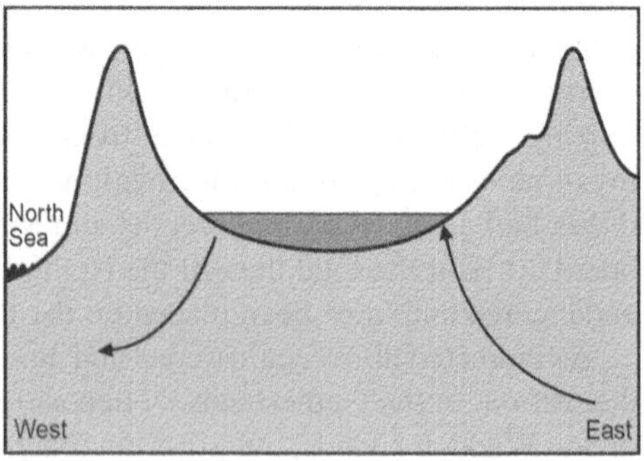

Fig. 15.2 Groundwater flow pattern in a wet dune slack. Calcareous and iron-rich groundwater is entering the dune slack on one side, proceeds as surface water and infiltrates again at the left end side (after Stuyfzand 1993)

oxygenation, but may lead to low availability of water. However, typical dune slack species are adapted to both anaerobic conditions and temporary desiccation. For instance, some early successional stage species such as *Littorella uniflora* and *Schoenus nigricans* are capable of radial oxygen loss (ROL), a mechanism of pumping oxygen into the soil. This oxygen pumping helps them to neutralize anoxic conditions around their roots, thus eliminating toxic concentration of reduced sulphide, iron, and manganese (Adema et al. 2005). Early-stage dune slack species have developed adaptations such as a very low nutrient demand (*L. uniflora*, *Centaurium pulchellum*, and *Radiola linoides*) or the ability to retain and recycle nutrients within their living and senescing tissues (*S. nigricans*). Later successional species such as *Carex nigra* and *Salix repens* are not capable of ROL. They can only win the competition with *Littorella* or *S. nigricans* when the nutrient availability increases. But *Littorella,* in particular is very successful at avoiding rapid accumulation of organic matter. Because of its low litter production and high mineralization conditions (due to ROL) this pioneer vegetation can dominate the vegetation for a long time. Adema et al. (2002) showed a case on the island of Texel in which *Littorella* was able to dominate the vegetation for more than 80 years.

The hydrology of dune slacks is dynamic and its effect on dune slack vegetation is diverse. Its effect depends on the nature of the soil and the source of water: groundwater, surface water or precipitation (Zunzunegui et al. 1998; Munoz-Reinoso 2001; Clarke and Sanitwong Na Ayuthaya 2010). Dune slack soils are usually calcareous because they often originated from recently deposited sand that is rich in shell materials. When dune slacks are formed on a beach plain that has a low initial lime (<0.3 %) content, they are acidic. When dune slacks are fed by groundwater that comes from deep layers, the calcium-rich water buffers the acidity of the soil, which keeps the accumulation of organic matter at a low level. However, in the past few decades, this has increased considerably in many dune slacks, owing to high atmospheric deposition, but also the lowering of groundwater levels in the surroundings of the slacks. Groundwater tables may have further dropped because of increased growth of vegetation in the whole dune area. Consequently, less groundwater is available for dune slacks, leading to a positive feedback loop, i.e., further accelerating succession in the dune slacks.

15.2 Approaches to Dune and Dune Slack Restoration

15.2.1 Mowing and Grazing

Mowing is a traditional management technique in dune slacks, preventing grasses, willows, and tree species to dominate the vegetation, but it does not prevent acidification when hydrological conditions are suboptimal. In calcareous dune areas, mowing is quite efficient at maintaining basiphilous pioneer stages. Grazing is often applied in dry dunes to counter grass and shrub encroachment. Accumulation of organic matter is not counteracted by grazing, but perhaps retarded, owing to a lower input of litter. The effect of grazing partly depends on the productivity and nutrient limitations of the vegetation. Grazing does not affect P availability, which is largely regulated by soil chemistry, but it may affect litter input and N cycling. In dune areas with P limitation, such as calcareous soils, or iron-rich soils with low organic matter content, productivity is already low and grazing may counteract grass encroachment. In N-limited dunes, such as in the Wadden islands, grazing may reduce net N mineralization and N availability to the vegetation, thus reducing grass encroachment. However, in partly decalcified middle dunes along the Holland coast with high availability of N and P, grazing cannot prevent further grass encroachment (Kooijman and Hedenäs 2009).

15.2.2 Sod Cutting and Rewetting

Sod cutting is a widely applied management tool to reduce nutrient availability in dune slacks, and is usually applied in combination with rewetting. Sod cutting includes removal of the organic A horizon, leaving the mineral C horizon intact. Also sod cutting has been practiced in dune areas in northwest Europe for a long time. Rewetting and restoration of hydrological systems consist of eliminating drainage ditches as much as possible, terminating groundwater abstraction facilities, and decreasing evaporation intensities by cutting pine plantations (Grootjans et al. 2002).

15.2.3 Initiating Aeolian Processes Again

As mentioned in the Sect.15.1, apart from changes in hydrology or high atmospheric deposition, a major problem in the Dutch dunes is the reduction of the natural regeneration potential by the lack of aeolian dynamics.

Over the years, experiments on different scales have been conducted to reactivate dune dynamics, varying from local blowouts to parabolic dunes and even foredune ridges (Arens and Geelen 2006). A survey of active blowouts in different dune zones suggests that even apparently stable reference vegetation in the neighborhood of the actual blowouts, may be affected (Table 15.1). In both calcareous and acid dunes, pH of the reference dune grasslands was higher than values characteristic of the particular dune zone. However, the results of reactivation experiments have generally been less successful than expected. In many cases, vegetation regrowth stabilized open sand rapidly after reactivation (Arens and Geelen 2006).

Table 15.1 Soil pH and plant characteristics of pioneer stages and reference dune grasslands in different dune zones

	Dune zone	Marram grass pioneer stage	Open pioneer vegetation	Reference dune grasslands	Older dune grasslands
pH-H_2O	H foredunes	8.4 (0.3)	8.1 (0.2)	7.8 (0.1)	7.4 (0.3)
	H middle dunes	8.0 (0.1)	7.8 (0.2)	7.5 (0.8)	5.0 (0.5)
	H hinter dunes	8.4 (0.2)	8.4 (0.3)	7.8 (0.3)	4.1 (0.1)
	Wadden islands	7.9 (0.6)	7.3 (0.4)	6.6 (1.2)	4.2 (0.3)
Living biomass ($g\ m^{-2}$)	H foredunes	259 (130)	60 (22)	236 (116)	225 (36)
	H middle dunes	250 (143)	49 (25)	189 (95)	220 (28)
	H hinter dunes	232 (248)	78 (24)	186 (87)	125 (35)
	Wadden islands	416 (248)	120 (50)	245 (99)	300 (350)
Foliar N:P ratio ($g\ g^{-1}$)	H fore dunes	10.1 (2.1)	11.0 (4.1)	13.8 (1.7)	16.0 (1.4)
	H middle dunes	12.1 (1.5)	10.2 (2.4)	15.8 (5.7)	11.5 (0.7)
	H hinter dunes	10.2 (1.9)	9.7 (3.5)	15.4 (1.5)	15.0 (1.4)
	Wadden islands	9.1 (2.4)	9.6 (1.3)	7.4 (1.6)	13.0 (2.8)

H foredunes calcareous foredunes along the Holland coast in Meijendel, *H middle dunes* partly decalcified middle dunes along the Holland coast in Zuid-Kennemerland, *H hinter dunes* acid dunes along the Holland coast in Meijendel, *Wadden islands* acid dunes of the Wadden island Terschellling
Mean values (and standard deviations)

15.3 Results of Restoration

15.3.1 Effect of Grazing on Dry Dune Vegetation

As indicated before, dry dune zones may differ in productivity, depending on the availability of N and P (Kooijman et al. 1998; Kooijman and Hedenäs 2009; Kooijman and Besse 2002). It is therefore not surprising that the effect of grazing also depends on the productivity and nutrient limitations of the vegetation (Ten Harkel and van der Meulen 1996).

In the foredunes along the Holland coast, grazing has a large and positive effect on the percentage of open vegetation. The soil is calcareous, the pH is high and the percentage of organic material is low. The availability of P is restricted by storage in insoluble calcium phosphates, and the availability of N is relatively low because of the high microbial demand and storage in the soil in stable organic matter. Consequently, the productivity is also low and grazing has a relatively large impact on the vegetation cover. The middle dunes along the Holland coast, however, are highly productive and dominated by tall grasses and shrubs. The soils are partly decalcified soils with pH values around 5, and all calcium phosphates have dissolved. Under such conditions, the productivity of the vegetation is barely restricted by grazing (Table 15.2). In the oldest dunes in the hinterland of the Holland dunes, and also in most of the dry dunes of the Wadden Sea islands, grazing has a marked

Table 15.2 Effects of grazing in various parts of Dutch dune areas

Dune zone	Organic matter	pH	Open pioneer vegetation (%)[a]	
			Grazed	Not grazed
H foredunes	Low	7.4 (0.3)	80	45
H middle dunes	Intermediate	5.0 (0.5)	13	9
H hinter dunes	High	4.1 (0.1)	75	55
Wadden islands	High	4.2 (0.3)	60	13

[a] Data on the effect of grazing were obtained from vegetation surveys between 2001 and 2008 (Holland coast) and between 1992 and 2008 (Wadden Sea islands)

influence on the area of open pioneer vegetation. Thus, grazing can be an efficient tool for increasing biodiversity in grass-encroached dune areas, but the youngest and oldest stages of the dune landscape have the best opportunities.

15.3.2 Rewetting and Sod Cutting in Dune Slacks

Three cases will be discussed with respect to dune slack restoration, where sod cutting and rewetting was carried out and where the results have been monitored for 12–18 years. Two examples are from the Dutch Wadden Sea islands (Koegelwieck, Terschelling and the Moksloot area, Texel) and one example from the Holland coast.

In the Koegelwieck valley on the island of Terschelling sod-cutting experiments have been present since 1954 and various stages of vegetation development are present in a chronosequence.

In the Moksloot area a groundwater abstraction facility has been terminated in 1993 and large scale sod-cutting has been applied to restore typical dune slack vegetation (Grootjans et al. 2001).

In the Holland dune area (Meinderswaal valley) sod cutting has been applied after the introduction of new groundwater abstraction techniques (see Box 15.1), which led to a considerable rewetting and restoration of the hydrological system.

15.3.2.1 Koegelwieck, Island of Terschelling

The dune slack in the Koegelwieck area is about 100 years old. It is a large secondary dune slack on the island of Terschelling, which was formed because of massive sand blowing in the early twentieth century and became vegetated around 1910. It is mainly fed by precipitation water, except during winter and early spring, when calcareous groundwater enters the slack from an inland dune massive

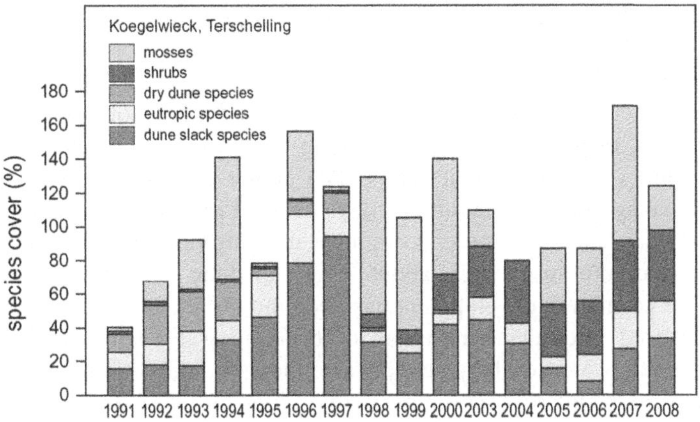

Fig. 15.3 Change in ecological groups since sod cutting in a calcareous slack (Koegelwieck, Terschelling) from 1991 to 2008. Total cover can be higher than 100 %, since species can form different vegetation layers that partly overlap

(Grootjans et al. 2002). The vegetation of the slack consists of acidic heathland and small willows with very low species diversity. Commonly encountered species are *Oxycoccus macrocarpa, S. repens*, and *Calamagrostis epigejos*. Various attempts to restore younger vegetation stages with typical dune slack species have been carried out. The oldest experiment is from 1954, the youngest is from 1995. In 1986 and 1991 parts of the slack have also undergone sod cutting. The activities of dune managers thus produced a chronosequence of five stages over a period of almost 100 years. However, we will only discuss an experiment in which the restoration started in 1986 (Fig. 15.3). Five years after sod cutting, vegetation cover was still less than 40 %, but target species, such as *S. nigricans* and various orchid species, were already present, while, for instance, the percentage cover of mosses was close to zero. However, 12 years after sod cutting the target community suddenly collapsed, which was associated with prolonged inundations during the summer. These summer inundations started to be regular phenomena and triggered a sudden expansion of moss cover and after 5 years an increase in willows (*S. repens*) and cranberry (*O. macrocarpa*) too.

These experiences show that history does not repeat itself when the hydrological conditions of the dune slacks have changed. In this case the following factors have altered the hydrological balance of the slacks: an increase in summer rainfall during the last decade and expansion of the coast owing to natural geomorphological processes. Both factors have led to a rise in surface water levels. Thus, precipitation water had stagnated in the dune complex and prevented calcareous groundwater reaching the slack, which consequently made the development of calcareous dune slack vegetation an impossible goal.

15.3.2.2 Moksloot Valley, Island of Texel (Wadden Sea)

The Moksloot valley is situated in the southwestern part of Texel. In 1993 the drinking water company decided to terminate the abstraction of groundwater in the Moksloot area, mainly because an adjoining camping site had polluted the well areas, because of leaking septic tanks. The nature conservation organization in charge of management of the dune valleys decided to implement sod cutting in a number of dune valleys in order to remove the eutrophic top soil and reset succession. In the autumn of 1993, about 100 ha of dune slack area underwent sod cutting. The vegetation development has been monitored since then. Groundwater levels increased after the termination of drinking water abstraction. Owing to increased groundwater discharge the pH also increased (Grootjans et al. 2001). We will discuss the results of restoration in two types of dunes slacks:

1. A surface water slack that receives both groundwater and surface water from upstream areas and which as a consequence has become a nesting place for numerous birds.
2. A groundwater-fed slack that only receives groundwater from an adjoining dune area and from precipitation water.

The monitoring results are shown in Fig. 15.4. During the first 5 years there was almost no difference in vegetation development, except that the total species richness in the surface water slack was higher than in the groundwater slack (Grootjans et al. 2001). After 5 years the calciphilous pioneer species increased considerably in the groundwater-fed slack, while in the surface water slack the eutrophic grassland and marsh species increased at the cost of the nutrient-poor pioneer species.

In the groundwater-fed slack the re-establishment of calciphilous target species was very slow, but stabilized after 10 years, while in the surface water slack development of the target community was fast, but it did not last long. These different responses after rewetting clearly show that rewetting with groundwater is more successful for the development of a typical dune slack community than rewetting with surface water. The latter type of water mobilizes nutrients from the organic matter, while iron-rich groundwater reduces the availability of phosphorous in the soil (Lamers et al. 2002). Furthermore, the large areas of surface water attracted many water birds, which increased the availability of nutrients considerably.

15.3.2.3 Meinderswaal Valley (Goeree, Holland Coast)

The third restoration project was carried out in a dune area that is also used for the production of drinking water. With the development of new techniques for this process for the large cities, parts of the dunes were no longer needed; thus, many

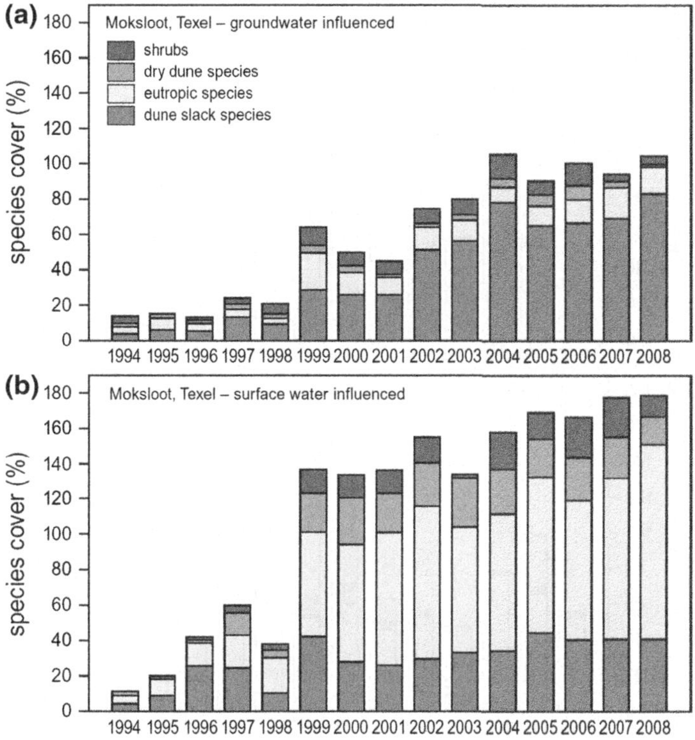

Fig. 15.4 a, b Change in ecological groups since sod cutting and the elimination of groundwater abstraction activities since 1993

restoration projects were subsequently carried out, including large-scale sod cutting in eutrophicated dune slacks. The dune soil in this part of the Dutch coast is very calcareous ($CaCO_3 > 2\ \%$) and the prospects of restoring the typical dune slack vegetation with many rare species were considered very favorable. We present one example of such a restoration project (Meinderswaal Valley), which underwent sod cutting in 1989 and was monitored until 2009. The vegetation was monitored at two sites. One site was situated on the valley flank where calcareous groundwater discharged, while another was situated slightly lower where the flooding frequency was higher. The restoration was very successful at the beginning. Typical dune slack species re-established after only 2 years, most frequently in the higher plot in the seepage zone. Within 5 years the target community with *S. nigricans* and many orchids re-emerged (Fig. 15.5). However, lower in the dune slack a eutrophic marsh community developed with *Phragmites australis, Juncus subnodulosus*, and *S. repens*, but few target species. In addition, the target community with *S. nigricans* did not last very long. After 10 years the calciphilous species disappeared again, and productive moss species with *P. australis* and

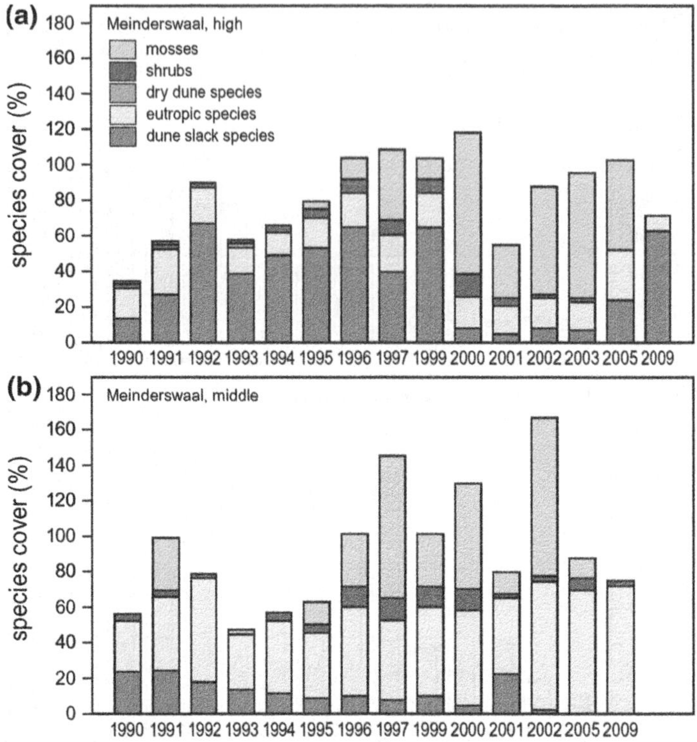

Fig. 15.5 a, b Change in ecological groups since sod cutting in a calcareous slack (Meinderswaal) from 1990 to 2009. Re-establishment of calciphilous target species was rapid in the groundwater discharge zone, but after 10 years the target community disappeared again at the lower side of the dune slack

J. subnodulosus took over. At the moist valley flank a vegetation with *S. nigricans* and many Red List species developed well.

It is likely that, despite the purification of the surface water to reduce the phosphate content, the phosphate availability in the dune slack itself was still too high to redevelop a stable calciphilous dune slack vegetation, such as the one on Texel, where a target community survived for over 80 years (Adema et al. 2002). It is possible that the past infiltration of phosphate-rich surface water has led to high phosphate loading of the sandy substrate, and that too much phosphate is being released from adsorbed fractions in the dune soil itself (Stuyfzand 1993).

From these three examples we may conclude that costly management measures, such as sod cutting, do not result in a stable target community in the dune slack if the hydrological conditions are not optimal. Phosphorous availability also appears to play a key role here.

15.4 Overall Effectiveness of Restoration Measures in Dutch Dune Areas

The case studies discussed above are some illustrations of well-monitored restoration projects over time periods exceeding 10 years. Long-term monitoring showed that restoration may lead to short-term successes, such as in the sod-cutting experiments in the dune slacks, but that long-term responses may be completely different. Unfortunately, most restoration projects are not monitored at all or only for a short time. Restoration of dry and wet dunes is not easy. In dry dunes, counteraction of grass encroachment by grazing is mainly successful in areas with low P availability and low productivity of the vegetation. Also, with respect to Red List species, a recent evaluation of 138 restoration projects, including both small areas of a few hectares to large areas of more than 100 ha, showed that restoration projects in dry dunes were generally less successful than those in wet dune slacks (Jansen et al. 2009). In dry dunes, a grazing regime generally led to more diverse vegetation, which is in accordance with earlier findings. However, only 12 Red List species (out of 66 for this habitat) in 11 projects responded positively to these measures (Jansen et al. 2009). Renewed dune formation is possibly a much more effective way to increase biodiversity in dry dunes again, because aeolian activity will transport calcareous sand to areas that have been acidified and it will probably also help to disperse seeds more effectively.

An evaluation of 126 restoration projects in wet dune slacks showed that sod cutting was most successful in increasing populations of Red List species (Jansen et al. 2009). More than 50 % of the Red List species (51 out of 95 for this habitat) responded well to the measures. Moreover, 43 of these species have improved so much that they could be removed from the list. Also, 10 of the endangered orchid species responded slowly but positively to the restoration measures during the period of observation (5–15 years). Long-term monitoring suggests that sod cutting in wet dune slacks may only be successful when hydrology has also improved, and that soils are not loaded with P because of the former input of P-rich surface water. Few species showed a negative response over a longer time span, which could be because of their adaptation to pioneer habitats. They appear to be successful initially, but disappear again as succession proceeds and organic matter accumulates in the habitat.

Jansen et al. (2009) showed that Red List species that were positively responding to restoration measures did not usually have long-living seed banks, especially not where seeds have been removed through the restoration measures. This indicates that dispersal mechanisms in these projects were more important than soil seed banks (Fig. 15.6), which is in accordance with Ozinga et al. (2009). This may be true for less endangered species as well.

In the Moksloot area, we noticed that species diversity increased very rapidly after sod cutting, although approximately 99 % of the soil seed banks had been lost because of top soil removal (Grootjans et al. 2001). This explains the relatively small contribution of species with long-living seeds.

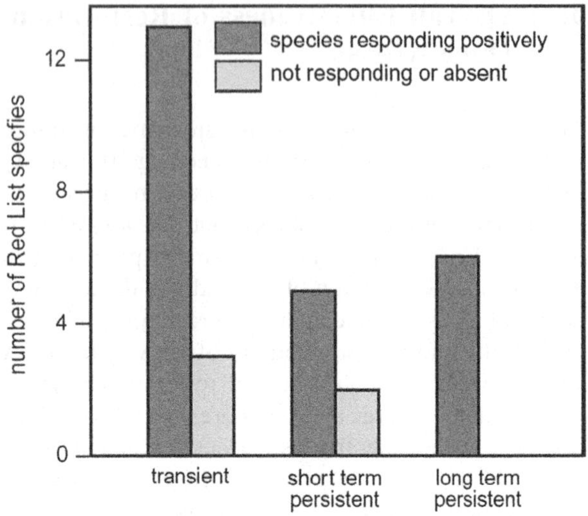

Fig. 15.6 Seed bank characteristics of Red List species that have responded positively in restoration projects compared with Red List species that did not respond positively or were not found at all. Transient means that the seeds die within a year. Long-term persistent means that the seeds can survive for more than 5 years in the soil. Short-term persistent means that the seed can survive in the soil for between 1 and 5 years. This figure shows that having long-living seed banks is no guarantee of the successful restoration of a population

Figure 15.7 indicates that Red List species that were dispersed by only one dispersal factor, for instance either wind or water, appeared to be more successful in increasing their populations after restoration measures than species that had multiple dispersal vectors. The latter species may have invested in the effectiveness of dispersal through more than one vector. However, they then depend highly on the configuration of the habitat in the landscape. When flooding is limited to small isolated valleys, then such species may only disperse effectively within that valley, but may not reach other valleys in that landscape.

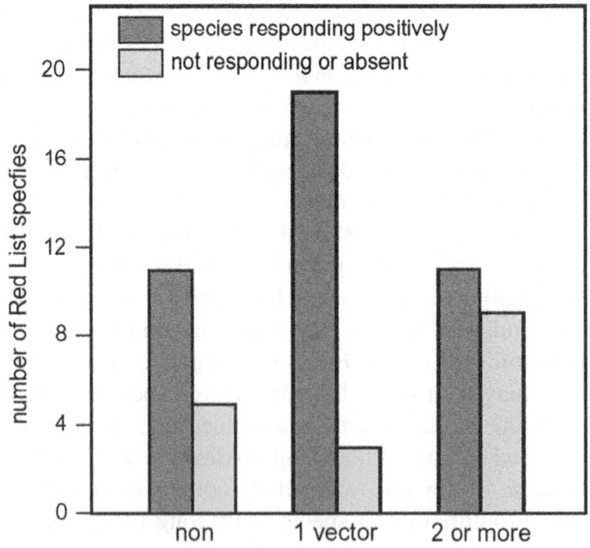

Fig. 15.7 Dispersal characteristics of Red List species that have responded positively in restoration projects compared with Red List species that did not respond positively or that were not found at all. This figure shows that species that respond positively to restoration measures are usually specialized in one dispersal factor, for instance, only wind dispersal or only water dispersal. Specialists in two or more vectors are also successful, but their success highly depends on the configuration of the landscape

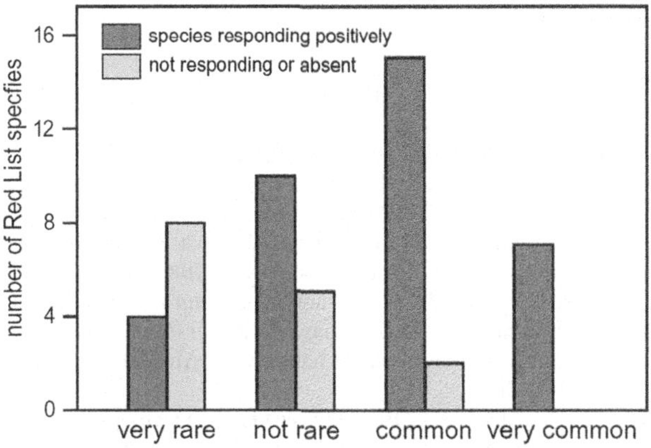

Fig. 15.8 Rarity of Red List species that have responded positively to restoration measures compared with Red List species that did not respond positively or were not found at all. This figure shows that Red List species that are very rare do not respond positively to restoration measures

Rarity as such also remains an important factor for successful restoration. Figure 15.8 clearly shows that if a Red List species is occurring in only a few localities, the chances that they will respond positively to restoration measures are low, most likely because they are not present or cannot reach the area.

15.5 Concluding Remarks

In summary, we have seen that the results of restoration projects differed considerably between wet and dry dune areas. Restoration projects were generally very successful in dune slacks, in particular after removing the top soil, although some projects were less sustainable than others, depending on the hydrological conditions. In the dry dunes the successes were very modest. Grazing appeared to be an efficient management tool for keeping the landscape open in young and in old successional stages. However, the grass encroachment and rapid growth of shrubs could not be stopped by only grazing, because of a high availability of phosphates and a high atmospheric deposition in the dunes areas. Grazing generally increased the biodiversity of the dry dunes, but very few endangered and rare species returned. This implies that additional measures have to be taken to conserve the characteristic species of dry dunes.

What we have learned from numerous restoration projects in our dune areas is that we cannot compensate for the ecological function of natural processes, such as intensive sand blowing, a regular supply of groundwater, and clean air. Some of the damage can be repaired, but usually only for a short period of time. The examples described here clearly show that conserving the natural functioning of a dune area is essential for the long-term conservation of our nature reserves. Dune areas that are destined to protect endangered species should be allowed to move more freely, within the limits of coastal protection for urban areas. In that way we can restore natural processes, which can sustain protected species naturally. If natural processes have to be restricted, then intensive management of the dune area is necessary in order to protect endangered species.

References

Adema EB, Grootjans AP, Petersen J, Grijpstra J (2002) Alternative stable states in a wet calcareous dune slack in the Netherlands. J Veg Sci 13:107–144

Adema EB, Van de Koppel J, Meyer HAJ, Grootjans AP (2005) Enhanced nitrogen loss may explain alternative stable states in dune slack succession. Oikos 109:374–386

Arens SM, Geelen LHWT (2006) Dune landscape rejuvenation by intended destabilisation in the Amsterdam Water Supply Dunes. J Coast Res 22:1094–1107

Bobbink R, Hornung M, Roelofs JGM (1998) The effects of air-borne nitrogen pollutants on species diversity in natural and semi-natural vegetation: a review. J Ecol 86:717–738

Clarke D, Sanitwong Na Ayuthaya S (2010) Predicted effects of climate change, vegetation and tree cover on dune slack habitats at Ainsdale on the Sefton Coast, UK. J Coast Cons 14:115–126

Gerlach A, Albers EA, Broedlin W (1994) Development of the nitrogen cycle in the soils of a coastal dune succession. Acta Bot Neerl 43:189–203

Grootjans AP, Grootjans AP, Everts FH, Fresco LF (2001) Restoration of wet dune slacks on the Dutch Wadden Sea Islands: recolonization after large-scale sod cutting. Restor Ecol 9:137–146

Grootjans AP, Geelen L, Jansen AJM, Lammerts EJ (2002) Restoration of coastal dune slacks. Hydrobiologia 478:181–203

Grootjans A, Adema E, Bekker R, Lammerts E (2004) Why coastal dune slacks sustain a high biodiversity. In: Martinez ML, Psuty NP (eds) Coastal dunes: ecology and conservation. Ecological studies, vol 117. Springer, Berlin, pp 85–101

Jansen AJM, Bekker RM, Bobbink R, Bouwman JH, Van Dobben H, Van Duinen GA, Wallis de Vries MF (2009) Efficiency of restoration measures in the framework of EGM for Red List species (1990–2009; in Dutch). Report DKI nr 2010/dk137O, Ede, NL, 222 pp

Jones MLM, Reynolds B, Brittain SA, Norris DA, Rhind PM, Jones RE (2006) Complex hydrological controls on wet dune slacks: the importance of local variability. Sci Total Environ 372:266–277

Kooijman AM, Besse M (2002) On the higher availability of N and P in lime-poor than in lime-rich coastal dunes in the Netherlands. J Ecol 90:394–403

Kooijman AM, Hedenäs L (2009) Changes in nutrient availability from calcareous to acid wetland habitats with closely related brown moss species: increase instead of decrease in N and P. Plant Soil 324:267–278

Kooijman AM, Dopheide J, Sevink J, Takken I, Verstraten JM (1998) Nutrient limitation and their implications on the effects of atmospheric deposition in coastal dunes: lime-poor and lime-rich sites in the Netherlands. J Ecol 86:511–526

Kooijman AM, Lubbers I, Van Til M (2009) Iron-rich dune grasslands: relations between soil organic matter and sorption of Fe and P. Environ Pollut 157:3158–3165

Lamers LPM, Smolders AJP, Roelofs JGM (2002) The restoration of fens in the Netherlands. Hydrobiologia 478:107–130

Lammerts EJ, Pegtel DM, Grootjans AP, Van der Veen A (1999) Nutrient limitation and vegetation change in a coastal dune slack. J Veg Sci 10:11–122

Munoz-Reinoso JC (2001) Vegetation changes and groundwater abstraction in SW Doñana, Spain. J Hydrol 242:197–209

Ozinga WA, Römermann C, Bekker RM, Prinzing A, Tamis WLM, Schaminée JHJ, Hennekens SM, Thompson K, Poschlod P, Kleyer M, Bakker JP, van Groenendael JM (2009) Dispersal failure contributes to plant losses in NW Europe. Ecol Lett 12:66–74

Provoost S, Laurence M, Jones M, Edmondson SE (2009) Changes in landscape and vegetation of coastal dunes in northwest Europe: a review. J Coast Cons 15:207–226

Remke E, Brouwer E, Kooijman AM, Blindow I, Esselink H, Roelofs JGM (2009) Even low to medium nitrogen deposition impacts vegetation of dry coastal dunes around the Baltic Sea. Environ Pollut 157:792–800

Sival FP, Strijkstra-Kalk M (1999) Atmospheric deposition of acidifying and eutrophicating substances in dune slacks. Water Air Soil Pollut 116:461–477

Stuyfzand PJ (1993) Hydrochemistry and hydrology of the coastal dune area of the Western Netherlands. PhD thesis, Vrije University, Amsterdam, published by KIWA, 366 pp

Stuyfzand PJ (1994) Behaviour of phosphate in eutrophic surface water upon artificial recharge in the Western Netherlands. In: ASCE (eds) Proceedings of the second international symposium on artificial recharge, Orlando, FL, 17–22 July

Ten Harkel MJ, Van der Meulen F (1996) Impact of grazing and atmospheric deposition on the vegetation of dry coastal dune grasslands. J Veg Sci 7:445–452

Van der Maarel E, Boot RGA, Van Dorp D, Rijntjes J (1985) Vegetation succession on the dunes near Oostvoorne, The Netherlands: a comparison of the vegetation in 1959 and 1980. Vegetatio 58:137–187

Van Dijk H, Grootjans AP (1993) Wet dune slacks: decline and new opportunities. Hydrobiologia 265:281–304

Zunzunegui M, Diaz Barradas M, García Novo F (1998) Vegetation fluctuation in Mediterranean dune ponds in relation to rainfall variation and water extraction. Appl Veg Sci 1:151–160

Chapter 16
Interdune Wetland Restoration in Central Veracruz, Mexico: Plant Diversity Recovery Mediated by the Hydroperiod

Hugo López-Rosas, Patricia Moreno-Casasola, Fabiola López-Barrera, Lorena E. Sánchez-Higueredo, Verónica E. Espejel-González and Judith Vázquez

16.1 Introduction

Wetlands are among the most productive ecosystems worldwide and provide critical habitats for the survival of a diverse group of organisms, ecological services, and economic benefits (Bobbink et al. 2006). For instance, even though wetlands only cover 6 % of the world's land surface, they contain about 12 % of the global carbon pool, playing an important role in the global carbon cycle, as stated by the International Panel on Climate Change (Erwin 2009). In spite of the high biodiversity and the services they provide, wetland ecosystems have suffered severe disturbances owing to their destruction and degradation as a result of a wide variety of human impacts. Worldwide, more than 50 % of the wetland resources have been lost because of human

P. Moreno-Casasola (✉) · F. López-Barrera · L. E. Sánchez-Higueredo
V. E. Espejel-González · J. Vázquez
Red de Ecología Funcional, Instituto de Ecología, A.C., Antigua carretera a Coatepec No. 351, El Haya 91070 Xalapa, VER, Mexico
e-mail: patricia.moreno@inecol.edu.mx

F. López-Barrera
e-mail: fabiola.lopez@inecol.edu.mx

L. E. Sánchez-Higueredo
e-mail: lorenaelisa@gmail.com

V. E. Espejel-González
e-mail: espejelveronica@gmail.com

J. Vázquez
e-mail: judith-s18@hotmail.com

H. López-Rosas
Estación El Carmen, Instituto de Ciencias del Mar y Limnología, UNAM, Km 9.5 carretera Carmen-Puerto Real 24157 cd. del carmen, Campeche, Mexico
e-mail: hugoloper@cmarl.unam.mx

M. L. Martínez et al. (eds.), *Restoration of Coastal Dunes*, Springer Series on Environmental Management, DOI: 10.1007/978-3-642-33445-0_16,

activities as drainage for agriculture, aquaculture, and pasturelands, floodplain reclamation, construction of flood control structures, excavation of peat for fuel and the modification and straightening of river channels (Bobbink et al. 2006).

Wetlands can be found in a great variety of geomorphological settings along floodplains, and in low areas or depressions in almost every environment. Dune slacks in temperate and tropical regions are usually formed in depressions and can harbor wetlands that remain flooded only during 2 or 3 months of the year (humid and wet slacks) or permanently, forming dune lakes. These important types of wetlands are also facing degradation and destruction because of expanding agriculture and population growth. Considering this scenario, restoration has been a priority.

Wetland restoration aims to restore lost biodiversity or provide services, such as flood-peak reduction and water quality improvement (Zedler 2000). Since the early 1970s, the science for successful wetland restoration has been developed and although we have made much progress on that front, the diversity of wetland types and their individual characteristics requires restoration remedies that are customized to particular situations (Erwin 2009). In general, the success of wetland restoration may depend on several factors, such as the presence of seed banks (Kelly et al. 2009), the hydrological regime (Bruland et al. 2003), the microtopography (Kurt et al. 2009), the forms and amounts of nutrients in the soil (Christopher et al. 2008), the habitat and matrix type (Roe et al. 2003), the landscape setting (Simenstad et al. 2006), disturbance regimes, and invasive species (Lyubov et al. 2009; Miller and Zedler 2003). Nevertheless, several authors agree that the effectiveness of the wetland restoration will depend mainly on the flow of the water through the system and the degree to which re-flooding occurs. Restoration of wetland structure, increased secondary productivity, and enhanced wildlife value can be achieved by subtle improvements in hydrology in former wetlands that have been converted to uplands, such as pasture (David 1999).

Hydrology is a primary filter for the recruitment of vegetation and propagule bank expression in herbaceous wetlands (Battaglia and Collins 2006; Flores-Verdugo et al. 2007). Germination of some species is inhibited by inundation (Kozlowski 2002), whereas others require it (Leck and Simpson 1995). Some studies have analyzed the differential responses of native and exotic plants to hydrological conditions in order to explain the dominance of some species over others (Miller and Zedler 2003; López Rosas 2007). The re-flooding of former agricultural areas is a relatively common strategy of wetland restoration. Ideally, under flooded conditions undesired species may present a combination of lower germination and physiological stress from flooding on seedlings that can lead to lower plant densities with increasing water depth. On the other hand, floating-leaved and submerged aquatics had higher cover; vegetation converged on simpler, less variable communities dominated by obligate wetland species. One gap in our understanding is the strength of the linkage between hydroperiod and vegetation expression and the thresholds in the changes in hydrology that affect vegetation composition or species abundances.

Despite the large amount of research on wetlands restoration, most of the information comes from temperate freshwater wetlands, and research on tropical interdune wetlands is limited. We present in this chapter the results of a 4-year study monitoring the effects of an induced hydroperiod change to promote the recovery of the vegetation in a tropical interdune wetland in central Veracruz, Mexico. The primary objective was to restore native wetland plant communities and control the invader, *Echinochloa pyramidalis* (antelope grass), an African grass introduced for cattle forage. A monitoring program was initiated in 2007 to assess the first phase of restoration, which involved blocking the drain of the wetland to increase the hydroperiod (i.e., the depth and duration of flooding). Distribution and density of the exotic grass, as well as desirable wetland plant assemblages, were examined relative to increasing hydroperiod and biomass destruction of the invading plant, and their response to the subsequent normalization of the flooding events. The first set of results of this study was published in López Rosas et al. (2010).

16.2 Methods

16.2.1 Study Site

The restoration was conducted in a dune slack dominated by herbaceous vegetation with the presence of some isolated trees. This marsh has an area of 2.6 ha and is within a private reserve (Coastal Research Center "La Mancha", Institute of Ecology A. C.) on the central coast of Veracruz, Gulf of Mexico (19°35′45″N, 96°23′05″W, Fig. 16.1). In this region the climate is warm sub-humid with summer rains, ranging from June to September and a period of relative drought (midsummer) in August and a dry season (January–May). The mean annual precipitation is 1,286 mm. Seventy-eight percent of the total annual rainfall occurs between June and September. The mean monthly temperature fluctuates between 21.1 °C in January and 27.3 °C in June.

The dune slack under restoration was modified in the early 1970s to create nine wetland farming plots (*chinampas*) to grow vegetables, covering an area of 729 m^2 (López Martínez 1985). To build *chinampas*, farmers need to raise the ground level to reduce flooding and build canals to maintain aquatic fauna and fish production. On the eastern site edge, five ponds for aquaculture were built by excavating the soil and sediments. This historical land use resulted in a microsite with high heterogeneity, with areas permanently inundated with freshwater saturation prevailing. Also, the vegetation of the region was modified by the introduction of African grasses, mainly *Panicum maximum*, *Cynodon dactylon*, and *E. pyramidalis*. The latter species has been widely introduced, behaving as an invasive species in wetlands. The site

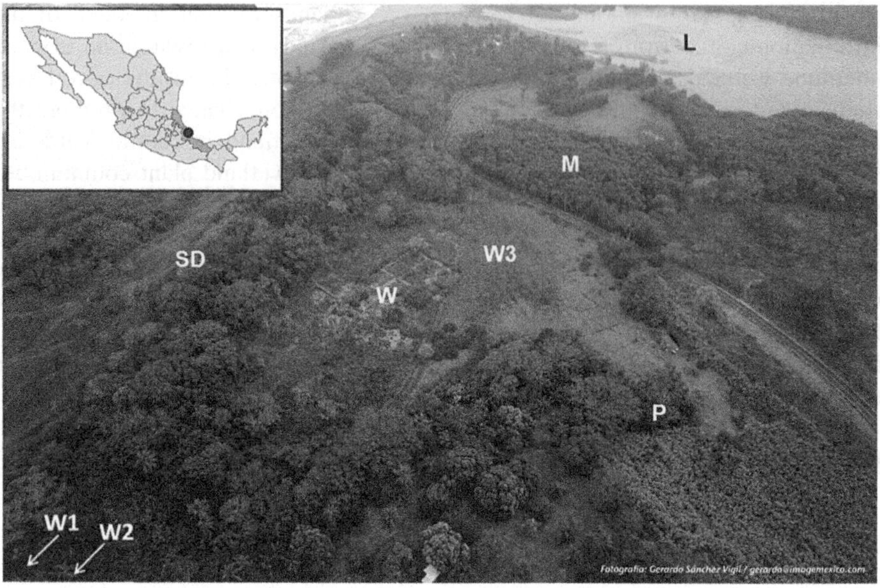

Fig. 16.1 Location of the study site in Veracruz, Southern Mexico. Dune systems and wetlands are indicated in the aerial photograph (2009, elaborated by R. Monroy). *SD* stabilized dunes, *P* old stabilized dunes with orchards, pasture, and sugar cane plantations, *W* wetland under restoration in the interdune, *M* mangrove, *L* coastal lagoon. Other types of wetlands on three sides surround the site. On the north side, there was originally a dune slack marsh that was seasonally flooded, but was subsequently excavated and converted into a permanent lake and is now used for watering cattle (*W1*). A forest swamp borders it (*W2*). Behind it there is sugar cane agriculture on old dunes. The wetland on the west side geomorphologically belongs to the same slack of the study site, but the owner uses it as pasture; thus, vegetation is dominated by the African grass *Echinochloa pyramidalis* and the southern cattail *T. domingensis* (*W3*). To the west the stabilized dune system continues, covered with grassland for cattle ranching. On the south side there is a road that separates the dune slack from a mangrove swamp and the coastal lagoon. An artificial pipe facilitates the water flow between these two systems

was used as a pasture for several years and abandoned 20 years ago; therefore, there was a significant increase in grass cover, height, and biomass. Before the introduction of the grass, the original physiognomy of this dune slack was herbaceous vegetation dominated by *Typha domingensis* and *Sagittaria lancifolia* (Novelo 1978). With the invasion of *E. pyramidalis*, the wetland vegetation was transformed and the grass became dominant. The original wetland plant species *S. lancifolia*, *Laportea mexicana*, and *T. domingensis* (López Rosas et al. 2005) remained in small patches. López Rosas (2007) performed experimental treatments to eradicate the grass and showed that human intervention would be needed to recuperate the original vegetation structure and composition of the dune slack.

Alteration of topography and the consequent change in the hydroperiod are two of the main causes of wetland degradation in the site (Flores-Verdugo et al. 2007). By compiling the site history use during the last 40 years and the consequences of the land use in terms of the topography and the hydrology of the area, it is possible to understand the invasion patterns of an African grass that had been introduced to neighboring ranches. In the study area, available water was not a limiting factor. Yetter's (2004) findings showed that water flowed all year, mainly as a subsuperficial flow, and that the direction was from the sugarcane fields and the dune lake, through the dune slack, toward the mangrove and lagoon (Fig. 16.1). This same author obtained the hydrographs for the site prior to the beginning of restoration activities.

16.2.2 Restoration Activities

Restoration activities in the dune slack were focused on increasing the flooding level and reducing the cover of *E. pyramidalis* (hereafter named *Echinochloa*). In the dry season, in areas dominated by *Echinochloa*, the grass layer was cut at ground level using a machete. In areas where *Echinochloa* was present but mixed with native vegetation, there was selective cutting and extraction of native plants to be relocated. As vegetation was cleared, patches dominated by *Dalbergia brownei* (hereafter named *Dalbergia*) were found on the eastern border. This is a native shrub that grows as a climbing shrub on isolated trees, covering large areas, creating shaded microhabitats, and replacing wetland plant species. The areas covered by *Dalbergia* were also cut at ground level. Where *Echinochloa* or *Dalbergia* re-sprouted, herbicide was sprayed locally. When the herbicide application was not enough to eliminate the sprouts, prescribed burning or covering with black plastic for 2–3 months were necessary. The sprouts of *Dalbergia* that still resisted were injected with herbicide on the basal portion of the sprouting branches (Fig. 16.2; Table 16.1).

At the beginning of the rainy season in 2007, the drain pipe leading to the mangroves was blocked and the wetland flooded for 16 months, ensuring that the grass stems, which had been cut, were drowned. The areas that were flooded but had a low proportion of *Echinochloa* in relation to desirable native species were covered with shade cloth to reduce the amount of light by 50 %, at a height of approximately 2 m above ground level. Every 2 months, teams of three or four people manually removed the re-sprouts of *Echinochloa*. After 16 months of permanent flooding, we unblocked the drain pipe to allow the flow of water into the mangrove. In elevated areas where flooding was shallow, the ground level was lowered around 30–40 cm with a shovel to ensure higher levels of flooding (Fig. 16.2).

Table 16.1 Main restoration activities performed to remove the exotic species from the site

Dominant vegetation	Treatment	Intervention	Location in the area
Sagittaria	Control (original wetland)	Subjected only to induced flooding	Central area of the restoration site
Echinochloa	Control (disturbed area)	Subjected only to induced flooding	Distributed over the whole site
Echinochloa–Typha	Cattail	Cutting plants to soil level, burning, subjected to prolonged flooding	Distributed over the whole site
Echinochloa–Sagittaria	Border	Cutting plants to soil level, reduction of light with shade cloth, subjected to prolonged flooding	Distributed toward the northern part of the site
Dalbergia	Mucal	Cutting plants to soil level, applying herbicide, burning, and subjected to prolonged flooding	Along the eastern border, drier and closer to the dune

16.2.3 Restoration Monitoring

Before starting the restoration actions, we located five distinct areas on the basis of wetland type (dominated by *Echinochloa*, by *T. domingensis* (hereafter named *Typha*) or by *S. lancifolia* (hereafter named *Sagittaria*)—see López Rosas et al. 2005) and two control areas (Table 16.1) where no management activities were performed; however, they were exposed to the induced flooding. In each of these areas we placed five 0.7 $\times$ 0.7-m quadrats to record vegetation changes as well as some environmental parameters through time (Table 16.1).

Each quadrat was monitored bimonthly. All species present in the quadrat were recorded and their cover (%) and height were estimated. Cover and height data were used to calculate the relative importance value (RIV) of each species in each quadrat (RIV of species Y = (percentage cover of species Y + relative height of species Y in relation to the tallest species)/2; the tallest species was *Typha* at 4 m). Along with the vegetation data, water level and pH, electric conductivity and salinity; and soil Eh, moisture and bulk density were measured.

16.3 Results

16.3.1 Hydroperiod

Figure 16.3a shows the hydroperiod for each of the five areas during the 36 months since the restoration began. All five sets of quadrats remained flooded for more than a year because of the blocking of the water flow, and this coincided

Fig. 16.2 Study site changes across time due to restoration activities. **a** Prior restoration. **b** Grass layer cutting. **c** Controlled burning to avoid re-sprouting of invasive species. **d** Selective extraction and translocation of native plants. **e** A black plastic cover was placed on the soil to decrease the light and increase the temperature to eliminate the regeneration of invasive species. **f** Microtopography manipulations, up to 30–40 cm of topsoil was removed in some areas. **g** Shading to reduce the exotic grass cover. **h** View of the restored wetland 12 months after initiation of the restoration process

with a year of high precipitation values (Fig. 16.3b). The quadrats that were originally dominated by *Dalbergia* remained dry for a longer period of time and show high variation in the water levels, followed by the quadrats dominated by *Echinochloa* and *Sagittaria*. The two control areas as well as the *Echinochloa*–*Typha*-dominated quadrats show prolonged flooding with relatively lower variation over time.

16.3.2 Plant Diversity

Both plant diversity and dominance changed over time. The richness increased from 14 to 19 plant species in March 2007 and April 2010 respectively. The Shannon diversity index (in Log_{10}) increased from 0.88 to 0.94. There is 47 % species similitude in the plant communities comparing time 0 (March 2007) and time 36 (April 2010).

Plant dominance showed high variation owing to restoration activities during the first year (Fig. 16.4). *Echinochloa* and *Dalbergia* were the dominant species when the restoration started. The RIV of *Echinochloa* decreased almost tenfold, from an average of 0.35 to 0.034, and remained very low until month 28. *Sagittaria* and *Typha*, two native wetland species were also abundant and interspersed in the area, as well as *Ipomoea tiliacea*, a creeper associated with emergent

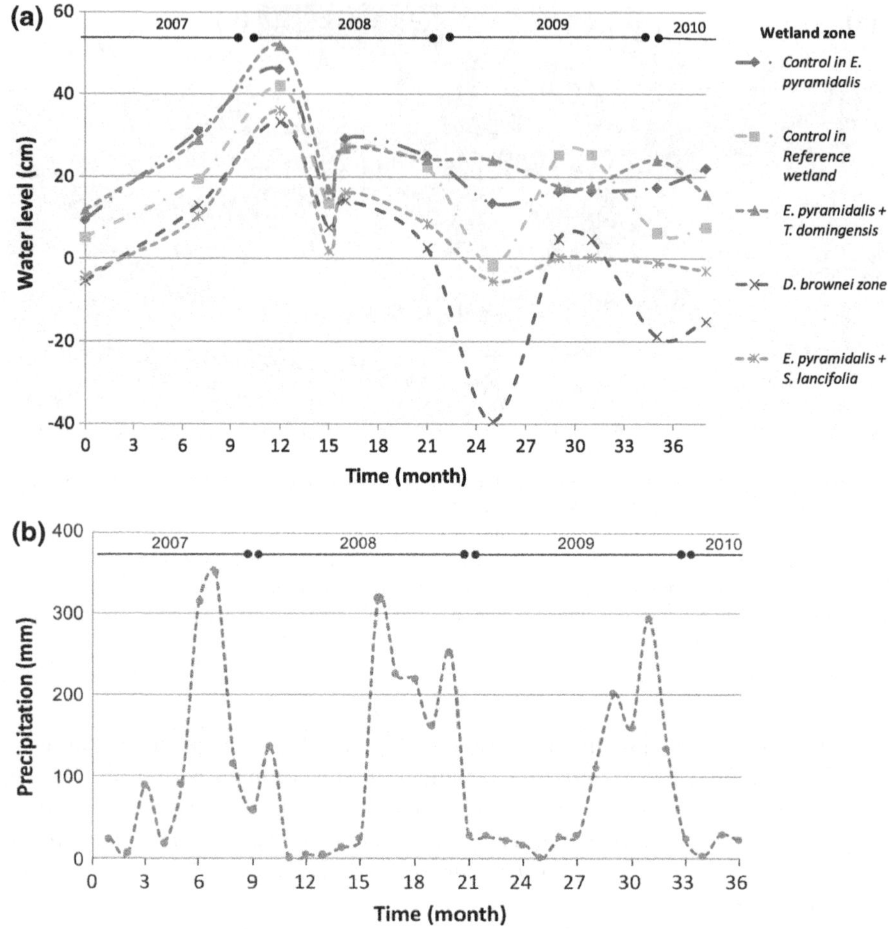

Fig. 16.3 **a** Hydroperiod changes across time from March 2007 (project start, time 0) to April 2010 (time 36). **b** Mean monthly rainfall during the study period obtained from La Mancha weather station

hydrophytes. One year later, in 2008, *Typha* had increased its RIV and has remained the dominant species over time. *Dalbergia* disappeared from the monitored zones, while *Mikania micrantha* and *Pontederia sagittata* (hereafter named *Pontederia*), both desirable species, increased six- and five-fold respectively. In several species the importance values fell, and this was when water was flushed out of the site. *Sagittaria* remained stable and by 2009, had increased its RIV values. In 2010 *Sagittaria* was the third dominant after *Typha* and *Pontederia*. *Echinochloa* was recorded again in some quadrats with low RIV values.

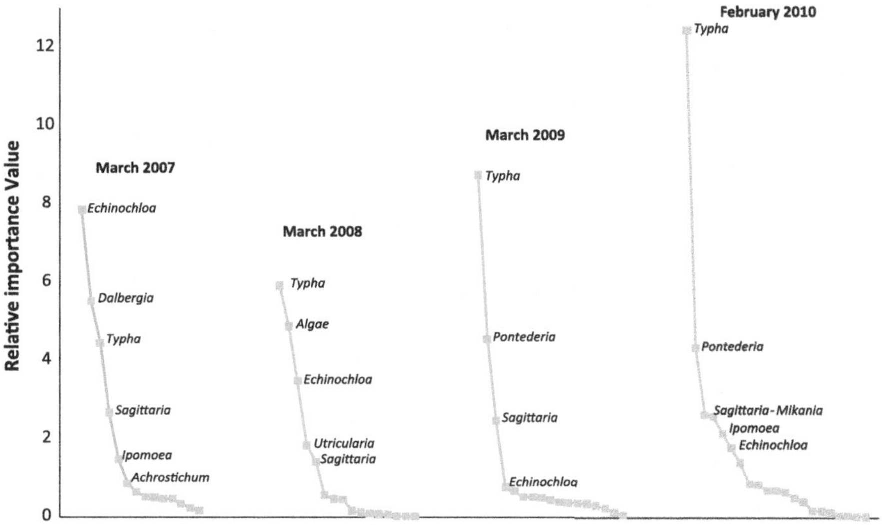

Fig. 16.4 Relative importance value (RIVs) of plant species in four restoration times (sum of RIVs obtained by species in 25 vegetation quadrats). Only generic names of species with higher RIVs are typed

The more abundant plant species in deep-water areas were the emergent cattail, with the presence of the free-floating common duckweed (*Lemna minor*), giant duckweed (*Spirodela polyrhiza*), water lettuce (*Pistia stratiotes*), and giant salvinia (*Salvinia molesta*).

16.3.3 Plant Composition and Microhabitat

Figure 16.5a shows a principal component analysis (PCA) ordination that was performed to represent the abiotic microhabitat variables related to the plant composition. The abiotic factors with major correlation values with the ordination solution were water level ($r = -0.67$) and redox potential (Eh; $r = 0.64$). The control plots dominated by *Echinochloa* and the *Dalbergia*-dominated plots are located toward the drier areas of the restoration site. Water level is highest in *Typha*-dominated areas. Figure 16.5b shows the same ordination with points connected, with vectors indicating the direction in which each quadrat "moves" from time 0 to time 36. The original *Dalbergia*-dominated plots (top right) moved to an area dominated by *Pontederia* (bottom right). The *Typha* and *Echinochloa* (top right) and the *Echinochloa*-dominated plots changed to an area dominated by *Typha* (center left). The control area dominated by native wetland species

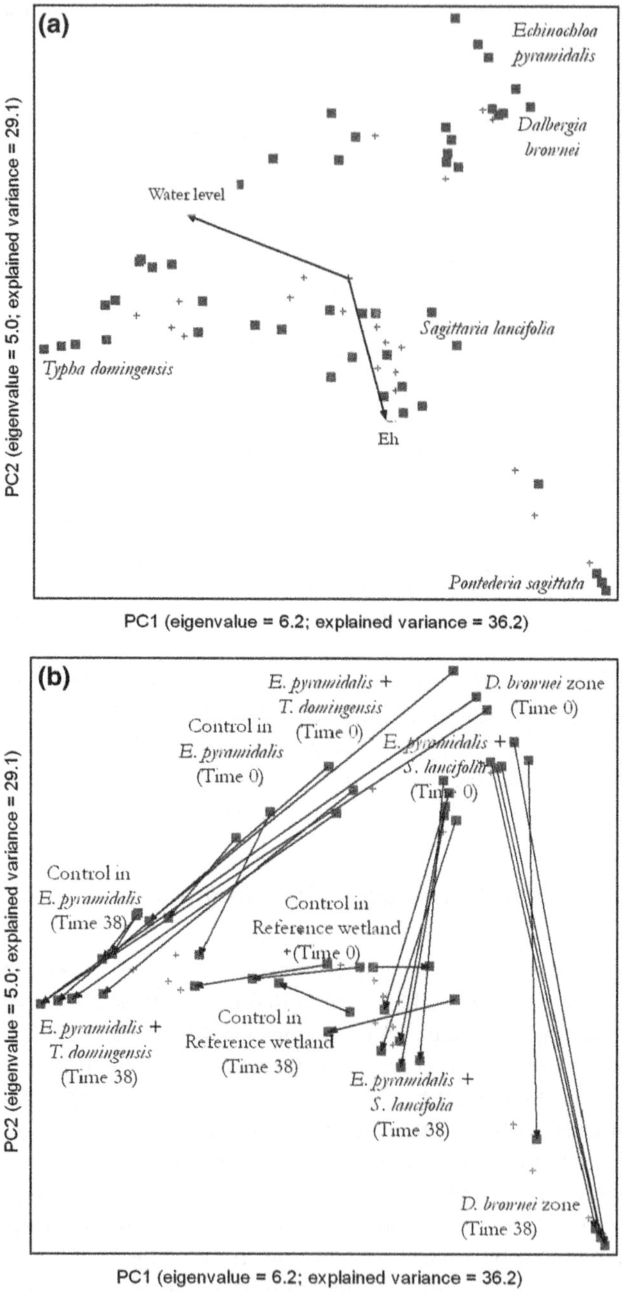
(a)
Echinochloa pyramidalis
Dalbergia brownei
Water level
Sagittaria lancifolia
Typha domingensis
Eh
Pontederia sagittata
PC2 (eigenvalue = 5.0; explained variance = 29.1)
PC1 (eigenvalue = 6.2; explained variance = 36.2)
(b)
E. pyramidalis + T. domingensis (Time 0)
D. brownei zone (Time 0)
Control in E. pyramidalis (Time 0)
E. pyramidalis + S. lancifolia (Time 0)
Control in E. pyramidalis (Time 38)
Control in Reference wetland (Time 0)
E. pyramidalis + T. domingensis (Time 38)
Control in Reference wetland (Time 38)
E. pyramidalis + S. lancifolia (Time 38)
D. brownei zone (Time 38)
PC2 (eigenvalue = 5.0; explained variance = 29.1)
PC1 (eigenvalue = 6.2; explained variance = 36.2)

◀ **Fig. 16.5** Principal components analysis (PCA) ordination of quadrats in five wetland areas and the RIVs of plant species in March 2007 (Time 0) and April 2010 (Time 36): **a** significant abiotic vectors were water level ($r = -0.67$) and Eh ($r = 0.64$), where the angle and length of vectors indicate the direction and strength of the relationships between abiotic factors and ordination scores. **b** Succession vectors that indicate the change in species composition of each wetland zone

remained in the central area with a slight movement toward the left, indicating the increased presence of *Typha*. The border site quadrats dominated by *Echinochloa* and *Sagittaria* (top center) changed to an area similar to the control area dominated by native species.

16.4 Discussion

16.4.1 Invasive Species Removal

This study showed that removing the topsoil (in higher ground) up to 30–40 cm and continuous flooding for 16 months decreased significantly the dominance of *Echinochloa* and increased the dominance of native wetland plant species. López Rosas (2007) experimentally exposed *Echinochloa, Sagittaria*, and *Typha* to prolonged inundation and at the end of the experiment the biomass was lower for *Echinochloa* compared with the native species (*Sagittaria* and *Typha).* The opposite trend was recorded when shorter flooding periods were applied to the same species. In this study, modifying the hydroperiod by flooding the area for 16 months after cutting the exotic grass prevented it from re-sprouting. Also, the prolonged flooding helped native species that are more tolerant to inundation to re-colonize the area successfully. However, long-term monitoring is needed, for instance, 2009 was a relatively dry year that induced the exotic grass re-sprouting in some favorable areas, considering the grass requirements. Shading the area (50–60 %) was also useful as there were patches of native species interspersed with the exotic species. Shading in controlled experiments (50 %), has been documented to be an efficient mechanism for decreasing the invader's biomass (López Rosas 2007).

Another species that had to be removed from the site was *D. brownei*, which is a native shrub/climber species that may be dominant in disturbed sites and covers isolated or wetland edge trees, also forming dense, spiny thickets. *Dalbergia* has been observed to increase its cover in the Reserve Cienaga del Fuerte, north of Veracruz, after hurricanes Dean and Lorenzo in 2007. There is no published information about the ecological response of this species to anthropogenic activities and it was relatively difficult to eliminate it from our study site, by performing several actions, such as cutting aerial biomass, burning, flooding, spraying, and injecting with herbicide. There is a need for further information about more effective treatment to control this species and eliminate its re-sprouting capacity.

Typha domingensis is a native wetland species whose dominance has been increasing throughout the restoration process at our study site. Rejmánková et al. (1995) and Johnson and Rejmánková (2005) found that marshes dominated by this species occupied areas with high levels of nitrogen and phosphorus. Soil phosphorus was the most important factor affecting *Typha* abundance; however, other dominant species, such as *Eleocharis* spp. and *Cladium jamaicense*, correlated negatively with soil phosphorus. The increasing dominance of *Typha* at our restoration site is a non-desired situation as *Typha* creates monodominated patches; therefore, the monitoring and selective extraction of this species will have to be addressed.

16.4.2 Plant Diversity Recovery

The species diversity recorded at the 4-year restored site is low compared with European dune slacks (Grootjans et al. 2004), but similar to diversity records in wetland dune lake vegetation (Peralta Peláez and Moreno-Casasola 2009) and in other freshwater marshes along the coastal plain of Veracruz (Moreno-Casasola et al. 2010). This similarity is probably the result of longer inundation periods, which have favored organic accumulation, as occurs in other wetlands. Few species are shared with those dominant (*Hydrocotyle bonariensis*, *Lippia nodiflora*, *Cyperus articulatus)* of wet slacks found in the same dune (Moreno-Casasola and Vázquez 1999), and in this study, they were present along the restoration process, but had a low importance value.

Moreno-Casasola et al. (2010) obtained hydrographs for different types of broad-leaved marshes in the state of Veracruz, which showed that flooding periods for *Sagittaria* and *Pontederia* floristic associations was around 10 months per year. At the restoration site, hydrographs for the different wetland types varied, but some of them (Fig. 16.3) showed that areas remained flooded all year round (both controls and the *Echinochloa–Typha* zone) or for several months (*Echinochloa*, *Sagittaria*, and *Dalbergia* zones). In other dune slacks of this same system, flooding lasts only for 2–3 months (Martínez et al. 1997). This may explain the difference in the species composition found, and the similarity to other wetlands along the floodplain.

Bakker et al. (2006) tested in a field experiment the effects of groundwater level rise with a combination of topsoil removal or mowing on the recovery of oligotrophic wet dune slack plants. They found that the re-wetting was crucial to nutrient availability. Plant traits related to dispersal, nutrient, and moisture regime were all significantly affected by re-wetting in the Kennemer dunes slacks in the Netherlands. The re-wetting did not coincide with traits of plant species of wet, oligotrophic conditions that were predominantly the target species of wetland restoration (van Bodegom et al. 2006). They concluded that the lack of re-development of target vegetation in restored wetlands deserves further analysis.

16.5 Conclusions

Restoration strategies should be based on the knowledge of the biology of the problematic species and the knowledge of the minimal ecological requirements of native species. In this study area, the restoration process is on track, three invasive species have been partially or totally eliminated, there is recovery of the hydroperiod compared with that of similar broadleaved marshes along the coastal plain, and a freshwater marsh has been created with a heterogeneous plant cover dominated by native species. Nevertheless, the wetland is not yet self-sustaining. Its small size increases the edge effect, the water flowing into it is enriched with nutrients that favor the mono-dominance of *Typha*, and next to the restored site there is grassland dominated by *Echinochloa*. In the short and mid-term, continuous monitoring and intervention are still needed. In the long term, consideration about climate changes and their effects on the hydroperiod will have to be taken into account to ensure the success of the restored marsh.

Acknowledgments This research was made possible with a grant given by Conacyt-SEP (106451) and CONABIO (HH024), and the support of the Instituto de Ecología A.C. (902-17). Thanks to the editors for their critical appraisal of this work.

References

Bakker C, Van Bodegom PM, Nelissen HJM, Ernst WHO, Aerts R (2006) Plant responses to rising water tables and nutrient management in calcareous dune slacks. Plant Ecol 185:19–28

Battaglia LL, Collins BS (2006) Linking hydroperiod and vegetation response in Carolina bay wetlands. Plant Ecol 184:173–185

Bobbink R, Beltman B, Verhoeven JTA, Whigham DF (eds) (2006) Wetlands: functioning, biodiversity conservation, and restoration. Springer, Berlin

Bruland GL, Hanchey MF, Richardson CJ (2003) Effects of agriculture and wetland restoration on hydrology, soils, and water quality of a Carolina bay complex. Wetlands Ecol Manag 11:141–156

Christopher DB, Danielle MA, Randall KK (2008) Evaluating hydroperiod response in restored Carolina bay wetlands using soil physicochemical properties. Restor Ecol 16:668–677

David PG (1999) Response of exotics to restored hydroperiod at Dupuis Reserve, Florida. Restor Ecol 7:407–410

Erwin KL (2009) Wetlands and global climate change: the role of wetland restoration in a changing world. Wetlands Ecol Manag 17:71–84

Flores-Verdugo F, Moreno-Casasola P, Agraz-Hernández CM, López-Rosas H, Benítez-Pardo D, Travieso-Bello AC (2007) La topografía y el hidroperiodo: dos factores que condicionan la restauración de los humedales costeros. Boletín de la Sociedad Botánica de México 80(Suppl):33–47

Grootjans AP, Adema EB, Bekker RM, Lammerts EJ (2004) Why young coastal dune slacks sustain a high biodiversity? In: Martínez ML, Psuty NP (eds) Coastal dunes, ecology and conservation ecological studies, vol 171. Springer, Berlin, pp 85–101

Johnson S, Rejmánková E (2005) Impacts of land use on nutrient distribution and vegetation composition of freshwater wetlands in northern Belize. Wetlands 25:89–100

Kelly PN, Kristin R, Andrew HB (2009) Rapid seed bank development in restored tidal freshwater wetlands. Restor Ecol 17:539–548

Kozlowski TT (2002) Physiological-ecological impacts of flooding on riparian forest ecosystems. Wetlands 22:550–561

Kurt FM, Changwoo A, Gregory BN (2009) The influence of microtopography on soil nutrients in created mitigation wetlands. Restor Ecol 17:641–651

Leck MA, Simpson RL (1995) Ten-year seed bank and vegetation dynamics of a tidal freshwater marsh. Am J Bot 82:1547–1557

López Martínez A (1985) Estudios de productividad en áreas de chinampas: El caso de la Estación de Investigaciones sobre Recursos Bióticos, "El Morro de La Mancha" Mpio. de Actopan, Ver. BS thesis, Colegio Superior de Agricultura Tropical, Tabasco, Mexico

López Rosas H (2007) Respuesta de un humedal transformado por la invasión de la gramínea exótica *Echinochloa pyramidalis* Hitchc. & A. Chase a los disturbios inducidos (cambios en el hidroperíodo, apertura de espacios y modificación de la intensidad lumínica). PhD thesis, Instituto de Ecología, A. C., Xalapa, Mexico

López Rosas H, Moreno-Casasola P, Mendelssohn IA (2005) Effects of an African grass invasion on vegetation, soil and interstitial water characteristics in a tropical freshwater marsh in La Mancha, Veracruz (Mexico). J Plant Interact 1:187–195

López Rosas H, López-Barrera F, Moreno-Casasola P, Aguirre-León G, Cázares-Hernández E, Sánchez-Higueredo L (2010) Indicators of recovery in a tropical freshwater marsh invaded by an African grass. Ecol Restor 28:324–332

Lyubov EB, Alexander YK, Dianna KP, Leah DC, David NH (2009) Wetland restoration and invasive species: apple snail (*Pomacea insularum*) feeding on native and invasive aquatic plants. Restor Ecol 17:433–440

Martínez ML, Moreno-Casasola P, Vázquez G (1997) Effects of disturbance by sand movement and inundation by water on tropical dune vegetation dynamics. Can J Bot 75:2005–2014

Miller RC, Zedler JB (2003) Responses of native and invasive wetland plants to hydroperiod and water depth. Plant Ecol 167:57–69

Moreno-Casasola P, Vázquez G (1999) The relationship between vegetation dynamics and water table in tropical dune slacks. J Veg Sci 10:515–524

Moreno-Casasola P, Cejudo-Espinosa E, Capistrán-Barradas A, Infante-Mata D, López-Rosas H, Castillo-Campos G, Pale-Pale J, Campos-Cascaredo A (2010) Composición florística, diversidad y ecología de humedales herbáceos emergentes en la planicie costera central de Veracruz, México. Boletín de la Sociedad Botánica de México 87:29–50

Novelo A (1978) La vegetación de la estación biológica El Morro de la Mancha, Veracruz. Biotica 3:9–23

Peralta Peláez LA, Moreno-Casasola P (2009) Composición florística y diversidad de la vegetación de humedales en los lagos interdunarios de Veracruz. Boletín de la Sociedad Botánica de México 85:89–101

Rejmánková E, Pope KO, Pohl MD, Rey Benayas JM (1995) Fresh-water wetland plant-communities of Northern Belize—implications for paleoecological studies of Maya wetland agriculture. Biotropica 27:28–36

Roe JH, Kingsbury BA, Herbert NR (2003) Wetland and upland use patterns in semi-aquatic snakes: implications for wetland conservation. Wetlands 23:1003–1014

Simenstad C, Reed D, Ford M (2006) When is restoration not? Incorporating landscape-scale processes to restore self-sustaining ecosystems in coastal wetland restoration. Ecol Eng 26:27–39

Van Bodegom PM, Grootjans AP, Sorrell BK, Bekker RM, Bakker C, Ozinga WA (2006) Plant traits in response to raising groundwater levels in wetland restoration: evidence from three case studies. Appl Veg Sci 9:251–260

Yetter JC (2004) Hydrology and geochemistry of freshwater wetlands on the Gulf Coast of Veracruz, Mexico. MS thesis, University of Waterloo

Zedler JB (2000) Progress in wetland restoration ecology. Nature 402:523–526 [illegible] (2009) Compression horizontale [illegible] Bulletin de la Société [illegible]

[illegible]

[illegible]

[illegible]

[illegible]

[illegible]

Part III
The Costs of Coastal Dune Restoration and Ecosystem Services

Chapter 17
The Value of Coastal Sand Dunes as a Measure to Plan an Optimal Policy for Invasive Plant Species: The Case of the *Acacia saligna* at the Nizzanim LTER Coastal Sand Dune Nature Reserve, Israel

David Lehrer, Nir Becker and Pua Kutiel (Bar)

17.1 Introduction

The introduction of *Acacia saligna* into Israel by the British Mandate and later on by the Jewish National Fund's (JNF) forestation department began at the start of the 20th century and continued for over 70 years. Because of its rapid growth rate over a broad ecological range and being a legume, it was chosen either for solving erosion and aeolian problems such as dune mobility, or as a fodder plant in semi-arid and arid regions. Since being planted in Israeli coastal sand dunes, *A. saligna* has spread rapidly at an annual growth rate of 2.92 % (Bar (Kutiel) et al. 2004). This has caused significant undesired changes, from the conservation point of view, such as dune stabilization, exclusion of psammophiles (sand-loving species—some of which endemic to the region) and replacement by generalists and species noncharacteristic to the dunes, and a significant reduction in the regional biodiversity as a whole (Mehta 2000; Cohen and Bar (Kutiel) 2004) High biodiversity is considered crucial for ecosystem sustainability (Tilman et al. 1996).

This research combines our ecological understanding of the spread of the invasive species *A. saligna* within the coastal sand dunes in Israel with known

D. Lehrer (✉)
The Arava Institute for Environmental Studies, 88840 Kibbutz Ketura, D.N. Hevel Eilot, Israel
e-mail: david.lehrer@arava.org

N. Becker
Department of Economics and Management, Tel Hai College, 12210 Upper Galilee, Israel
e-mail: nbecker@telhai.ac.il

P. Kutiel (Bar)
Department of Geography & Environmental Development, Ben-Gurion University, P.O.B. 653 84105 Beer Sheva, Israel
e-mail: kutiel@bgu.ac.il

M. L. Martínez et al. (eds.), *Restoration of Coastal Dunes*, Springer Series on Environmental Management, DOI: 10.1007/978-3-642-33445-0_17,

economic tools in order to answer two fundamental questions: Does the benefit of slowing or preventing its spread outweigh the costs? What is the most cost effective method to control the spread?

17.2 *Acacia saligna*: An Invasive Species

17.2.1 Background

Acacia saligna is a small tree indigenous to Australia and known around the world as an invasive species owing to its rapid growth rate and ability to competitively exclude native species (Cohen and Bar (Kutiel) 2004). It was first brought to Israel by the British in the early part of the twentieth century for forestation, soil conservation, and stabilization of sand dunes to protect roads, agriculture, and settlements (Bar (Kutiel) et al. 2004; Cohen and Bar (Kutiel) 2004). In the early 1960s, the JNF planted *A. saligna* in the Nizzanim coastal sand dunes as a means of stabilizing shifting dunes. *Acacia saligna* has spread quickly into native habitats by transforming the nature of the soil (such as increasing organic matter and nitrate content) and the hydrology (such as decreasing water infiltration rate and the ground water level) to better suit its own ecological needs (Mehta 2000). Likewise, it out-competes dwarf native plants through shade and an enormous seed bank that may stay dormant for 50 years (Mehta 2000).

17.2.2 The Impact of Acacia saligna in Israel

The transformation of an organism into an invasive species and its spread is a stochastic process. "A species must advance through several stages before being considered invasive: (1) it must be imported to an area where it is not native; (2) it must be introduced into the wild; (3) it must become established with a self-sustaining population; (4) it must be a pest which means that it must trigger costs to humans or ecosystems that outweigh any attendant benefits" (Shogren and Tschirhart 2004, p. 2). *Acacia saligna* has clearly advanced through stages 1–3 within Israel (Bar (Kutiel) et al. 2004; Cohen and Bar (Kutiel) 2004) and therefore its invasion is seen as costly to both humans and ecosystems in that country.

The presence of *A. saligna* in the coastal sand dunes in Israel has a negative effect on the characteristic native vegetation as well as on the fauna (Cohen and Bar (Kutiel) 2004; Manor et al. 2007). It is obvious mainly at the inter-dune depressions and at the aeolianites (local type of sandstones), where it is expressed in cover, richness, and composition. The negative effects of *A. saligna* on the plant composition (abundance and richness) are expressed chiefly by reducing the psammophyle species; nearly 100 % of the species in the non-invaded mobile dunes are psammophytes, while less than 51 % of the species in the invaded dunes are psammophytes. Moreover, the composition of the psammophyte species at the invaded sites differed from that on the mobile dunes. Instead of psammophytes, such as the native *Echinops philistaeus* (endemic to the region), *Polygonum palaestinum*, and *Silene villosa* (a frequent annual), that are characteristic to mobile sand dunes, there were psammophyte plants such as *Rumex pictus*, *Hormuzakia aggregate*, and *Daucus glaber* (all annuals) that are typical of stabilized dunes. Most of the herbaceous species found in the invaded sites are *opportunist* and *generalist* species. These two categories of species generally favor disturbed habitats rich with available nutrients and upper (0–30 cm) soil moisture. The overall impact of Acacia on the native richness and composition of plants is expressed in the EVI (ecological value index: an index that is the sum for the entire species *s* in a certain habitat of the product of the conservation importance value of species *i*, ranked from 5 to −3, and its abundance) of the various habitat types, which significantly decreases compared with that of similar un-invaded habitats (Cohen and Bar (Kutiel) 2005). Similar trends were found for small mammals, such as field mice (*Apodemus mystacinus*) and rats (*Ratus rates*), which replace the psammophyle gerbil species characteristic to the Israeli coastal dunes (Manor et al. 2007). There is little doubt that the invasion of *A. saligna* has unfavorable consequences for native ecosystems and, thus, human intervention and management control are necessary.

17.2.3 Nizzanim LTER Coastal Sand Dune Nature Reserve and its Ecological Value

The Nizzanim long-term ecosystem research (LTER) Coastal Sand Dune Nature Reserve is located in the southern part of Israel's Mediterranean coast. It covers 2,000 ha and is the only large, open, and minimally disturbed area along the coast. The reserve consists of various types of dunes with characteristic assemblages of plants and animals. The most unique feature of this area is the fact that it is an arid ecosystem inhabited by plants and animals originating from the Sahara and the Arab deserts, and is located within the Mediterranean region of Israel. The nature reserve is currently covered by thinly dispersed flora and fauna thriving on sand dunes; however, this is slowly being replaced by a dense cover of the invasive species, *A. saligna*.

Owing to the rapid increase in Israel's population since its establishment in 1948, and the resultant acceleration of building and development along Israel's coastal plain, the sand dune areas have been reduced to a third of their original size (Bar (Kutiel) et al. 2004). Israel's coastal dunes are significant habitats to a variety of endemic species, including species that are specifically adapted to shifting sand dunes such as the *Meriones sacramenti*, and the largest lizard in Israel, *Varanus grizeum*. Many of these species are listed as vulnerable or endangered (Kutiel 2001).

17.3 Using a Contingent Value Methodology Survey

Policy makers need to decide whether or not the damage caused by invasive species, such as *A. saligna*, justifies the expense of controlling them. Ecologists must be able to express the benefits of the expenditure on management and conservation of ecological resources in terms of goods and services to the public (Richardson and van Wilgen 2004).

The most common tools used to estimate the value of public goods, such as nature reserves, open spaces, and biodiversity, are the contingent valuation method (CVM) and the travel cost method (Maille and Mendelsohn 1993; Carson 2000; Hackett 2000; Jakobsson and Dragun 2001; Carson et al. 2003; Becker et al. 2007).

Contingent valuation methodology is a survey method that asks the respondent to express how much he or she is willing to pay for a nonmarket environmental good or service. This method is used to determine how people value environmental goods and services through direct questioning and the benefits are therefore measured directly. The assumption is that individuals have hidden preferences (detected from their behavior), which they will reveal when questioned about their preferences. This is the only method available to estimate non-use or passive use value (sometimes referred to as existence value, first mentioned by Krutilla 1967) (Carson 2000; Jakobsson and Dragun 2001; Carson et al. 2003).

The basic problem confronted by the research was the difficulty in explaining the complexity of biodiversity loss to the unaware public so that informed answers could be collected. If the respondents to a CVM survey are unaware of the importance of the goods that they are being asked to value, they may not reveal a strong preference and the good may be undervalued (Carson et al. 2000; Christie et al. 2006).

17.3.1 Contingent Value Methodology Survey Design

In order to determine the economic benefit of controlling *A. saligna* or the economic damage caused to the biodiversity of the coastal sand dunes at Nizzanim Nature Reserve by the invasion of the species, a CVM survey of 404 Israelis was conducted. There were two groups of respondents, from which an overall response rate of 80 % was achieved. One group consisted of 113 visitors to the Nizzanim LTER Coastal Sand Dune Nature Reserve. The second group consisted of 291 respondents from all over Israel. The purpose of surveying these two groups was to compare the answers of users and non-users.

Owing to incomplete surveys and "protest" answers (a protest answer is a zero value given by the respondent that indicates non-acceptance of the constructed scenario); the results of 103 surveys were removed from the data, leaving 301 surveys for analysis.

17.3.2 Components of the Contingent Value Survey

A CVM survey must include several components, which are listed below (Carson 2000; Carson et al. 2000).

1. This section provided a one-page introduction and background information, explaining the context in which the decision must be made and gave a detailed but brief description of the environmental good.
2. This section presented the respondent with three scenarios accompanied by photographs of the Nizzanim LTER Coastal Sand Dune Nature Reserve. The three photographs represented three scenarios: total elimination of the invasive species, containment of the spread of the invasive species, and non-intervention, resulting in 90 % coverage of the nature reserve by the invading species within 20 years.
3. This section of the survey briefly described how the environmental good will be provided through a public fund dedicated to the prevention of the spread of *A. saligna* at the Nizzanim LTER Coastal Sand Dune Nature Reserve. The first question in this section (question 1) asks:

   Under the assumption that a designated public fund is created for the purpose of financing programs to prevent the spread of the Blue Acacia tree on the Nizzanim Nature Reserve and the restoration of the beach, what is the maximum amount of money that you would be willing to pay as an contribution to this fund - In order to preserve the present situation, I would be willing to contribute from my own pocket once a year ...(payment card)

   The second question in this section (question 2) asks:

   Under the assumption that a designated public fund is created for the purpose of financing programs to prevent the spread of the Blue Acacia tree on the Nizzanim Nature Reserve and the restoration of the beach, what is the maximum amount of money that you would be willing to pay as an contribution to this fund - In order to improve the present situation, eliminate the Blue Acacia tree at the Nizzanim Nature Reserve and restore the reserve,

   I would be wiling to contribute from my own pocket once a year ...(payment card)

   After each question, an identical "payment card" was presented with numerical values in New Israeli Shekels (NIS)[1] which when converted into US$ ranging from $0.00 to $33.34 increasing in increments of $1.11 and options to write another amount or to claim to not have enough information to answer the question.

4. This section of the CVM survey debriefs the respondents by asking them to reveal the reasons for their answers. Respondents were asked to choose the statement that best explains their preferences. These answers indicated whether the respondent attributed use or non-use value to the environmental good or alternatively was registering a protest bid or a legitimate zero response.

[1] Throughout the paper the ongoing exchange rate is 5.1 NIS = 1 Euro, 4.5 NIS = 1 US$.

5. This section asked the respondent to provide demographic information, such as gender, age, family status, etc. These answers enabled a regression analysis of the data, which in turn allowed the researchers to test the results of the survey against expected adherence to economic theory. The answers are also used to compare the demographics of the sample with the demographics of the general population of Israel.

17.4 Analysis of the CVM Survey Results

17.4.1 Descriptive Statistics

Table 17.1 summarizes the demographic make-up of the survey respondents and compares their demographic background with that of the population of Israel in general. All national statistics were taken from the Israeli Government's Central Bureau of Statistics. The comparison showed that the respondents were a representative sample of the national population in almost all categories except for gender. The survey sample had a lower percentage of women than the general population of Israel. Because gender may have had an impact on the mean willingness to pay (WTP), an adjustment was made in the final valuation of the nature reserve in order to account for this deviation from the general population statistics. We found that on the WTP question on species elimination (question 2)

Table 17.1 Comparison of survey demographic data with national data

	Demographics	National (%)	Surveyed (%)
Gender	Male	49	54
	Female	51	46
Origin	Israel	72	89
	Abroad	28	11
Age	18–25	19	33
	26–35	22	29
	36–45	17	23
	45–55	16	12
	Over 55	25	3
Family status	Married	43	42
	Not married	57	58
Average persons per household		3.35	3.7
Education	Elementary	2.2	0
	High school	33	37
	Vocational	17	18
	Academic	44	47
Income	Under average	30	30
	Average	40	40
	Above average	30	30

women in our survey were willing to pay $2.84 more than men. Because the percentage of women in our survey was lower than the percentage of women in the general Israeli population, we recalculated the total WTP in question 2 based on the product of this additional WTP for women and the actual percentage of women in the population. This adjustment resulted in a slightly higher WTP value for question 2 than in the original results.

17.4.2 The Mean Willingness to Pay

Table 17.2 summarizes the mean and median WTP of Israelis to contain or eradicate the invasion of *A. saligna* at the Nizzanim LTER Coastal Sand Dune Nature Reserve. The results show that the mean WTP of the Israeli public to prevent the spread of *A. saligna* at the nature reserve is $8.50 with a standard deviation of 9.70 (question 1). The Israeli public's mean WTP to prevent the spread of *A. saligna* at the nature reserve and to restore the reserve was $8.92 with a standard deviation of 9.95 (question 2).

The mean WTP for question 1 was $8.50 and for question 2 was $8.92, a difference of $0.42. However, the paired t test results, which test whether the mean of the difference between the WTP in question 1 and the WTP in question 2 is significantly different than zero, were inconclusive. The mean difference ($M = 0.42$, SD = 5.2, $N = 301$) was not significantly greater than zero, t (299) = −0.309, two-tailed $p = 0.165$, not providing strong evidence of conformity to economic theory. A 95 % confidence interval for the mean WTP difference between questions 1 and 2 is (1.00, −0.18) which includes the zero value.

17.4.3 Use and Non-use Value

This section of the survey contained debriefing questions. The respondents were asked to select from among eight possible explanations the one best describing the reasons for their answers to the previous section. These explanations were placed into three categories: use value, non-use value and zero value.

Use value is the value that the respondent places on an environmental good because he or she receives direct benefit, such as recreational benefit, health benefit, the use of a natural resource for consumption like clean water or timber, or the option to use this resource in the future (Carson 2000; Carson et al. 2000, 2003).

Table 17.2 Willingness to pay (WTP) descriptive statistics

US$	Value	Mean	SD	Median
Question 1	WTP for contamination	8.50	9.70	4.44
Question 2	WTP for eradication	8.92	9.95	4.44

Table 17.3 Expressed mean use and non-use value

US$	Value	Use	Non-use	Zero	Total
Question 1	Mean WTP for containment	1.87	5.29	1.34	8.50
Question 2	Mean WTP for eradication	2.07	5.57	1.28	8.92

Non-use value is the value that the respondent places on an environmental good, even though he or she receives no direct benefit. Non-use values include bequest value for future generations, as well as the satisfaction of knowing that a resource is there, called existence value (Carson 2000; Carson et al. 2000, 2003).

Zero-value is the value a respondent places on an environmental good when he or she does not feel that it is their job to pay for the environmental good, or when he or she sees no value in preserving or protecting the environmental good.

Table 17.3 presents the mean WTP for questions 1 and 2 broken down into use value, non-use value, and zero-value amounts.

It is important to note that while the Nizzanim Coastal Sand Dune Nature Reserve is an important visiting area, and although 25 % of the surveys were implemented in and around the Nature Reserve, the majority of the value of the mean amount of WTP to preserve and protect the nature reserve is non-use value.

17.4.4 Regression Analysis of the Demographic Variables

Multiple regressions were performed on the two WTP questions posed to the respondents at the site and within the general population sample. The purpose of running a regression analysis on the demographic variables was to validate whether the predictors found in the analysis were consistent with economic theory and previous CVM surveys. The results are presented in Table 17.4. As can be seen from the table, income and membership of green organizations were found to be significant variables consistently throughout the four regressions (two sample types and two questions).

The level of income indicated significance as a predictor of WTP in questions 1 and 2. In both questions the positive coefficient indicated a positive correlation between level of income and the average WTP to prevent the spread of invasive species or eliminate it altogether at the Nizzanim LTER Coastal Sand Dune Nature Reserve. These results are consistent with previous findings of other CVM studies, as well as general economic theory (Carson et al. 2000; Jakobsson and Dragun 2001; Amirnejad et al. 2006).

Membership in an environmental organization also indicated significance as a predictor of WTP in questions 1 and 2. In both questions the positive coefficient indicated a positive correlation between membership in an environmental organization and the average willingness of members of environmental organizations to pay more than nonmembers. These results are consistent with previous findings

Table 17.4 Regression results

	General population		Visitors	
Question type → Variable ↓	Q.1	Q.2	Q.1	Q.2
Gender	9.126 (6.249)	11.263 (5.867)*	6.926 (9.160)	19.494 (10.852)
Age	0.429 (0.361)	0.163 (0.339)	−0.298 (0.572)	−0.581 (0.678)
Country of birth	3.589 (10.724)	6.522 (10.068)	−14.646 (13.459)	−23.805 (15.994)
Family status	−1.279 (2.869)	−2.774 (2.694)	−4.698 (4.560)	−4.017 (5.402)
Number of children	−3.723 (9.237)	−7.287 (8.671)	−19.002 (22.987)	−21.587 (27.232)
Town of residence	−0.029 (0.053)	−0.004 (0.049)	−0.445 (0.153)*	−0.549 (0.181)*
Membership in green organization	26.347 (12.319)*	36.649 (11.565)*	2.429 (15.487)	4.881 (1.347)*
Education	−0.682 (3.559)	−0.325 (3.341)	6.378 (6.296)	5.751 (7.458)
Income	8.820 (4.651)*	10.495 (4.366)*	11.831 (7.213)*	13.140 (98.544)*
Constant	19.188 (22.204)	12.976 (20.846)	60.367 (34.646)*	76.665 (41.044)*
R^2	0.176	0.224	0.433	0.165

Standard error in parenthesis
*Indicates significance at 10 % level

of other CVM studies (Carson et al. 2000; Jakobsson and Dragun 2001; Amirnejad et al. 2005). Interestingly, it can be seen that within the visitors' group, the additional WTP relative to the general population is smaller. This might be explained by the fact that environmental commitment is higher within nonmembers (and hence the small difference), provided they visit the given site.

17.5 Benefit Cost Analysis

17.5.1 Estimating the Economic Value of the Nizzanim LTER Coastal Sand Dune Nature Reserve

According to the Israeli Government's Central Bureau of Statistics (2007) Annual Report, there were 1,968,300 households in Israel in 2006. If we assume that the survey sample is a representative sample of the households in Israel, the total value to the Israeli public of containing the spread of *A. saligna* at the Nizzanim LTER Coastal Sand Dune Nature Reserve will be equal to the mean WTP from question 1 multiplied by the number of households in Israel, which comes to US$16,725,827. The total value of eradication of *A. saligna* at the site and restoring the nature reserve will be equal to the mean WTP from question 2 multiplied by the number of households in Israel, which comes to US$18,806,166.

In order to exercise maximum caution in the benefit–cost analysis, we chose to use the mean use value of the visitors to the Nizzanim LTER Coastal Sand Dune Nature Reserve, estimated by the Nature Reserve Authority to be 200,000 visitors a year. Although this calculation indicates a much lower economic value of the environmental goods than the other methods, it can be argued that the calculation expresses the use value of the nature reserve to those Israelis who are actual users of the environmental goods. By this method, the value of containing the spread of *A. saligna* at the Nizzanim LTER Coastal Sand Dune Nature Reserve will be equal to the visitor's mean WTP by using the value from question 1 ($2.25) multiplied by the number of annual visitors to the nature reserve, which comes to US$450,000. The value of eradicating *A. saligna* at the Nizzanim LTER Coastal Sand Dune Nature Reserve will be equal to the visitor's mean WTP by using the value from question 2 ($2.85) multiplied by the number of annual visitors to the nature reserve, which comes to US$570,000.

17.5.2 Cost

The cost estimates for the containment and eradication of the spread of *A. saligna* at the Nizzanim LTER Coastal Sand Dune Nature Reserve can be narrowed down

Table 17.5 Cost of control at Nizzanim Nature Reserve

Treatment	Description	Annual cost of containment (US$)	Total cost of elimination (US$)
A	Cutting and clearing, spraying of stumps and sprouts	40,151	401,556
B	Cutting and burning, spraying of stumps and sprouts	38,962	389,556
C	Cutting and clearing, solar sterilization—natural irrigation	33,167	331,111
D	Uprooting, solar sterilization—natural irrigation	22,333	223,333
E	Uprooting, spraying stumps and sprouts	19,540	195,333

to five major treatments. Table 17.5 presents the different treatments along with their annual costs.

The total costs per acre of each treatment type ranges from $1,600 for Treatment A to $780 for Treatment E. It is presently estimated that *A. saligna* has established itself in an area of approximately 250 acres out of a total of about 5,000 acres for the entire nature reserve (Cohen and Bar (Kutiel) 2004). In order to completely eradicate *A. saligna* from the Nizzanim LTER Coastal Sand Dune Nature Reserve, a one-time investment would have to be made of between US$195,000 and $400,000 (see Table 17.5) depending on which method of treatment was chosen. Treatments A and B require a small reapplication of herbicides. This was taken into consideration in the costs of the treatments as the net present value of the cost of the repeat application over 20 years at a 5 % interest rate.

The current rate of invasion is approximately 25 acres per year. The expectation is that within 20 years, 90 % of the Nizzanim LTER Coastal Sand Dune Nature Reserve will have been invaded by *A. saligna*. Therefore, in order to contain the spread of the invasive species by maintaining the current level of invasion, an investment would need to be made of between US$19,500 and US$40,000, depending on the treatment level.

17.5.3 Analysis

The annual cost of containing the spread of the *A. saligna* at the Nizzanim LTER Coastal Sand Dune Nature Reserve is between US$19,500 and US$40,000. The one-time cost of eradicating the invasive species completely from the reserve is between US$195,000 and US$400,000 (see Table 17.5). In order to exercise maximum caution in the estimation of the economic value of the elimination of the *A. saligna* species, we chose to compare the total value derived from the mean use value of the visitors to the nature reserve with the cost of containment or elimination of the invasive species.

It is our conclusion that policy makers can justify a one-time expense of US$195,000–US$400,000 in order to provide a benefit to the visitors of the Nizzanim LTER Coastal Sand Dune Nature Reserve, which is conservatively valued at approximately US$570,000. On the other hand, comparing the cost of containment, approximately US$19,500–US$40,000 annually (see Table 17.6), with the annualized benefit of approximately US$450,000 derived from the mean use value of the visitors to the nature reserve gives a less straightforward answer. The annual payment for an NPV of US$450,000 is US$35,708 (at a 5 % interest rate over 20 years). This is a slightly smaller annual benefit than the annual cost of containment using Treatments A or B, is slightly higher than the cost of Treatment C, and considerably higher than the annual costs of Treatments D and E (see Table 17.6). We conclude that if policy makers choose to contain the spread and not eliminate the invasive species, there is only an economic justification for such actions if one of the methods, Treatments C through E, is deemed effective and used. The annual benefit from containment (US$35,708) is less than the annual costs of treatments A and B (US$40,151 and US$38,962 respectively) resulting in a net loss (see Table 17.6).

17.6 Discussion

17.6.1 Economic Theory

In order to validate the results of the CVM survey, they must conform to economic theory. Carson et al. (2000) suggested two tests for conformity to economic theory:

1. The number of respondents who are willing to pay for an environmental good should fall as the price increases.
2. The respondents should be willing to pay more for a higher quality or larger amount of the environmental good.

Figure 17.1 demonstrates that as the hypothetical price of controlling the spread of *A. saligna* in the Nizzanim LTER Coastal Sand Dune Nature Reserve increases, the number of respondents willing to pay for that environmental good will fall.

The CVM survey asked respondents their WTP for two different environmental goods, containment of the spread and eradication of the species from the nature reserve. The intention of presenting respondents with these two environmental goods was to test whether the respondents were willing to pay for a higher quality of the environmental good. The results of this study show that the mean contribution that respondents were willing to pay for eradicating *A. saligna* from the Nizzanim LTER Coastal Sand Dune Nature Reserve was higher ($8.92) than the mean contribution that the respondents were willing to pay for the conservation of the current situation or stopping further spread of *A. saligna* ($8.50). However, this difference was not found to be statistically significant. This result raises the

Table 17.6 Summary of benefit–cost results

Eradication of *A. saligna*			
Benefits	Use value (US$)	Annual number of visitors	Total economic value (US$)
WTP of visitors at the nature reserve	2.85	200,000	570,000
Cost			
Treatment		Total cost (US$)	Net benefit (US$)
A		401,556	168,444
B		389,556	180,444
C		331,111	238,889
D		223,333	346,667
E		195,333	374,667
Control of *A. saligna*			
Benefits	Use value (US$)	Number of visitors	Net benefit (US$)
WTP of visitors at the nature reserve	2.25	200,000	450,000
Annualized value (20 years, 5 %)			35,708
Annual cost			
Treatment		Annual cost (US$)	Annual net benefit (US$)
A		40,151	−4,443
B		38,962	−3,254
C		33,167	2,541
D		22,333	13,375
E		19,540	16,168

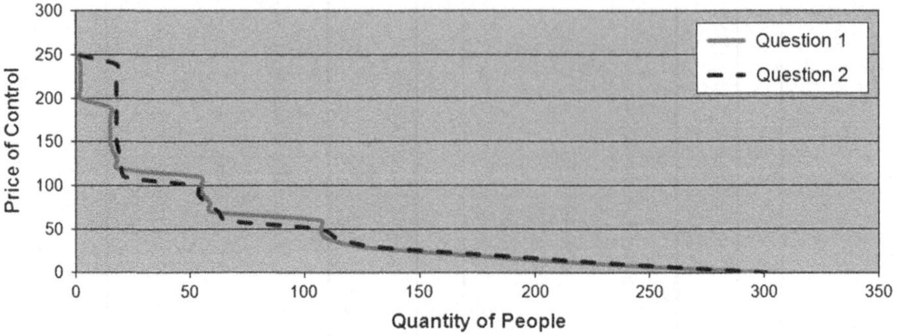

Fig. 17.1 Demand for invasive species control—demand curve for the environmental good. The demand decreases as the price of the good increases for both questions

question of embedding or sensitivity to scope, which is discussed in the Report of the NOAA Panel on Contingent Valuation by Arrow et al. (1993) as well as in other articles dealing with CVM (Carson 2000; Carson et al. 2000, 2003). The concern is that the respondents are stating WTP without regard for the amount or scope of the environmental good being offered. A larger mean difference was found in the WTP of visitors to the nature reserve to eradicate the species compared with just containment. This gives some support to the idea that, among visitors to the nature reserve at least, the difference between containment and eradication was clear. The implication that they were willing to pay more for the better environmental good is consistent with economic theory.

In addition to these two basic tests of conformity to economic theory, regression analysis indicates two predictors of WTP, consistent with the results of previous CVM studies: income level and membership in environmental organizations. The regression analysis revealed that a higher level of income and membership in environmental organizations correlated positively with a higher WTP (see Sect. 14.4.4).

17.6.2 Use and Non-use Value

Contingent valuation is the only methodology that allows us to estimate non-use or passive use value (sometimes referred to as existence value, first mentioned by Krutilla 1967) (Carson 2000; Jakobsson and Dragun 2001; Carson et al. 2003). Our study reveals that both for visitors to the Nizzanim LTER Coastal Sand Dune Nature Reserve and the general population, the non-use or existence value of containment or eradication of the invading species is greater than the use value. This seems to indicate that the Israeli public, both visitors and non-visitors to the nature reserve, recognize the value of conserving the nature reserve for its own sake and not simply for its utility to human beings.

17.6.3 Benefit–Cost Analysis

The benefit–cost analysis compared the value that the public placed on eliminating the spread of *A. saligna* in the nature reserve with the estimated cost of controlling the spread. Estimating the total economic value using the mean use value of visitors to the nature reserve of eradication of the invasive species we arrived at a benefit of approximately US$570,000. The estimated one-time cost of controlling the spread of *A. saligna* ranged between US$195,000 and US$400,000 (Table 17.6). Thus, we can conclude that the value of the coastal sand dune biodiversity is higher than the cost of *A. saligna* eradication. On the other hand, the annual net benefit of containment of the invasive species depended on the method of containment chosen.

References

Amirnejad H, Khalilian S, Assareh M, Ahmadian M (2006) Estimating the existence value of north forests of Iran by using a contingent valuation method. Ecol Econ 58:665–675

Arrow K, Solow R, Portney PR, Leamer EE, Radner R, Schuman H (1993) Report of the NOAA Panel on Contingent Valuation. NOAA, January 11th, 1993

Bar (Kutiel) P, Cohen O, Shoshany M (2004) The invasion rate of the alien species *A. saligna* within the coastal sand dune habitats in Israel. Isr J Plant Sci 52:115–124

Becker N, Choresh Y, Inbar M, Bahat O (2007) Combining TCM and CVM of endangered species: estimation of the marginal value of vultures in the presence of species—visitors' interaction. In: Kontoleon A, Pascual U, Swanson T (eds) Biodiversity economics: issues, methodologies and applications. Cambridge University Press, Cambridge, pp 313–342

Carson RT (2000) Contingent valuation: a user's guide. Environ Sci Tech 34(8):1413–1418

Carson RT, Flores NE, Meade NF (2000) Contingent valuation: controversies and evidence. Environ Resource Econ 19(2):173–310

Carson RT, Mitchell RC, Hanemann M, Kopp RJ, Presser S, Ruud PA (2003) Contingent valuation and lost passive use: damages from the Exxon Valdez oil spill. Environ Resource Econ 25(3):257–286

Christie M, Hanley N, Warren J, Murphy K, Wright R, Hyde T (2006) Valuing the diversity of biodiversity. Ecol Econ 58:304–317

Cohen O, Bar (Kutiel) P (2004) The invasion rate of *A. saligna* and its impact on protected Mediterranean coastal habitats. Proceedings of 10th MEDECOS conference: 25th April–1st May, 2004, Rhodes, Greece

Cohen O, Bar (Kutiel) P (2005) Effect of invasive alien plant—*Acacia saligna*—on natural vegetation of coastal sand ecosystems. Forest 7:11–17 (Hebrew, with English abstract)

Hackett SC (2000) The recreational economic value of the eastern Trinity Alps wilderness. School of Business and Economics, Humboldt State University, Arcata

Israeli Government's Central Bureau of Statistics (2007) Annual report. http://www1.cbs.gov.il/reader/shnatonenew_site.htm. Accessed 9 August 2007

Jakobsson KM, Dragun AK (2001) The worth of a possum: valuing species with the contingent valuation method. Environ Resource Econ 19(3):211–277

Krutilla J (1967) Conservation reconsidered. Am Econ Rev 57(4):777–786

Kutiel P (2001) Conservation and management of the Mediterranean coastal sand dunes in Israel. J Coast Conservat 7:183–192

Maille P, Mendelsohn R (1993) Valuing ecotourism in Madagascar. J Environ Manag 38:213–218

Manor R, Cohen O, Saltz D (2007) Homogenization and invasiveness of commensal species in Mediterranean afforested landscapes. Biol Invas 10:507–515

Mehta S (2000) The Invasion of South African Fynbos by an Australian immigrant: the story of *A. saligna*. http://hort.agri.umn.edu/h5015/rrr.htm. Accessed 15 July 2007

Richardson D, van Wilgen B (2004) Invasive alien plants in South Africa: how well do we understand the ecological impacts? S Afr J Sci 100(1):45–52

Shogren JF, Tschirhart J (2004) Integrating ecology and economics to address bioinvasions. Ecol Econ 52(3):267–271

Tilman D, Wedin D, Knops J (1996) Productivity and sustainability influenced by biodiversity in grassland ecosystems. Nature 379(6567):718–720

Chapter 18
The Coasts and Their Costs

O. Pérez-Maqueo, M. L. Martínez, D. Lithgow, G. Mendoza-González, R. A. Feagin and J. B. Gallego-Fernández

18.1 Introduction

Coastal dunes are unique ecosystems with considerable ecological, social, and economic relevance (Martínez et al. 2004a, b). Ecologically, they are unique because they include species with specific tolerance to burial by sand and salinity (Maun 2009; Gallego-Fernández and Martínez 2011). These species share their functional responses to the peculiar environment of the coastal dunes and yet their phylogeny is very diverse, providing a clear example of evolutionary convergence in both plants and animals. They are also the preferred site to test the theory of successional dynamics (Martínez et al. 2001; Feagin et al. 2005; Isermann 2011).

O. Pérez-Maqueo (✉)
Red de Ambiente y Sustentabilidad, Instituto de Ecología, A.C., Antigua carretera a Coatepec No. 351, El Haya 91070 Xalapa, Veracruz, Mexico
e-mail: octavio.maqueo@inecol.edu.mx

M. L. Martínez · D. Lithgow · G. Mendoza-González
Red de Ecología Funcional, Instituto de Ecología, A.C., Antigua carretera a Coatepec No. 351, El Haya 91070 Xalapa, Veracruz, Mexico
e-mail: marisa.martinez@inecol.edu.mx

R. A. Feagin
Spatial Sciences Laboratory, Department of Ecosystem Science and Management, Texas A&M University, College Station, TX 77845, USA
e-mail: feaginr@tamu.edu

J. B. Gallego-Fernández
Departamento de Biología Vegetal y Ecología, Universidad de Sevilla, Ap.1095 41080 Sevilla, Spain
e-mail: galfer@us.es

M. L. Martínez et al. (eds.), *Restoration of Coastal Dunes*, Springer Series on Environmental Management, DOI: 10.1007/978-3-642-33445-0_18,

In fact, ecological theory on plant succession was first described in coastal dunes a century ago (Cowles 1899), and there is still ongoing research on how community dynamics works in these ecosystems.

Socially, a large fraction of the global human population lives at or near the coast (Martínez et al. 2007); many cities are also located here, and have developed on top of dunes. Twenty-one of the 33 megacities (>10 million inhabitants) of the world are located at the coast and population growth trends indicate that here, human population will continue increasing at a faster rate than inland (Martínez et al. 2007). This human preference results in urban development and the construction of infrastructure, which transforms and even destroys coastal dunes.

Economically, tourism, one of the largest industries in the world, takes place predominantly at the coast, especially on sandy coasts, yielding millions of dollars in revenue. Besides these direct economic benefits, coastal dunes are also highly relevant to humans, since they offer many ecosystem services to society, such as the protection of human structures from erosion as well as the impact of hurricanes and storms by providing sediments, a physical barrier, or resistant vegetation (Nordstrom and Jackson, Chap. 2, this volume). Protection, in particular, will become increasingly relevant because of the encroaching human population at the coast, and also because of the risks associated with global climate change such as an increasing sea level and storminess (Webster et al. 2005).

Other additional ecosystem services provided by sandy coasts and coastal dunes include: sand and minerals for extraction, retention and purification of ground water, raw materials, plants with pharmaceutical use, food for primary and higher trophic level consumers, habitats and refuge for plants and animals, recreation, education, scientific research, cultural and environmental heritage (Bell and Leeworthy 1990; King 1995; Lubke and Avis 1998; Arens et al. 2001; Peterson and Lipcius 2003; Everard et al. 2010; Nordstrom and Jackson, Chap. 2, this volume).

In spite of the evident relevance of coastal dunes, they are threatened by human encroachment throughout the world. As they are lost, the ecosystem services provided by them will be lost too. Because the economic value of the ecosystem services has seldom been assessed, the ecological and socio-economic consequences of their destruction remain largely unknown. Large amounts of money have already been invested in coastal dune restoration and, since destruction continues, restoration needs (and their costs) will also increase. Given this scenario, it is relevant to analyze what would be the best option for our beaches and coastal dunes: to restore? To conserve these ecosystems and manage their ecosystem services? Should we sustain the status quo because of immediate revenues? An important question to answer is: what are the costs of our coasts? What are the costs of restoring beaches and coastal dunes, and what are the economic benefits derived from ecosystem services provided by them?

In this chapter we aim to achieve three goals: first, we performed a literature search to assess a proxy estimate of the economic value of the beach and coastal dunes in terms of their ecosystem services. Second, we used the information that was available on the costs of restoration projects described in this book, which have been performed on beaches and coastal dunes in different countries: New

Zealand (Hesp and Hilton, Chap. 5, this volume); Italy (Acosta et al., Chap. 12, this volume); USA (Pickart, Chap. 10, this volume; Feagin, Chap. 6, this volume); Netherlands (Arens et al., Chap. 7, this volume; Grootjans et al., Chap. 15, this volume); Denmark (Vestergaard, Chap. 4, this volume), and Spain (Muñoz-Reinoso et al., Chap. 9, this volume). We also gathered additional information in order to have a larger set of data with which we could analyze the costs of coastal dune restoration. Finally, we carried out an analysis of the conservation benefits in terms of ecosystem services, versus the costs of restoring beaches and coastal dunes, to determine the best management option for our sandy coasts.

18.2 Ecosystem Services

Natural ecosystems provide a variety of direct and indirect services and intangible benefits to humans and other living organisms (Costanza et al. 1997), and because of their relevance to society, these ecosystem services, goods and their economic value have become a focus of interest for scientists, policy makers, and stakeholders over the last decade (Troy and Wilson 2006). The provision of ecosystem services is directly related to the functionality of natural ecosystems upon which ecological processes and ecosystem structures depend (de Groot et al. 2002). Thus, the better preserved natural ecosystems are, the more ecosystem services they can provide to society (Balvanera et al. 2001).

To gather information on the ecosystem services provided by the beach and coastal dunes, we performed a literature search including the databases environmental valuation reference inventory (EVRI), Envalue, and ecosystem services database (ESD) (McComb et al. 2006) and applicable gray literature (government statistics and graduate theses) (Shuang 2007; Mendoza-González 2009; Lithgow-Serrano 2007). Based on these databases and the published literature we determined specific ecosystem services from the beach and coastal dunes. Using the Millennium Ecosystem Assessment ecosystem services classification (supporting, provisioning, regulatory, and cultural) the ecosystem services provided by the beach and coastal dunes can be grouped as shown in Table 18.1.

Table 18.1 Summary of ecosystem services provided by coastal dunes to society (modified from Everard et al. 2010)

Supporting	Provisioning	Regulatory	Cultural/aesthetic
Soil formation	Drinking water	Storm/flood/erosion protection	Recreation/tourism
Primary production	Food	Climate regulation	Aesthetic value
Nutrient cycling	Fiber and fuels	Water storage	Cultural heritage
Water cycling	Genetic resources	Pest/disease regulation	Spiritual
Photosynthesis	Medicine	Pest control	Art
Carbon storage	Ornamental	Water purification	Social relations
Habitat/diversity	Mineral extraction	Pollination	Education/science

The above-mentioned ecosystem services have not been studied in as much detail as is needed. For instance, the literature on ecosystem services (ISI Web of Science January 2012) displayed 3,070 publications that mention ecosystem services in the title, abstract, or key words. Of these, 271 focus on the coast, and this includes the beach, coastal dunes, and any other coastal ecosystem such as wetlands, marshes and mangrove forests. Less than 30 studies referred to ecosystem services of coastal dunes and the beach.

We located a total of only 18 studies where ecosystem services from the beach and coastal dunes have been analyzed and their economic value has been calculated (Table 18.2). Most of them (11; 55 %) focus on the beach; 6 (30 %) deal exclusively with coastal dunes and only 2 (10 %) with the beach and coastal dunes as an integrated system. We found only one study (5 %) that took place in wet slacks. In this set of studies, the ecosystem services that have been studied on the beach and coastal dunes are: aesthetic/recreational/cultural (cultural); disturbance/prevention/protection (regulatory), and carbon sequestration (supporting; Table 18.2). To our knowledge, no ecosystem services that deal with provisioning have been studied, so far, for the beach and coastal dunes, probably because here, some ecosystem services (aesthetic, protection) are more heavily exploited (and apparently considered more relevant) than others (pollination, water cycle, nutrient cycle, and others mentioned in Table 18.1). However, the above does not mean that these are the only ecosystem services of coastal ecosystems, but that a handful of them are of greater interest to humans than the others. In addition, it is interesting to note that it is necessary to consider if losing these ecosystem services after land use change (urbanization, for example, instead of tourism) is compensated for by the value gain. Such a valuation is currently being performed.

The economic values of ecosystem services that we found in the literature were estimated in different years, in different countries, and with a variety of methodologies. This heterogeneity in how economic values were assessed made it necessary to standardize them in order to be able to compare studies. Thus, the economic values estimated for ecosystem services from different ecosystems were adjusted to US$ currency using the consumer price index (CPI) and the purchasing power parity (PPP) for 2010, obtained from US government statistics (US Department of Labor). We thus adjusted the original values estimated in the 18 studies we used (Table 18.2), to US dollars (2010), using the following formula (Envalue 2007):

$$\text{ESV} = \frac{(\text{Value}/\text{CPI}) \times 100}{\text{PPP}} \times \text{USA PPP}$$

where:

Value is the value in the original year in the original currency
CPI is an index of inflation of the source data, with a base year in 2010
PPP is the PPP between the original currency and US$ in 2010

Table 18.2 Ecosystem services and their calculated economic value estimated in different countries. Valuation methods are: WTP= willingness to pay; HP = hedonic pricing; CV =contingent valuation; TC = travel cost. All estimated costs are standardized to USD per ha per year in 2010

Ecosystem	Ecosystem service	Reference	Country	GDP Per capita /year 2010	GINI Index/ year	Method	Year of study	CPI	PPP country studied 2010	US(2010) /ha/year
Beach	Aesthetic and Recreational	Edwards and Gable 1991	USA	47,198.50	45(2007)	HP	2004	96.72	1	$55
Beach	Aesthetic and Recreational	Kline and Swallow 1998	USA	47,198.50	45(2007)	TC	2004	96.72	1	$15,838
Beach	Aesthetic and Recreational	Silberman et al. 1992	USA	47,198.50	45(2007)	CV	2004	96.72	1	$8,653
Beach	Aesthetic and Recreational	Taylor and Smith 2000	USA	47,198.50	45(2007)	HP	2004	96.72	1	$303
Average										**$6,212**
Beach	Recreation	Lindsay et al. 1992.	USA	47,198.50	45(2007)	WTP	1992	71.85	1	$43
Average										**$43**
Dunes	Recreation	Mendoza-González et al. 2012	MEXICO	9,123.41	46.05 (2004)	WTP	2011	103.2	7.951476	$12,192
Dunes	Recreation	Mendoza-González et al. 2012	CHILE	12,431.03	54.92 (2003)	WTP	2011	103.2	403.1928	$8,388
Average										**$10,290**
Beach	Cultural and Spiritual	Taylor and Smith 2000	USA	47,198.50	45(2007)	HP	2004	96.72	1	$10
Average										**$10**
Wetslack	Carbon sequestration	Jones et al. 2010	UK	36,143.94	34(2005)		2010	114.5	0.65151	$4
Average										**$4**
Drydune	Carbon sequestration	Jones et al. 2010	UK	36,143.94	34(2005)		2010	114.5	0.65151	$3

(continued)

Table 18.2 (continued)

Ecosystem	Ecosystem service	Reference	Country	GDP Per capita /year 2010	GINI Index/ year	Method	Year of study	CPI	PPP country studied 2010	US(2010) /ha/year
Average										**$3**
Beach	Disturbance prevention	Pearsons and Powell 2001	USA	47,198.50	45(2007)	HP	2004	96.72	1	$8,709
Beach	Disturbance prevention	Pompe and Rinehart 1995	USA	47,198.50	45(2007)	HP	2004	96.72	1	$14,117
Average										**$11,413**
Dunes & Shingles	protection	Jones et al. 2010	UK	36,143.94	34(2005)	Avoided costs	2010	114.5	0.65151	5272
Beach and Dunes	Protection/ Erosion	Scottish Natural Heritage 2000	UK	36,143.94	34(2005)	Replacement cost	2000	93.06	0.65151	19424
Average										**$12,348**
Dunes	Protection	Mendoza-González et al. 2012	MEXICO	9,123.41	46.05 (2004)	Replacement cost	2011	103.2	7.951476	65743
Dunes	protection	Pye et al. 2007	UK	36,143.94	34(2005)		2007	104.7	0.65151	6661
Average										**$36,202**
Beach, Dune & Sand plains	All	Alves et al. 2009	Portugal	21,504.81	38.5(2007)	Tranfer value	2000	85.55	0.633975	$27,812
Average										**$27,812**
Beach	All		USA	47,198.50	45(2007)	WTP	2004	96.72	1	$17,642
Average										**$17,642**

There are limited data available that estimate the economic values of ecosystem services from the beach and coastal dunes, and those data do not cover many of the possible ecosystem services in detail. We only found economic values for beach and dune ecosystem services from Chile, Mexico, Portugal, the UK, and the USA, although most of the data acquired came from the UK and the USA. When we compare the list of ecosystem services that are recognized as being supplied by the beach and coastal dunes (Table 18.1) with the list of ecosystem services that have been studied in more detail and their economic values calculated, it becomes evident that supporting ecosystem services are generally overlooked in the literature, and other provisioning and regulatory ecosystem services are also ignored.

The estimated economic values of ecosystem services such as recreational, cultural, and aesthetic have been studied most, with a total of 8. In this case, most of them have focused on the beach and only 2 were performed on coastal dunes. It is interesting to note that the economic value of the ecosystem service of protection against storms, hurricanes, and floods had a relatively low number of estimates (a total of 6), even though it seems likely that it would be a very important ecosystem service. Most of these studies have focused on coastal dunes (4) and 2 on the beach. Finally, carbon sequestration is an ecosystem service that has been studied occasionally in coastal dunes and slacks (Table 18.2).

In the economic estimates that we gathered, it is obvious that the economic values of the costs of protection were high both for the beach and the coastal dunes, although those for the dunes were higher. Slowly but surely, stakeholders and the government are becoming more aware of the relevance of coastal protection and are willing to invest more in coastal dune restoration in order to gain protection against natural hazards (storm, storm surges, flooding, erosion). Aesthetic and recreational ecosystem services are also considered to be very valuable to the beach (aesthetic and recreational) and the dunes (recreational). In fact, Mendoza-González (2009) found that tourists were willing to pay for higher hotel prices as long as they were closer to the beach (recreational) and had access to the scenic beauty of the coast (hotel rooms looking at the ocean versus not looking at the ocean; aesthetic value). Of course, further studies are necessary to obtain a better understanding of the economic benefits that society receives from the ecosystem services provided by the beach and coastal dunes.

18.3 The Costs of Restoration Efforts

Although the set of examples on restoration efforts presented in this book were performed in several countries, the economic costs of such actions were not always readily available. Usually, scientists are asked to monitor the effectiveness of restoration actions and do not deal with the socioeconomic part of restoration. In our case, we were only able to gather information from European countries and for the USA. This is probably because the information from the funding agencies (local and federal governments) is not always available to the public.

The activities involved in coastal dune restoration include a wide array of actions, such as use of machinery, planting, fencing, elimination of exotics, remobilization of stabilized dunes, geotubes, sod cutting, and even tree felling (Table 18.3). The goals of these activities are not as diverse, and mostly include protection of endangered or native species; restoring mobility and native flora; recreation, protection, and aesthetic. Indeed, the costs involved in these actions are equally variable. In the data we gathered, we noticed that removal of unwanted vegetation by means of sod cutting, tree felling, and elimination of exotics was, no doubt, amongst the most expensive coastal dune restoration actions. Likewise, creating an artificial dune and planting it with native vegetation was a very expensive activity in Denmark (US$77,393/ha), but it seems to have been successful (Table 18.3) (Vestergaard, Chap. 4, this volume), with high annual maintenance costs (US$2,229/ha; Table 18.3). Coastal dune re-mobilization is also a difficult and expensive task and needs to be repeated every few years (Arens et al., Chap. 7, this volume).

Although the information is still not readily available, and restoration actions are scattered in different countries and take place with different intensities, millions of dollars are already being spent on this activity. For instance, in The Netherlands, there have been more than 100 restoration projects over the last 20 years that, overall, have cost from 10 to 20 million Euros (Grootjans, personal communication). In Spain, more than one million Euros were spent on recovering the native maritime juniper woodlands in Andalusia in a project that lasted 5 years (Muñoz-Reinoso et al., Chap. 9, this volume).

It is also interesting to notice that funding for restoration actions almost always comes from local and federal governments, and usually restoration is performed in national parks, nature reserves, and on public beaches (Table 18.3). An exception to this were the activities that were carried out in Galveston, Texas (USA), where private owners and stakeholders paid for beach and coastal dune restoration as an investment in the protection of their properties against the recurring impacts of hurricanes. In this case, the owners felt that the original dune and swale structure had protected their homes from incurring much greater expense during Hurricane Ike. Moreover, residents said that the reconstructed dunes also enhanced the aesthetic value of their property and buffered them from public intrusion onto their property (Feagin, Chap. 6, this volume; Table 18.3).

Besides the government and stakeholders being interested in paying for coastal dune restoration programs, the public, in general, seems to be gradually becoming more and more interested in these actions (Table 18.3). For instance, volunteer work for either planting dune vegetation (Grootjans et al., Chap. 15, this volume) or removing invasive species (Pickart, Chap. 10, this volume) has played a key role in achieving the goals of restoration in the Netherlands and the USA respectively. Additionally, in Israel, Lehrer et al. (Chap. 17, this volume) found that the public was willing to pay US$10.06 (38 NIS) a year for the containment efforts of invasive species and US$10.63 (40.12 NIS) for its elimination from a national park. Certainly, as the society becomes more involved in the conservation or restoration of natural ecosystems in general (not only the beach and coastal

Table 3 Example of the costs of restoration projects performed in different countries

Action/ mechanism	Goals of restoration actions	Country	Area/ volume restored	Year	Reference	Funding Agency	CPI	PPP country studied 2010	$US(2010)	Units
Removing pine trees	Protect native Maritime Juniper Woodlands in Andalusia	Spain	50 ha	2002-2006	Muñoz-Reinoso et al.	Government	103.517792	0.719016464	1,545,628	USD/ ha
Creating artificial dune; planting native vegetation	Recreation and protection	Denmark	500 ha	1978	Vestergaard	Government	33.5947391	7.959994283	77,393	USD/ ha
Elimination of exotics	Restore mobility and diversity of native flora	USA	327 ha	1998	Pickart	Government/ volunteers	83.46917	1	65,293	USD/ ha
Sod cutting	Restore mobility and diversity of native flora	Netherlands	6 ha	last 20 years	Grootjans et al.	Government	107.939317	0.838259729	29,414	USD/ ha
Restoration (grazing, sod cutting, rewetting, re-mobilization)	Restore mobility and diversity of native flora	Netherlands	100 ha	1993	Grootjans et al.	Government	75.9065759	0.838259729	12,053	USD/ ha
Yearly maintenance costs of created dune	Recreation and protection	Denmark	500 ha	1986	Vestergaard	Government	62.7957965	7.959994283	2,229	USD/ ha
Re-mobilization/, monitoring	Restore mobility and diversity of native flora	New Zealand	150m shoreline	2011	Hesp and Hilton	Government	115.252411	1	1,123	USD/ m

(continued)

Table 3 (continued)

Action/ mechanism	Goals of restoration actions	Country	Area/ volume restored	Year	Reference	Funding Agency	CPI	PPP country studied 2010	$US(2010)	Units
Fencing	Plant diversity	Italy	7,000 m2	2007	Acosta et al.	Government	103.928931	0.811506578	962	USD/ ha
Vegetation	Aesthetic	USA	2,373 m	2007	Feagin	Stakeholders/ Government	106.170642	1	275	USD/ m
Geotubes	Storm protection	USA	2,373 m	1999	Feagin	Stakeholders/ Government	85.2954982	1	247	USD/ m
Sand movement/ vegetation	Storm protection	USA	2,373 m shoreline	2009	Feagin	Stakeholders/ Government	109.854662	1	99	USD/ m
Consultant fee, heavy machinery, planting, fencing,	Restore mobility and diversity of native flora	New Zealand	150m shoreline	2000	Hesp and Hilton	Government	88.4352555	1	58	USD/ m
Re-mobilization	Restore mobility and diversity of native flora	Netherlands	3,000 m3	2010	Arens et al.	Government	107.939317	0.838259729	9	USD/ m3
Planting	Aesthetic	USA	17,830 m3	1999	Feagin	Stakeholders/ Government	85.2954982	1	2	USD/ m3

dunes) these activities will become more effective. To achieve this, environmental education becomes a keystone.

18.4 Future Perspectives

The beach and coastal dunes provide society with a wide array of ecosystem services, including supporting, provisioning, regulatory and cultural/aesthetic (Table 18.1). Protection, recreation, and aesthetic are the ones most frequently studied and their economic values have been calculated, while many other ecosystem services remain largely overlooked (Tables 18.1, 18.2) (Everard et al. 2010). Needless to say, it is obvious that many of the goals of restoration actions are also aimed at recovering ecosystem services such as storm protection, recreation, and aesthetic (Table 18.3). An additional goal of restoration that is not directly related to publically recognized ecosystem services is recovering the natural dynamics of the beach and coastal dunes. That is, many restoration projects developed on coastal dunes have been aimed at restoring the mobility and diversity of native flora. Vegetation removal may seem counter-intuitive to what is expected by restoration, but, in the case of coastal dunes and beaches, recovering their natural dynamics is a very important goal (Martínez et al., Chap. 20, this volume). Upon recovering the natural functionality of these ecosystems, it will be possible to recover ecosystem services.

Our findings show that the economic value of ecosystem services provided by coastal dunes is quite high. That is, the economic benefits that society receives because of the natural functioning of the beach and coastal dunes are very high, especially protection, recreation, and aesthetic. The porous structure and substrate mobility of coastal dunes and beaches absorbs and dissipates wave energy, which helps to protect the coasts and inland infrastructure with minimal human intervention necessary (as long as natural dynamics are allowed to operate) (Everard et al. 2010). Through this capacity of absorbing and dissipating wave energy, the costs of hard engineering solutions are reduced. In addition to the above, a dynamic beach and coastal dune stores sand and provides new sediments that re-enter the marine sediment transport system, which can later nourish beaches after erosion events. Finally, as the natural dynamics of coastal dunes is maintained, native species will be able to survive and grow, and the scenic beauty of the beach/dune system is maintained. In brief, in well-preserved coastal dunes and beaches, the potential benefits to society are maximized, since ecosystem services are maintained under optimal conditions.

The paradox is that, because of the strong interest in these ecosystems, coasts are often over-exploited in order to obtain short-term benefits from recreation and their natural beauty. But, as we exploit these ecosystem services, we degrade them: the landscape is not as beautiful; beaches are flat and deserted; a coast with urban infrastructure is not capable of offering protection. That is, our short-term actions and interests negatively affect the ecosystem services that we value the most.

A logical consequence is that the funds required for restoration actions are large, and will probably increase as the human impact on the coasts increases.

Why are we moving toward this dead-end situation? Two explanations can be set out to interpret this difficult situation and suggest possible ways out: ecosystem services are not considered in decision-making processes, and some ecosystem services are over-exploited, resulting in their loss or degradation together with other ecosystem services that are also very relevant to society, although they may not be directly appreciated.

18.4.1 Ecosystem Services are not Considered in the Decision-Making Processes

Most of the ecosystem services provided by the beach and coastal dunes are not considered in the decision-making process. The result of this incomplete assessment of the use of natural resources is that coastal ecosystems are being fragmented and degraded, on the basis of short-term financial gain rather than their long-term value to society (de Groot 2006). A consequence of this myopic vision is that ecosystem services (even those considered most valuable) are spoiled and even lost. This problem is derived from the fact that ecosystem services are assigned little or no value in the cost–benefit analysis of development projects, because of the absence of market mechanisms. That is, many ecosystem services that the beach and coastal dunes provide to society are usually overlooked, even though the importance of these systems is increasingly being acknowledged. Nevertheless, the societal value of natural ecosystems (specifically the beach and coastal dunes) is often underappreciated (Everard et al. 2010).

One potential solution to the currently inadequate decision-making process is to recognize the societal value of the beach and coastal dunes, including the full range of benefits that they confer to society besides recreational, aesthetic, and protective (see Table 18.1). The estimates of the economic value of these ecosystem services constitute a useful tool that can help with a better-informed process, because marginal costs can also be considered. For instance, before making the decision to destroy coastal dunes and build an urban infrastructure on top of them (houses, hotels, roads), it would be useful to consider the ecological, social, and economic costs of losing the protection from a dynamic beach and its mobile dunes. In this case we would use the ecosystem service value calculated for coastal protection and incorporate it in the cost–benefit analyses of development projects. Here, it is worth highlighting that calculating the economic costs of ecosystem services does not mean that nature and its dynamics are for sale. These calculations are only intended to be used as a tool to aid the decision-making processes.

18.4.2 Over-Exploitation of a Few Ecosystem Services at the Expense of Others

Often, one or a few ecosystem services provided by natural ecosystems are over-exploited and then the provision of other ecosystem services is reduced. This is known as a trade-off across ecosystem services (Rodríguez et al. 2006). These trade-offs can arise as a result of explicit management choices, but can also occur without any awareness that a trade-off is taking place. Trade-offs between ecosystem services can occur both across space (when one ecosystem service is used in one location, another is depleted somewhere else) (Pérez-Maqueo et al., submitted for publication) and over time (when the exploitation of one ecosystem service negatively affects another ecosystem service in the future) (Mendoza-González et al. 2012). Finally, some groups of ecosystem services are more frequently chosen over others. For instance, Rodriguez et al. (2006) found that provisioning and regulating services are frequently preferred over supporting and cultural services.

In the case of the beach and coastal dunes, direct explicit choices are made in which the recreational, aesthetic, and protective functions are generally exploited, at the expense of many other ecosystem services. That is, regulatory (storm protection) and cultural (recreational, aesthetic) services are the most frequently used ecosystem services in the beach and coastal dunes. These trade-offs occur in space and time. Spatially, when for instance the scenic beauty of the coast is chosen, then urban infrastructure is developed and the potential for protection is lost along with all the ecosystem services listed in Table 18.1. Urbanized coasts can become "deserts" with no natural plants or animals, and with a flat topography; these systems are non-functional and offer zero ecosystem services, except for aesthetics and recreation, although in a degraded system, even these are diminished. When the beach and coastal dunes are damaged or lost, ecosystem services for the future are also lost. Obviously, this is not the best possible scenario for the future. Ideally, biodiversity and dynamism should be preserved, which in turn would help maintain ecosystem services (Balvanera et al. 2001). Successful management policies need to consider ecosystem service trade-offs on different spatial and temporal scales (Rodríguez et al. 2006) and should aim to minimize the effects of trade-offs.

18.5 Conclusions

Beaches and coastal dunes provide many ecosystem services to society of which a few are excessively over-exploited. Such over-exploitation results in a degraded and even lost natural ecosystem and, consequently, valuable ecosystem services are also lost as well. In consequence, restoration efforts are becoming increasingly necessary at increasingly high costs.

A better alternative is to preserve these (and any other) natural ecosystems and use the many ecosystem services in a more rational manner, instead of over-exploiting a few of them. By preserving natural ecosystems, they remain dynamic and functional, and ecosystem services are preserved too (Balvanera et al. 2001). This would be a win–win situation yielding important economic benefits, improved human well-being, and better-preserved natural ecosystems. Because of the crucial role that sandy beaches and coastal dunes play in human society (recreational, aesthetic and protective) plus the multiple beneficial services that they provide to society and, on top of the inherent reasons for protection, our focus on these ecosystem services need to be prioritized and far more recognized than at present if we want to continue to receive the benefits from them. In brief, before investing in restoration (which is not always 100 % successful, but is usually very expensive) natural ecosystems should be preserved in their own integrity, and also because of the many benefits to our society.

Acknowledgments We are very grateful to all the authors of the chapters in this book who kindly provided us with information of the economic costs of the restoration efforts in which they have participated: Patrick Hesp, Mike Hilton, Alicia Acosta, Andrea Pickart, Bas Arens, Ab Grootjans, Peter Vestergaard, Rusty Feagin, and José Carlos Muñoz Reinoso.

References

Acosta ATR, Jucker T, Prisco I, Santoro R (2012) Passive recovery of Mediterranean coastal dunes following limitations to human trampling. In: Martínez ML, Gallego-Fernández JB, Hesp PA (eds) Restoration of coastal dunes, Chap. 12, this volume. Springer, Berlin

Alves F, Roebeling P, Pinto P, Batista P (2009) Valuing ecosystem services losses from coastal erosion using a benefit transfer approach: a case study for the central Portuguese coast. J Coastal Res 56:1169–1173

Arens SM, Jungerius PD, van der Meulen F (2001) Coastal dunes. In: Warren A, French JR (eds) Habitat conservation: managing the physical environment. Wiley, London

Arens SM, Slings QL, Geelen LHWT, Van der Hagen HGJM (2012) Restoration of dune mobility in the Netherlands. In: Martínez ML, Gallego-Fernández JB, Hesp PA (eds) Restoration of coastal dunes, Chap. 7, this volume. Springer, Berlin

Balvanera P, Daily GC, Ehrlich PR, Ricketts TH, Bailey SA, Kark S, Kremen C, Pereira H (2001) Conserving biodiversity and ecosystem services. Science 291:2047

Bell FW, Leeworthy VR (1990) Recreational demand by tourists for saltwater beach days. J Environ Econ Manag 18(3):189

Costanza R, D'Arge R, de Groot R, Farber S, Grasso M, Hannon B, Limburg K, Naeem S, O'Neill RV, Paruelo J, Raskin RG, Sutton P, van den belt M (1997) The value of the world's ecosystem services and natural capital. Nature 387:253–260

Cowles HC (1899) The ecological relations of vegetation on the sand dunes of Lake Michigan. Bot Gaz 27:95–117

De Groot RS, Wilson MA, Boumans RMJ (2002) A typology for the classification, description and valuation of ecosystem functions, goods and services. Ecol Econ 41(3):393–408

De Groot R (2006) Function-analysis and valuation as a tool to assess land-use conflicts inplanning for sustainable, multi-functional landscapes. Landscape and Urban Plan 75:175–186

Edwards SF, Gable FJ (1991) Estimating the value of beach recreation from property values: an exploration with comparisons to nourishment costs. Ocean Shorel Manag 15:37–55

Envalue (2007) New South Wales Environment Protection Authority. http://www.epa.nsw.gov.au/envalue
Everard M, Jones L, Watts B (2010) Have we neglected the societal importance of sand dunes? An ecosystem services perspective. Aquatic Conserv Mar Freshw Ecosyst 20:476–487
Feagin R (2012) Foredune restoration before and after hurricanes: inevitable destruction, certain reconstruction. In: Martínez ML, Gallego-Fernández JB, Hesp PA (eds) Restoration of coastal dunes, Chap. 6, this volume. Springer, Berlin
Feagin RA, Wu XB, Smeins FE, Whisenant SG, Grant WE (2005) Individual versus community level processes and pattern formation in a model of sand dune plant succession. Ecol Model 183(4):435–449
Gallego-Fernández JB, Martínez ML (2011) Environmental filtering and plant functional types on Mexican foredunes along the Gulf of Mexico. Ecoscience 18(1):52–62
Grootjans A, Dullo BW, Kooijman A, Bekker R, Aggenbach C (2012) Restoration of dune vegetation in the Netherlands. In: Martínez ML, Gallego-Fernández JB, Hesp PA (eds) Restoration of coastal dunes, Chap. 15, this volume. Springer, Berlin
Hesp PA, Hilton MJ (2012) Restoration of foredunes and transgressive dunefields—case studies from New Zealand. In: Martínez ML, Gallego-Fernández JB, Hesp PA (eds) Restoration of coastal dunes, Chap. 5, this volume. Springer, Berlin
Isermann M (2011) Patterns in species diversity during succession of coastal dunes. J Coastal Res 27(4):661–671
Jones L, Angus S, Cooper A, Doody P, Everard M, Garbutt A, Gilchrist P, Hansom J, Nicholls R, Pye K, Ravenscroft N, Rees S, Rhind P, Whitehouse A (2010) Chapter 11—coastal margin habitats. National Ecosystem Assessment
King OH (1995) Estimating the value of marine resources: a marine recreation case. Ocean Coast Manag 27(1–2):129
Kline JD, Swallow SK (1998) The demand for local access to coastal recreation in southern New England. Coastal Manag 26(3):177–190
Lehrer D, Becker N, Bar (Kutiel) P (2012) The value of coastal sand dunes as a measure to plan an optimal policy for invasive plant species: the case of the *Acacia saligna* at the Nizzanim LTER Coastal Sand Dune Nature Reserve, Israel. In: Martínez ML, Gallego-Fernández JB, Hesp PA (eds) Restoration of coastal dunes, Chap. 17, this volume. Springer, Berlin
Lindsay BE, Halstead JM, Tupper HC, Vaske JJ (1992) Factors influencing the willingness to pay for coastal beach protection. Coastal Manag 20:291–302
Lithgow-Serrano AD (2007) Los servicios ambientales de las costas del estado de Veracruz. UAM Xochimilco, México
Lubke RA, Avis AM (1998) A review of the concepts and application of rehabilitation following heavy mineral dune mining. Mar Pollut Bull 37:546–557
Martínez ML, Vázquez G, Sánchez-Colón S (2001) Spatial and temporal dynamics during primary succession on tropical coastal sand dunes. J Veg Sci 12:361–372
Martínez ML, Maun AM, Psuty NP (2004a) The fragility and conservation of the world's coastal dunes: geomorphological, ecological and socioeconomic perspectives. In: Martínez ML, Psuty N (eds) Coastal dunes: ecology and conservation. Springer, Berlin, pp 355–370
Martínez ML, Psuty N, Lubke R (2004b) A perspective on coastal dunes. In: Martínez ML, Psuty N (eds) Coastal dunes: ecology and conservation. Springer, Berlin, pp 3–10
Martínez ML, Intralawan A, Vázquez G, Pérez-Maqueo O, Sutton P, Landgrave R (2007) The coasts of our world: ecological, economic and social importance. Ecol Econ 63:254–272
Martínez ML, Hesp PA, Gallego-Fernández JB (2012) Coastal dune restoration: trends and perspectives. In: Martínez ML, Gallego-Fernández JB, Hesp PA (eds) Coastal dune restoration, Chap. 20, this volume. Springer, Berlin
Maun MA (2009) The biology of coastal sand dunes. Oxford University Press, Oxford 265 pp
McComb G, Lantz V, Nash K, Rittmaster R (2006) International valuation databases: overview, methods and operational issues. Ecol Econ 60(2):461–472

Mendoza González G (2009) Análisis del cambio de uso del suelo y valoración de los servicios ecosistémicos en tres sitios turísticos costeros del estado de Veracruz. Instituto de Ecología, A.C., Xalapa

Mendoza-González G, Martínez ML, Lithgow D, Pérez-Maqueo O, Simonin P (2012) Land use change and its effects on the value of ecosystem services along the coast of the Gulf of Mexico. Ecological Economics 82: 23–32

Muñoz-Reinoso JC, Saavedra-Azqueta C, Redondo-Morales I (2012) Restoration of Andalusian coastal juniper woodlands. In: Martínez ML, Gallego-Fernández JB, Hesp PA (eds) Restoration of coastal dunes, Chap. 9, this volume. Springer, Berlin

Nordstrom KF, Jackson NL (2012) Foredune restoration in urban settings. In: Martínez ML, Gallego-Fernández JB, Hesp PA (eds) Restoration of coastal dunes, Chap. 2, this volume. Springer, Berlin

Pearsons GR, Powell M (2001) Measuring the cost of beach retreat. Coastal Manag 29:91–103

Peterson CH, Lipcius RN (2003) Conceptual progress towards predicting quantitative ecosystem benefits of ecological restorations. Mar Ecol Prog Ser 264:297–307

Pickart AJ (2012) Dune restoration over two decades at the Lanphere and Ma-le'l Dunes in northern California. In: Martínez ML, Gallego-Fernández JB, Hesp PA (eds) Restoration of coastal dunes, Chap. 10, this volume. Springer, Berlin

Pompe JJ, Rinehart JR (1995) Beach quality and the enhancement of recreational property-values. J Leis Res 27(2):143–153

Pye K, Saye SE, Blott SJ (2007) Sand dune processes and management for flood and coastal defence. Parts 1 to 5. Joint DEFRA/EA flood and coastal erosion risk management R & D programme, R & D Technical Report FD1 1302/TR

Rodríguez JP, Beard TD Jr, Bennett EM, Cumming GS, Cork S, Agard J, Dobson AP, Peterson GD (2006) Trade-offs across space, time, and ecosystem services. Ecol Soc 11(1):28

Scottish Natural Heritage (2000) A guide to managing coastal erosion in beach/dune systems. http://www.snh.org.uk/publications/on-line/heritagemanagement/erosion/appendix_1.2.shtml

Silberman J, Gerlowski DA, Williams NA (1992) Estimating existence value for users and nonusers of New Jersey beaches. Land Econ 68(2):225–236

Shuang L (2007) Valuing ecosystem services: an ecological economic approach. Doctor of philosophy specializing in natural resources. PhD Thesis, The University of Vermont, Burlington

Taylor LO, Smith VK (2000) Environmental amenities as a source of market power. Land Econ 76(4):550–568

Troy A, Wilson MA (2006) Mapping ecosystem services: practical challenges and opportunities in linking GIS and value transfer. Ecol Econ 60:435–449

Vestergaard P (2012) Natural plant diversity development on a man-made dune system. In: Martínez ML, Gallego-Fernández JB, Hesp PA (eds) Restoration of coastal dunes, Chap. 4, this volume. Springer, Berlin

Webster P, Holland J, Curry GJ, Chang HR (2005) Changes in tropical cyclone number, duration, and intensity in a warming environment. Science 309:1844–1846

Part IV
Conclusions

Chapter 19
Multicriteria Analysis to Implement Actions Leading to Coastal Dune Restoration

Debora Lithgow, M. Luisa Martínez and Juan B. Gallego-Fernández

19.1 Introduction

Some 41 % of the human population lives within 100 km of the coast, and according to Povh (2000), three-quarters of the world's population will be living within 60 km of the coastline by 2020. Human overpopulation on the coast leads to major land use change, overexploitation of resources and the loss and degradation of natural ecosystems. This intensive use of the coast has resulted in coastal and estuarine ecosystems being the most threatened natural systems worldwide (Barbier et al. 2011). Among these ecosystems, sandy beaches and dunes are notable for both their wide distribution and the high degree of threat they face (Fig. 19.1).

Beaches and coastal dunes are distributed from the tropics to the polar circles (Martínez et al. 2004) and can range from the immediate coastline to several kilometers inland. This spatial distribution creates a wide variety of niches and thus a diverse flora and fauna. Besides featuring high biodiversity, coastal beaches and dunes are notable for the socioeconomic functions they fulfill (Everard et al. 2010). These functions include recreation, scenic beauty, pollutant capture, ability to act as reservoirs of water and sand, as well as coastal protection against extreme

D. Lithgow · M. L. Martínez (✉)
Instituto de Ecología, A.C., Red de Ecología Funcional, Carretera Antigua a Coatepec no. 351, El Haya Ver. 91070 Xalapa, Mexico
e-mail: marisa.martinez@inecol.edu.mx

D. Lithgow
e-mail: debora.lithgow@gmail.com

J. B. Gallego-Fernández
Departamento de Biología Vegetal y Ecología, Universidad de Sevilla, Ap.1095 - 41080, Sevilla, Spain
e-mail: galfer@us.es

M. L. Martínez et al. (eds.), *Restoration of Coastal Dunes*, Springer Series on Environmental Management, DOI: 10.1007/978-3-642-33445-0_19,

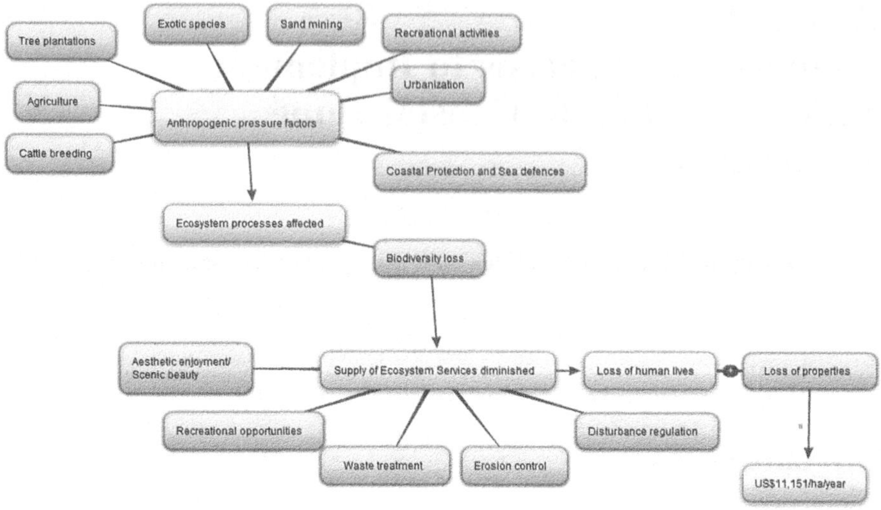

Fig. 19.1 Anthropogenic stress factors on sandy coasts and their effects on system destruction, alteration of biodiversity and reduction of ecosystem services. The value of material losses (US$11,151/ha/year) refers to sandy beaches with dunes in Mexico and was calculated based on Mendoza-González et al. 2012

climatic events (i.e., floods, hurricanes, and storms). The loss of these protection services is particularly important in view of growing human and economic losses derived from the impact of storms and hurricanes. These losses have led to a re-evaluation of the biological and socioeconomic importance of these ecosystems, and have highlighted the need to restore the natural capital they represent. This capital can only be recovered through programs of restoration (Ley-Vega et al. 2007) and by considering the local and regional scales of the processes acting on coastal ecosystem dynamics (Gallego-Fernández et al. 2011).

Sometimes, however, the state of degradation is so severe and extensive that restoration is no longer possible, and rehabilitation efforts become the most viable alternative. When it comes to restoring coastal dunes and sandy beaches, therefore, a choice between restoration and rehabilitation presents itself. In this study, the following definitions are adopted for these terms: Restoration is the set of actions taken for the re-establishment of the processes and features that allow the system to function naturally without subsequent human intervention (SER 2004). Rehabilitation, on the other hand, refers to the action taken to recover certain processes or features, since complete restoration is impossible. In this case, the system functions and is maintained with the help of human actions.

Where should these actions be implemented? What criteria should be taken into account when making such decisions? This chapter aims to establish the variables needed to identify sandy beaches with dunes where restoration actions are required and, furthermore, where such actions may be performed with a high probability of success. To this end, ecological, geomorphological, and socioeconomic criteria are all taken into consideration.

19.2 Methods

The decision regarding whether a beach should be restored, or even whether it is possible to carry out restoration actions, is a complex one, involving many diverse ecological, geomorphological, and socio-economic aspects. For this reason, there is no common scale by which to by which one can compare sites directly. This variety of factors also makes it virtually impossible for a single specialist to select the most representative and effective values. To solve both of these problems, this study utilized an Analytic Hierarchy Process (AHP) along with the contributions of a multidisciplinary expert panel. Together, these instruments allowed identification and weighting of the factors that determine the restoration potential of a beach with dunes.

19.2.1 Factor Selection

An extensive literature review was conducted, collating all the ecological, geomorphological, and socioeconomic information considered relevant to the success or failure of remedial action. From this review, a list of approximately 100 factors was compiled. This list was evaluated by three experts in restoration of beaches and coastal dunes, who eliminated redundant criteria and selected only those considered most representative. These representative factors were structured hierarchically and grouped according to criteria, subcriteria, and indicators for subsequent weighting.

19.2.2 Weighting of Criteria, Subcriteria and Indicators

The criteria, subcriteria, and indicators selected by the experts were integrated into a square matrix for weighting. Pairwise comparison allowed each criterion to be weighted, in relation to all the others, in an efficient manner. This comparison was carried out using a scale from 1 to 4 according to the degree of importance of the criterion (Table 19.1; modified from Saaty 1980).

Each of the pairwise comparison matrices was summarized, obtaining the total of each column of the matrix (Fig. 19.2). Each value of the matrix was then divided by the resulting sum total obtained in each column corresponding to that value. The result of this division is a normalized pairwise comparison matrix. The arithmetic average was subsequently calculated with the values from each row of the normalized matrix, producing a matrix with the priorities of each criterion.

Table 19.1 Scale used in the pairwise comparison by the expert panel

Intensity of importance	Definition	Explanation
1	Equal	Both evaluated elements contribute equally to the objective
2	Moderate	The experience and judgment slightly favor one of the elements over the other
3	Strong	The experience and judgment strongly favor one of the elements over the other
4	Very strong	The evaluated variable is strongly favored and its dominance can be demonstrated in practice

The scale was modified from that recommended by Saaty (1980)

Step 1

	Factors that cause stress	Degree of degradation	Elements that assist prioritization	Facilitators
Factors that cause stress	1	3	3	½
Degree of degradation		1	4	¼
Facilitators			1	1
Elements that assist prioritization				1

Step 2

	Factors that cause stress	Degree of degradation	Elements that assist prioritization	Facilitators
Factors that cause stress	1	3	3	½
Degree of degradation	1/3	1	4	¼
Facilitators	1/3	¼	1	1
Elements that assist prioritization	2	4	1	1
Sum	3 2/3	8 1/4	9	2 3/4

Step 3

	Factors that cause stress	Degree of degradation	Elements that assist prioritization	Facilitators
Factors that cause stress	0.27	0.36	0.33	0.18
Degree of degradation	0.09	0.12	0.44	0.09
Facilitators	0.09	0.03	0.11	0.36
Elements that assist prioritization	0.55	0.48	0.11	0.36

Step 4

$$W=1/4 \begin{pmatrix} 0.27+0.36+0.33+0.18 \\ 0.09+0.12+0.44+0.09 \\ 0.09+0.03+0.11+0.36 \\ 0.55+0.48+0.11+0.36 \end{pmatrix} = \begin{pmatrix} 0.29 \\ 0.19 \\ 0.15 \\ 0.38 \end{pmatrix} \begin{matrix} \text{Factors that cause stress} \\ \text{Degree of degradation} \\ \text{Facilitators} \\ \text{Elements that assist prioritization} \end{matrix}$$

Fig. 19.2 Simplified example for obtaining the relative weighting of each of the criteria. Step 1 shows the average matrix provided by the experts; this matrix was completed with the inverse of the values obtained by the experts, and was then added to each of the columns (step 2). Each of the values of the matrix was then divided by the corresponding value of each column (step 3). Finally, the average of each of the rows and the final relative weighting of each criterion were obtained (step 4)

In order to calculate the eigenvalues that allowed us to prioritize (weight) and the eigenvectors used to determine the consistency index of each evaluated matrix, we used the formula $B * w = \lambda * w$ where B is the reciprocal matrix of pairwise comparisons, λ is the maximum eigenvalue of B, and w is the eigenvector corresponding to λ.

Once the relative importance of each criteria and subcriteria was estimated, and in order to determine the stability of the system and assess the quality of the judgments made by each expert, the consistency ratio (CR) was calculated. The CR is the ratio of error committed by the expert to the random error. Twelve experts were each asked to contest an evaluation matrix, but ultimately it was only possible to take eight into account. The matrices that were chosen were those with less than 10 % inconsistency, i.e., those with 10 % or fewer incorrect answers. Matrices with greater "inconsistency" values (5 cases) were returned to the experts for review. In 4 cases, the matrices were not corrected on time and therefore had to be eliminated, resulting in the total of 8 considered here. The 8 "consistent" matrices were averaged and the relative importance calculated of each of the criteria, subcriteria, and indicators as described above.

19.3 Results

19.3.1 Selection Criteria

Based on the literature, four criteria, 10 subcriteria, and 38 indicators were extracted. The four criteria evaluated are listed in Table 19.2 and the subcriteria and indicators for each criterion can be seen in Fig. 19.3.

19.3.2 Weighting of Criteria and Field of Expertise

Although at first the panel was composed of equal numbers of experts for each of the areas covered, not all of them completed the process. This produced a slight imbalance, resulting in a greater number of experts in some areas than others. For example, there were 5 ecologists (2 functional ecologists, 3 experts in the restoration of beaches and coastal dunes), 2 geomorphologists, and only 1 anthropologist. This was partially offset by the fact that some of the ecologists had experience with geomorphological aspects and with work in human communities. Each of the members of the panel had expertise in a particular area and their opinion was unique; even those belonging to the same disciplines had had different experiences so that their views also differed; nevertheless, it was possible to group their decisions under the different criteria with different relative importance.

Table 19.3 shows the five most important factors identified by each group of experts and their weighting, normalized for easy comparison. It can be seen how 3 of the 4 groups of experts identified the extraction of sand as being the determining factor for achieving the restoration of a dune. However, there were wide differences among the other choices: Ecologists considered the biotic elements of the system as determinants, but this opinion was not shared by other groups.

Table 19.2 Criteria that influence the possibilities of restoration of dunes and beaches according to the expert panel

Criteria	Explanation
1. Degree of degradation	Identifies the existence as well as degree of degradation at a determined site. This criterion only takes into account anthropogenic degradation factors
2. Factors that cause stress	Include external and internal stressors that are having an influence on the system evaluated
3. Facilitators	Evaluate the existence of key biotic and abiotic elements that promote or are necessary for successful restoration
4. Elements that assist prioritization	Take into account priority elements from an anthropocentric point of view. The elements evaluated could attract public attention and investments needed to support the restoration project

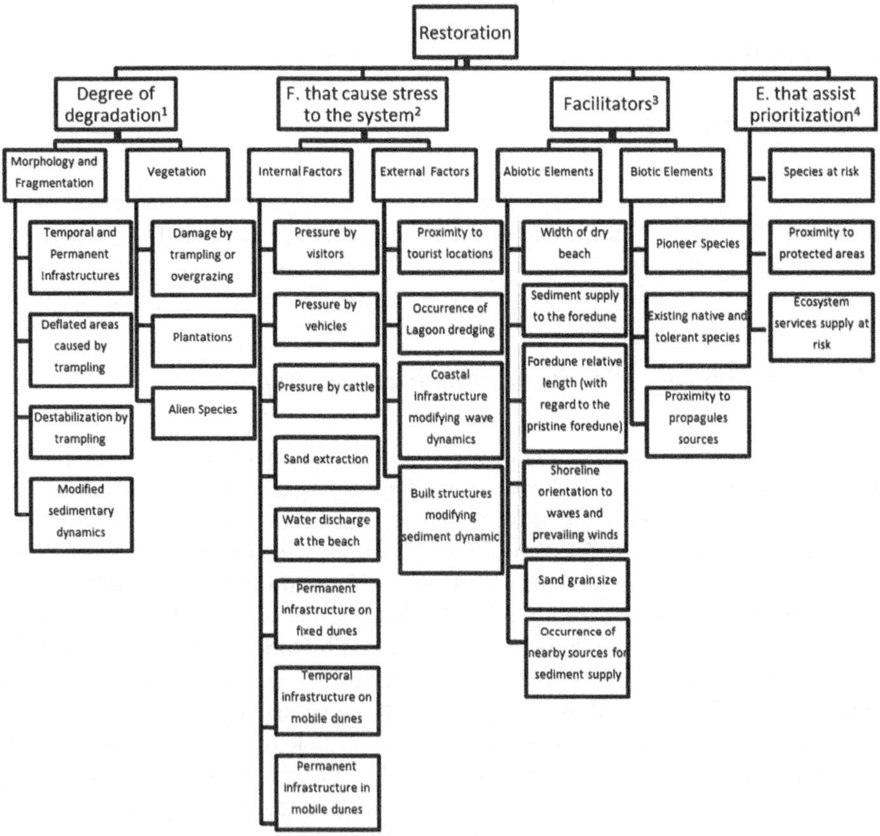

Fig. 19.3 Criteria, subcriteria, and indicators identified as relevant to the selection of beaches and coastal dunes where restoration actions are needed. Superscripts refer to the definitions for each of the criteria presented in Table 19.2

Table 19.3 Strongest indicators and weighting (W) according to the different groups of experts: ecologists; ecologists who were experts on dunes and beaches; anthropologists; and geomorphologists

Ecologists	W	Ecologist experts on beaches and dunes	W	Anthropologists	W	Geomorphologists	W
Existence of sand mining[a]	24.24	Existence of sand mining[a]	22.56	Presence of hotels[c]	26.13	Existence of sand mining[a]	24.93
Availability of pioneer species	21.35	Presence of infrastructure on mobile dunes[d]	20.68	Number of visitors per year	23.03	Sediment supply to the foredune	20.37
Presence of native and tolerant species	19.21	Permanent infrastructure on fixed dunes	20.39	Presence of human interests or properties at risk	17.75	Occurrence of nearby sources for sediment supply [b]	19.60
Presence of endemic species	17.96	Presence of hotels[c]	18.78	Recreational activities	17.74	Presence of infrastructure on mobile dunes[d]	19.48
Presence of species at risk	17.24	Occurrence of nearby sources for sediment supply[b]	17.59	Proximity to tourist locations	15.34	Built structures modifying sediment dynamic	15.62

[a,b,c,d] Similarities between groups of experts

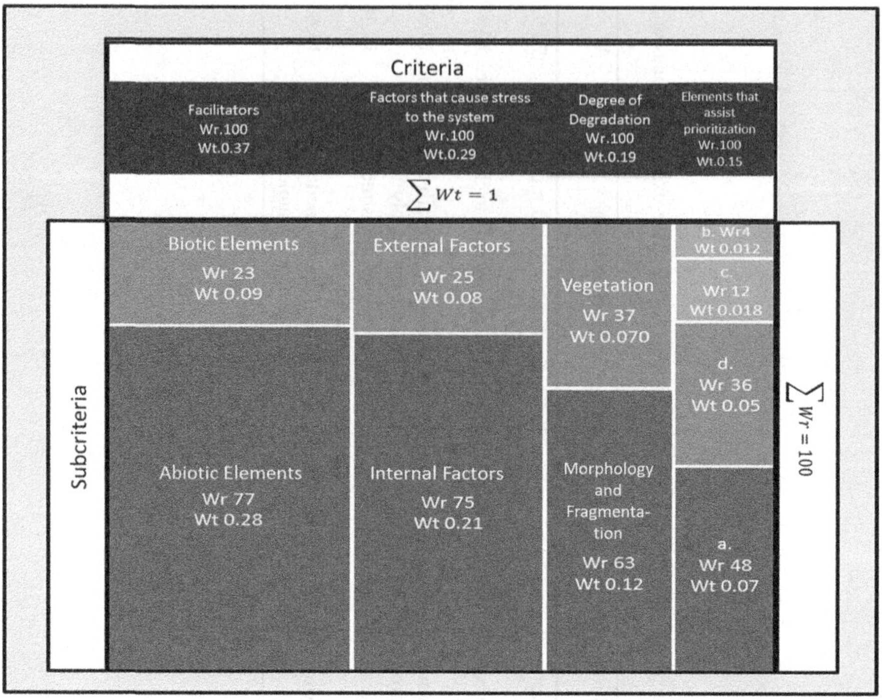

Fig. 19.4 Total weighting (Wt) and relative weighting (Wr) of the criteria and subcriteria evaluated. Four criteria and 10 subcriteria were observed, with their respective Wt and Wr. Greatest importance (Wt 0.37) is given to the criterion "facilitators", which comprises two subcriteria "biotic elements" and "abiotic elements". On the other hand, the criterion "Elements that assist prioritization" received the lowest score (Wt 0.15). This criterion includes four subcriteria: **a** endemic and/or high priority flora and fauna; **b** protected areas; **c** cultural sites; and **d** environmental services

On the other hand, ecologists who were expert in dunes and the geomorphologists chose three common elements: the occurrence of sand extraction, the presence of urbanization in mobile dunes, and the existence of nearby sources of sediment. The anthropologist shared a choice with the ecologists who were dune experts; the presence of hotels or urban infrastructure was an important factor in the achievement of a successful restoration of a sandy beach.

19.3.3 Mean Relative Importance

The averaging of the relative importance given to each criterion by each of the experts draws on the experience of all and incorporates all the different points of view.

The total weighting (Wt) of the criteria and subcriteria, taking into account the opinions of the entire panel of experts, is shown in Fig. 19.4. Facilitators, such as the existence of nearby sediment sources and the presence of burial-tolerant species, constituted the criterion given the highest weighting (0.37) followed by factors that cause stress to the system, such as the presence of infrastructure that alters the wind and sediment dynamics (0.29). The degree of degradation of the system obtained a value of 0.19. Finally, elements that assist prioritization, from the anthropogenic perspective, achieved the lowest total weighting (0.15).

19.3.4 Importance of Degradation and Stress Factors

Much of the anthropic degradation of the world's sandy coasts occurs because management is focused on maximizing the recreational experience of users of the beaches (Diefenderfer et al. 2009; McLachlan and Brown 2001).

In Mexico, tourism is an important economic activity and is the third largest source of revenue, after oil and manufacturing (Juárez et al. 2008). However, coastal management focused on tourism often involves ecologically damaging actions, such as the artificial filling of beaches (Speybroeck et al. 2006), the cleaning of beaches (Llewellyn and Shackley 1996; Dugan et al. 2003), dune destruction by the construction of tourism infrastructure (Nordstrom 2000), noise and light pollution (Bird et al. 2004; Longcore and Rich 2004), and the presence of vehicles on the beach.

Added to this, in many parts of the world, including Mexico, camping and management activities also have a severe impact on the beach and dune vegetation, but are not always identified as factors that must be reduced (Luckenbach and Bury 1983; Hockings and Twyford 1997; Groom et al. 2007). This may be because of the economic benefits that these activities bring to coastal communities. Overall, the direct impacts of recreational activities emerge as significant environmental problems that are degrading these natural ecosystems (Schlacher et al. 2008).

While reducing tourism along the sandy coasts is not an economically viable measure, improved management can certainly be promoted. One common example of poor management is the practice of permitting parking, and even driving of vehicles directly on the sandy beach. This activity causes damage such as the destabilization and destruction of embryonic dunes (Anders and Leatherman 1987; Kutiel et al. 1999; Priskin 2003; Schlacher and Thompson 2008), reduction of vegetation diversity and cover (Luckenbach and Bury 1983; Rickard et al. 1994; Groom et al. 2007; Gachuz 2009), and often injures or kills beach fauna, including endangered vertebrates, such as certain turtles and birds (Hosier et al. 1981; Buick and Paton 1989; Williams et al. 2004). Furthermore, the physical compaction of the sand caused by the presence and movement of vehicles is very difficult and costly to reverse.

Sand extraction was an indicator that was highly valued by the panel of experts, reflecting the widespread concern regarding mining in coastal areas. This activity creates a "lunar" landscape, depletes reserves of sand and impedes the natural response of the beach to storm events, increasing the rate of erosion of the shoreline and destroying archeological sites near the coast (Pilkey et al. 2007). Other small-scale and routine activities can dramatically affect these ecosystems: For example, beach cleaning (Poinar 1977; Llewellyn and Shackley 1996; Engelhard and Withers 1997; Dugan et al. 2003; Davenport and Davenport 2006), which, in addition to removing waste, also eliminates propagules and nutrients, disturbs organisms that inhabit the beach and flattens or changes the texture of the sand, exposing more of the surface to the erosive effect of wind. When such cleaning is conducted with machinery, the damage is intensified because of the associated physical compaction of the sand.

19.3.5 Factors Favoring Restoration and Criteria that Permit Prioritization of Restoration Requirements

Facilitators, i.e., those elements that facilitate restoration actions, constituted the criterion most highly valued by the experts. This criterion comprises two subcriteria: biotic and abiotic elements. Of these, the subcriterion "abiotic elements" was given the highest weighting. This reflects the fact that, when restoration activities are performed under adverse geomorphological conditions, they have little chance of success (Anfuso et al. 2000; Gallego-Fernández et al. 2003). For example, eroded beaches usually feature breakwaters, which, while contributing to the accumulation of sediment at their immediate location, create an imbalance in the processes of erosion accumulation and material transport, leading to modification of the coastline (Moreno-Casasola et al. 2006). Furthermore, experience worldwide has shown that the construction of fixed structures is both expensive and counterproductive in terms of erosion prevention, while beach nourishment can have adverse effects on the biota (Finkl and Walker 2004; Nelson 1993a, b; Bishop et al. 2006; Peterson et al. 2006; Speybroeck et al. 2006). On the other hand, the subcriterion "biotic elements" received a lower weighting. This may be because, although considered valuable elements for the system, several methods exist for the reintroduction of pioneer species from nearby fragments as well as for cultivating these plants under greenhouse conditions (e.g., Ley-Vega et al. 2007) and these methods have already proved to be relatively successful. It is also important to remember that, within these environments, abiotic elements dominate the system: if they do not function properly, then the biotic elements are strongly affected.

The criteria relating to elements that assist the prioritization of beach and dune restoration received the lowest weighting. This may be due to the difficulty of measuring the indicators considered here. For example, the presence of endangered species and the frequency of floods are indicators that are extremely sensitive to a lack of reliable information. Some of the indicators presented here have been previously used to assess the vulnerability of coastal dunes and beaches in Spain, Portugal, the UK, Chile, and the Gulf of Mexico (Williams et al. 1993, 1994, 2001; Pereira et al. 2000; García-Mora et al. 2001; Martínez et al. 2006;

Castro-Avaria 2004; Muñoz-Vallés et al. 2011).These indicators have shown that there are common factors that cause degradation and vulnerability on all the world's sandy beaches. These similarities can serve as a basis for developing easily accessible and applicable tools to help make informed decisions regarding which beaches and dunes should undergo restoration. An example of such a tool would be a checklist using all the criteria that were considered relevant by the expert panel and the AHP. A such a checklist has, in fact, already been developed to assess coastal dune vulnerability (Williams et al. 2001). Indeed, multcriteria checklists are multicriteria checklists are powerful tools since they organize and integrate a large amount of information by using a relatively reduced set of indicators. In our case, and as a follow-up to the multicriteria analysis shown in this chapter, we are already working on a pondered multicriteria checklist, "BEFORE" (BEach and FOredune Restoration Index), that will use the most relevant variables diagnosed by the expert panel. These variables will help to determine whether a beach–foredune system needs to be restored and the chances of restoration actions having favorable results (Lithgow Serrano 2010).

In summary, this study recognizes the existence of multiple criteria for determining the success or failure of actions to restore a beach and its adjacent dunes. These criteria were derived from an intensive literature review and by consultation with recognized experts. One way in which the multicriteria analysis performed in this study differs from that conducted in the other dune vulnerability studies described above is that, in many of those cases, all the evaluated criteria were given equal importance. In contrast, the present study confers a relative importance upon each criterion. One advantage of this system of relative weighting of the criteria is that it gives a clear indication of how each criterion contributes to the goal.

19.4 Conclusions

The restoration of degraded ecosystems is one of the most pressing environmental problems of this century (Lubbert and Chu 2001; Crowley and Ahearne 2002; Burger et al. 2007). Despite the fact that all restoration efforts are well intentioned, many projects continue to fail in reaching the goals that originally motivated the restoration (Kentula et al. 1992). This is due primarily to the difficulty of achieving ecosystem restoration because of the presence of complex interactions intrinsic to natural systems. However, the likelihood of a successful restoration can be increased if concurrent science-based and socially acceptable decisions are taken (Hopfensperger et al. 2007). One of the first steps necessary for restoration activities is therefore a full evaluation of the structure and function of ecosystems, their state of degradation, and the potential for eliminating or minimizing the degradation factors.

Integration of multiple criteria is of vital importance to the choice of sites for restoration. This is because the resilience of a system depends on the interaction between geomorphological and ecological components (van der Meulen and

Salman 1993) as well as human factors. However, this relationship does not occur in linear and equal proportions; thus, weighting of these components is necessary. Weighting provides a realistic view of the influence of each factor on the restoration viability of a beach and its adjacent dunes.

This chapter shows how multi-criteria analysis (MCA) can be used effectively to integrate ecological, geomorphological, and social criteria for the selection of restorable beaches and dunes. MCA can be carried out using multiple methods; however, the analytic hierarchy process (AHP) was adopted because it helps to reduce subjectivity and to simplify complex problems, such as that posed by the choice of beach and dune restoration. In addition, AHP allows the weighting of each of the criteria considered and can be used in conjunction with other instruments, such as the panel of experts. The combination of the two instruments allowed us to consider criteria of very different natures in order to increase the reliability of our findings.

The interdisciplinary panel of experts was used because the evaluation of such widely differing criteria went beyond the knowledge of one person. Furthermore, consideration of only one opinion, or of the opinion of experts from one area of knowledge, would have increased professional bias and taken the process further from reaching a solution closer to reality. In this context, Bulleri (2006) warns that disciplinary segregation has reduced our potential to understand the relationships between human activities and changes in processes and the ecology of the coast.

The AHP has a strict scientific basis, such that it represents an objective way of approaching decision making based on quantifiable components. In the best case scenario, the decision taken can be the most satisfactory, while in the worst, the least unsatisfactory (Hammond et al. 2001). That is to say, although the "correct" answer cannot be guaranteed, the use of AHP does ensure that the decision taken is based on a subtle analysis that summarizes the relevant information and utilizes the knowledge and experience of the experts who participated in the evaluations.

Acknowledgments We could not have written this chapter without the invaluable help of Dr Fabiola López, Dr Rusty Feagin, Dr Giorgio Anfuso, Dr Octavio Pérez- Maqueo, Gilberto Binnqüist MSc, Astrid Wojtarowski MSS, Dr José María Rey Benayas, Dr Jorge López Portillo, Dr A.P. Grootjans, Dr Bas Arens, Dr Norbert Psuty, and Dr Patrick Hesp who were generous enough to share with us their expertise and experience, as well as provide us with information and enriching comments at different stages of this research. For this we are deeply grateful. We would also like to express our gratitude to Keith MacMillan for translating the document. Finally, the Instituto de Ecología, A.C (INECOL) and the Consejo Nacional de Ciencia y Tecnología (CONACyT) are acknowledged for funding this research, which is part of a wider project.

References

Anders F, Leatherman S (1987) Disturbance of beach sediment by off-road vehicles. Environ Geol Water Sci 9:183–189

Anfuso G, Gracia F, Andres J, Sanchez F, Del Rio L, López- Aguago F (2000) Depth of disturbance in mesotidal beaches during a single tidal cycle. J Coast Res 16(2):446–457

Barbier E, Hacker S, Kennedy C, Koch E, Stier A, Silliman B (2011) The value of estuarine and coastal ecosystem services. Ecol Monogr 81(2):169–193

Bird B, Branch L, Miller D (2004) Effects of coastal lighting on foraging behavior of beach mice. Conserv Biol 18:1435–1439

Bishop M, Peterson C, Summerson H, Lenihan H, Grabowski J (2006) Deposition and long-shore transport of dredge spoils to nourish beaches: impacts on benthic infauna of an Ebb-Tidal Delta. J Coast Res 22:530–546

Buick A, Paton D (1989) Impact of off-road vehicles on the nesting success of Hooded Plovers Charadrius rubricollis in the Coorong region of South Australia. Emu 89:159–172

Bulleri F (2006) Is it time for urban ecology to include the marine realm? Trends Ecol Evol 21:658–659

Burger J, Gochfeld M, Powers C (2007) Integrating long-term stewardship goals into the remediation process: natural resource damages and the department of energy. J Environ Manag 82:189–199

Castro-Avaria C (2004) El índice de vulnerabilidad de dunas litorales: un instrumento para la gestión. Revista Geográfica de Chila Terra Australis 49:89–113

Crowley K, Ahearne J (2002) Managing the environmental legacy of U.S. nuclear-weapons production. Am Sci 90:514–523

Davenport J, Davenport J (2006) The impact of tourism and personal leisure transport on coastal environments; a review. Estuar Coast Shelf Sci 67:280–292

Diefenderfer H, Sobocinski K, Thom R, May C, Borde A, Southard S, Vavrinec J, Sather N (2009) Multiscale analysis of restoration priorities for marine shoreline planning. Environ Manag 44:712–731

Dugan J, Hubbard D, McCrary M, Pierson M (2003) The response of macrofauna communities and shorebirds to macrophyte wrack subsidies on exposed sandy beaches of southern California. Estuar Coast Shelf Sci 58:25–40

Engelhard T, Withers K (1997) Biological effects of mechanical beach raking in the upper intertidal zone on Padre Island National Seashore, Texas. Technical report TAMU-CC-9706-CCS. Padre Island National Seashore, Corpus Christi, TX, 41

Everard M, Jones L, Watts B (2010) Have we neglected the societal importance of sand dunes? An ecosystem services perspective. Aquat Conserv Marine Freshw Ecosyst 20:476–487

Finkl C, Walker H (2004) Beach nourishment. In: Schwartz M (ed) The encyclopedia of coastal science. Kluwer Academic, Dordrecht, pp 37–54

Gachuz S (2009) Reecuperación de la vegetación de las dunas costeras de La Mancha, Veracruz, después de un disturbio por caminatas. Tesis de Licenciatura. Instituto de Ecología A.C., y Universidad Autónoma Metropolitana (UAM-X), 43 pp

Gallego-Fernández J, García-Mora M, Ley C (2003) Restauración de Ecosistemas Costeros. In: Rey Benayas, JM (ed) Restauración de ecosistemas en ambientes mediterráneos. Posibilidades y limitaciones, pp 157–172

Gallego-Fernández JB, Sánchez IA, Ley C (2011) Restoration of isolated and small coastal sand dunes on the rocky coast of northern Spain. Ecol Eng 37:1822–1832

García-Mora M, Gallego J, Williams A, García Novo F (2001) A coastal dune vulnerability classification: SW Iberian Peninsula case study. J Coast Res 17:802–811

Groom J, McKinney L, Ball L, Winchell C (2007) Quantifying off-highway vehicle impacts on density and survival of a threatened dune-endemic plant. Biol Conserv 135:119–134

Hammond J, Keeney R, Raiffa H (2001) Decisiones inteligentes: guía practica para tomar mejores decisiones. Editorial Norma, Bogotá

Hockings M, Twyford K (1997) Assessment and management of beach camping within Fraser Island World Heritage Area, South East Queensland. Aust J Environ Manag 4:25–39

Hopfensperger K, Engelhardt K, Seagle S (2007) Ecological feasibility studies in restoration decision making. Environ Manag 39:843–852

Hosier P, Kochhar M, Thayer V (1981) Off-road vehicle and pedestrian track effects on the sea-approach of hatchling loggerhead turtles. Environ Conserv 8:158–161

Juárez I, Nava M, Gallardo F, Cruz y J, Fajersson P (2008) Potencial para turismo alternativo del municipio de Paso de Ovejas, Veracruz. Trop Subtrop Agroecosyst 8:199–208

Kentula ME, Brooks RP, Gwin SE, Holland CC, Sherman AD, Sifneos JC (1992) An approach to improving decision making in wetland restoration and creation: Hairston A (ed): U.S. Environmental Protection Agency, Environmental Research Laboratory, Corvallis, Oregon. US. 151p

Kutiel P, Zhevelev H, Harrison R (1999) The effect of recreational impacts on soil and vegetation of stabilised coastal dunes in the Sharon Park, Israel. Ocean Coast Manag 42:1041–1060

Ley-Vega J, Gallego-Fernández J, Vidal C (2007) Manual de Restauración de dunas costeras. Ed. Ministerio del Medio Ambiente. Dirección General de Costas. 258 pp

Lithgow Serrano AD (2010) Diseño de estrategias para la restauración de dunas costeras en el estado de Veracruz. Xalapa, Ver. México. INECOL

Llewellyn P, Shackley S (1996) The effects of mechanical beach-cleaning on invertebrate populations. British Wildl 7:147–155

Longcore T, Rich C (2004) Ecological light pollution. Front Ecol Environ 2:191–198

Lubbert R, Chu T (2001) Challenges to cleaning up formerly used defense sites in the twenty-first century. Remediation 11:19–31

Luckenbach R, Bury B (1983) Effects of off-road vehicles on the biota of the Algodones Dunes, Imperial County, California. J Appl Ecol 20:265–286

McLachlan A, Brown A (2001) The ecology of sandy shores. Publicaciones Elsevier, Amsterdam, 357 pp

Mendoza-González G, Martínez ML, Lithgow D, Pérez-Maqueo O, and Simonin P (2012) Land use change and its effects on the value of ecosystem services along the coast of the Gulf of Mexico. Ecological Economics 82:23–32

Martínez M, Psuty N, Lubke R (2004) A perspective on coastal dunes. In: Martínez M, Psuty N (eds) Coastal dunes: ecology and conservation, Ecological studies 171. Springer, Berlin, pp 3–10

Martínez M, Gallego-Fernández J, García-Franco J, Moctezuma C, Jiménez C (2006) Assessment of coastal dune vulnerability to natural and anthropogenic disturbances along the Gulf of Mexico. Environ Conserv 33(2):109–117

Moreno-Casasola P, Peresbarbosa E, Travieso-Bello A (2006) Estrategia para el Manejo Costero Integral. El Enfoque Municipal. Instituto de Ecología A.C. Xalapa, Veracruz, México

Muñoz-Vallés S, Gallego-Fernández J, Dellafiore C (2011) Dune vulnerability in relation with tourism pressure in central Gulf of Cádiz (SW Spain), a case study. J Coast Restor 27:243–251

Nelson W (1993a) Beach restoration in the southeastern US: environmental effects and biological monitoring. Ocean Coast Manag 19:157–182

Nelson W (1993b) Beach-inlet ecosystems of south-eastern Florida: a review of ecological research needs and management issues. J Coast Res 18:257–266

Nordstrom K (2000) Beaches and dunes on developed coasts. Cambridge University Press, Cambridge, 338 pp

Pereira A, Laranjeira M, Neves M (2000) A resilience checklist to evaluate coastal dune vulnerability. Coast Manag 102(1):309–318

Peterson C, Bishop M, Johnson G, D'Anna L, Manning L (2006) Exploiting beach filling as an unaffordable experiment: benthic intertidal impacts propagating upwards to shore birds. J Exp Mar Biol Ecol 338:205–221

Pilkey O, Young R, Kelley J, Griffith A (2007) Mining of coastal sand: a critical environmental and economic problem for Morocco. Reporte para coastal care ORG. http://coastalcare.org/wp-content/pdf/morocco_mining.pdf

Poinar G (1977) Observations on the kelp fly, Coelopa vanduzeei Cresson, in southern California. Pan-Pacific Entomol 53:81–86

Povh D (2000) Economic instruments for sustainable development in the Mediterranean region. Responsible coastal zone management. Period Biol 102(1):407–412

Priskin J (2003) Physical impacts of four-wheel drive related tourism and recreation in a semi-arid, natural environment. Ocean Coast Manag 46:127–155

Rickard C, McLachlan A, Kerley G (1994) The effects of vehicular and pedestrian traffic on dune vegetation in South Africa. Ocean Coast Manag 23:225–247

Saaty T (1980) The analytic hierarchical process. McGraw Hill, New York

Schlacher T, Thompson L (2008) Physical impacts caused by off-road vehicles (ORVs) to sandy beaches: spatial quantification of car tracks on an Australian barrier island. J Coastal Res 24:234–242

Schlacher T, Schoeman D, Dugan J, Lastra M, Jones A, Scapini F, McLachlan A (2008) Sandy beach ecosystems: key features, management challenges, climate change impacts and sampling issues. Mar Ecol 29:70–90

Society of Ecological Restoration (SER) International (2004) Principios de SER Internacional sobre restauración ecológica. http://www.ser.org/pdf/REV_Spanish_Primer.pdf

Speybroeck J, Bonte D, Courtens W, Gheskiere T, Grootaert P, Maelfait J, Mathys M, Provoost S, Sabbe K, Stienen W, Van Lancker V, Vincx M, Degraer S (2006) Beach nourishment: an ecologically sound coastal defence alternative? A review. Aquat Conserv Marine Freshw Ecosyst 16:419–435

van der Meulen F, Salman S (1993) Management of Mediterranean coastal dunes. In: Ozhan E (ed) MedCoast,'93. METU, Ankara

Williams A, Davies P, Curr R, Koh A, Bodéré J, Hallegouet B, Meur C, Yoni C (1993) A checklist assessment of dune vulnerability and protection in Devon and Cornwall, UK. In: Magoon OT (ed) Coastal zone'93. American Society of Civil Engineering, New York

Williams A, Davies P, Alveirinho-Dias J, Pereira A, García-Mora M, Tejada M (1994) A reevaluation of dune vulnerability checklist parameters. Gaia 8:179–182

Williams A, Alveirinho-Dias J, García-Novo F, García-Mora M, Curr R, Pereira A (2001) Integrated coastal dune management: checklists. Cont Shelf Res 21:1937–1960

Williams J, Ward V, Underhill L (2004) Waders respond quickly and positively to the banning of off-road vehicles from beaches in South Africa. Wader Study Group Bull 104:79–81

Chapter 20
Coastal Dune Restoration: Trends and Perspectives

M. Luisa Martínez, Patrick A. Hesp and Juan B. Gallego-Fernández

20.1 Introduction

Sandy beaches and coastal dunes suffer the intense impacts of human activities. Because of their foremost locations on the coast, they are preferred sites for urban and maritime development, destinations for tourists, and the location of many other human activities that take place on the beach or on coastal dunes. Thus, over the years (but especially during the last few decades) many of the previously natural dunescapes have been lost to urban, tourism, mining and industrial developments. Furthermore, a recurring problem of many coastal dune systems is over-stabilization, which is mostly the result of human actions (Hesp 1991; Nordstrom et al. 2000; Hanson et al. 2002; Arens et al. 2004; Everard et al. 2010; Hesp and Hilton 2012).

In brief, coastal dunes throughout the world are being rapidly lost and degraded; thus, it is evident that there is an urgent need to preserve the natural and valuable coastal dune remnants and, as much as possible, restore those that have been degraded.

M. L. Martínez (✉)
Instituto de Ecología, A.C., Red de Ecología Funcional,
Carretera Antigua a Coatepec no. 351, El Haya 91070 Xalapa, VER, Mexico
e-mail: marisa.martinez@inecol.edu.mx

P. A. Hesp
School of the Environment, Faculty of Science and Engineering, Finders University,
GPO Box 2100, Adelaide, SA 5001, Australia
e-mail: Patrick.hesp@flinders.edu.au

J. B. Gallego-Fernández
Departamento de Biología Vegetal y Ecología, Universidad de Sevilla,
Ap. 1095 41080 Sevilla, Spain
e-mail: galfer@us.es

M. L. Martínez et al. (eds.), *Restoration of Coastal Dunes*, Springer Series on Environmental Management, DOI: 10.1007/978-3-642-33445-0_20,

20.2 How can a Coastal Dune be Restored?

The society for ecological restoration (SER) defines restoration as "the process of assisting the recovery of an ecosystem that has been degraded, damaged or destroyed. It is an intentional activity that initiates or accelerates ecosystem recovery with respect to its health (functional processes), integrity (species composition and community structure), and sustainability (resistance to disturbance and resilience)" (http//www.ser.org/content/guidelines_ecological_restoration.asp). Certainly, this is a broad concept that applies to many situations and ecosystems, but it is not easy to use when it comes to coastal dune restoration, because recovery or restoration of coastal dune ecosystems may include a wide array of situations.

What can be considered to be the "recovery" of a coastal dune ecosystem? Does it mean recovering former landform types or geomorphological units? Does it mean recovering mobile stages or to stabilizing mobile dunes and recovering dune vegetation? Does it refer to restoring the environmental heterogeneity? Or recovering the many successional stages that can be found in a single system, adjacent to one another? Does it refer to the dynamic geomorphology, the vegetation, or both? Certainly, it seems that obtaining a clear definition of coastal dune restoration is not a simple task.

Because coastal dunes consist of very dynamic landforms and dynamic communities, heterogeneous and diverse (Martínez et al., Chap. 1, this volume), their restoration is anything but a simple and straightforward activity. There is no single way to restore a coastal dune. The commonly used definition of ecological restoration does not seem to apply to these environments, because there are many scenarios of recovery that can be looked for in a coastal dune restoration project. That is, the health, integrity, and sustainability of coastal dunes can refer to many situations.

The compendium of restoration projects gathered in this book is an example of the complexity and variety of coastal dune restoration activities that are taking place in a variety of countries from different continents. These activities include many contrasting actions, such as the creation of artificial dunes, the destruction of artificial dunes and berms, and the creation of more natural ones, the recovery after mine extraction activities, the reactivation of stabilized dunes, or the stabilization of active dunes, and the elimination of invasive species. All of them are attempts to recover functional processes and ecosystem integrity.

A summary of the different restoration activities that have taken place at coastal dunes in various countries and that are described in this book follows (Table 20.1). From this set of examples it is obvious that the restoration of coastal dunes has many goals, facets, and mechanisms.

20.2.1 Restoring Foredunes

Foredunes are exposed to many different disturbing events, such as the invasion of exotic species, artificial stabilization, nitrogen deposition, urbanization, storm surges, storms and hurricanes, human trampling, and 2WD and 4WD vehicle

Table 20.1 Summary of the restoration actions shown in this book. Order of authors follows chapter order

Chapter	Authors	Country	Disturbance	Goals for restoration	Morphology	Restoration actions	Costs	Funding agency
2	Nordstrom and Jackson	USA	Erosion by storm surges, human activities	Protection and recreation, habitat, refuge	Foredunes	This is a review of different methods used for coastal dune restoration	No data	Government
3	Psuty and Silveira	USA	Human and natural	Restore, recover sediment dynamics, maintenance of the processes	Foredunes	Passive restoration with no human intervention; allow dune inland migration; some human intervention with bulldozer, beach scraping and sand transportation	No data	Government; public parks
4	Vestergaard	Denmark	Human activities	Protection and recreation	Foredunes	Creating artificial dune; planting native vegetation	US$ 2,229/ha	Government; public parks
5	Hesp and Hilton	New Zealand	Grazing, planting exotics, reshaping, land use change, urbanization	Restore morpho-ecological states in a metropolitan context; restore natural dynamics of transgressive dunes stabilized with marram grass	Foredunes and transgressive dunes	Foredunes re-shaped with bulldozer; exotics sprayed with herbicide; native plants were planted; destabilization after eliminating marram; natural dynamics recovered	US$ 1,123/m	Government; public parks
6	Feagin	USA	Hurricanes	Protection of human infrastructure, aesthetic value; restoration after hurricane Ike	Foredunes	Three restoration projects: geotubes; reintroduction of *Uniola paniculata*; backhoes and tractors.	US$ 275/m	Stake-holders, government
7	Arens et al.	Netherlands	Stabilization leads to loss of biodiversity	Restoring dune mobility and increasing biodiversity	Foredunes; dunefield; parabolic dunes	Sod-cutting; restoration of aeolian processes; engineering; archeological research (to prevent damage to archeological heritage)	US$ 9/m^3	Government

(continued)

Table 20.1 (continued)

Chapter	Authors	Country	Disturbance	Goals for restoration	Morphology	Restoration actions	Costs	Funding agency
8	Rhind et al.	Wales, UK	Overstabilization, loss of herbivores	Restoring dune mobility and increasing biodiversity	Foredunes and dunefields	Deep ploughing; this is a review of different methods	No data	–
9	Muñoz-Reinoso et al.	Spain	Overstabilization, biodiversity loss, urbanization, tourist pressure	Recovery of Juniperus woodland	Dunefield	320 ha, 70,000 saplings were planted; pine tree clearings; hand removal of *Carpobrotus edulis*	No data	Government
10	Pickart	USA	Invasive exotic	Restoring biotic and abiotic processes	Foredune–blowout–parabolic dune complex	Hand removal of invasive species *Ammophila arenaria* and *Carpobrotus chilensis*	US$ 65,293/ha	Government; public parks
11	Kutiel	Israel	Stabilization leads to biodiversity loss	Restoring dune mobility and increasing biodiversity	Dunefield	Uprooting woody vegetation with bulldozer	No data	Nature reserve
12	Acosta et al.	Italy	Human trampling	Vegetation recovery	Foredunes	Fencing to avoid human trampling; natural recovery of vegetation	US$ 962/ha	Government
13	Lubke	South Africa	Mining for minerals: total destruction	Recreate previously existing natural ecosystems	Dunefield	Building dunes, planting plants and then spontaneous restoration	No data	Private (mining company)
14	Moreno-Casasola et al.	Mexico	Urbanization	Stabilization; preventing sand from affecting the city	Dunefield; parabolic dunes	Planting with *Casuarina trees*	No data	Government; port authorities
15	Grootjans et al.	Netherlands	Overstabilization leads to biodiversity loss; planting of pine forests	Recover hydrological regime and aeolian dynamics; dune vegetation	Foredunes; slacks and dunefields	Mowing and grazing; sod-cutting, re-wetting	US$ 12,053/ha	Water Company

(continued)

Table 20.1 (continued)

Chapter	Authors	Country	Disturbance	Goals for restoration	Morphology	Restoration actions	Costs	Funding agency
16	Lopez Rosas et al.	Mexico	Invasion of exotic species	Restoring flooding regime and recovering biodiversity	Slacks	Ploughing; digging	No data	Government; nature reserve
17	Lehrer et al.	Israel	Expansion of exotic invasive species	Eliminating invasive species and recovering diversity	Dunefield	Cutting, clearing, burning, spraying, uprooting, solar sterilization	US$ 640,917	Government; LTER
18	Pérez-Maqueo et al.	No specific country	Any kind	Any goal	Foredunes and coastal dunes	This is a review of the costs of restoration and evaluation of ecosystem services	Wide range of costs	Different sources
19	Lithgow et al.	Mexico	Any kind	Any goal	Foredunes and coastal dunes	Multicriteria analysis to improve planning, implementation and monitoring of coastal dune restoration	No data	Different sources

activity (Nordstrom and Jackson, Chap. 2, this volume; Psuty and Silveira, Chap. 3, this volume; Hesp and Hilton, Chap. 5, this volume; Acosta et al., Chap. 12, this volume; Wilcock and Carter 1977; Nordstrom et al. 2000; Miller et al. 2003; Hesp 2002, 2011). As a consequence, the goals of foredune restoration are equally diverse and include beach nourishment; the recovery of morpho-ecological states; the recovery of sediment dynamics; the restoration of natural vegetation and endemic species; and the creation of habitat refugia. Frequently, however, foredunes are restored for the protection of human infrastructure, recreation, and aesthetic values (Nordstrom and Jackson, Chap. 2, this volume; Vestergaard, Chap. 4, this volume; Feagin, Chap. 6, this volume; Peterson et al. 2000; López and Marcomini 2006; Nordstrom et al. 2000; Muñoz-Pérez et al. 2001). In general, the actions required to guarantee an adequate sediment supply to reach the location of pioneer of vegetation, accumulate, and help create the natural topographic features of fordunes. (Psuty and Silveira, Chap. 3, this volume; Webb et al. 2000; Miller et al. 2003).

The activities required to achieve these goals are very different, and range from "soft" measures, such as the modification of aeolian processes and sediment dynamics by using sand fences and/or vegetation planting (Nordstrom and Jackson, Chap. 2, this volume; Psuty and Silveira, Chap. 3, this volume; Nordstrom et al. 2002) as well as eliminating exotics by hand (Pickart, Chap. 10, this volume), through more intense methods, such as the eradication of exotics by means of herbicides (Hesp and Hilton, Chap. 5, this volume; Feagin, Chap. 6, this volume; Hilton et al. 2009), to "hard" methods, which include the use of geotubes to strengthen the foredune foundation and utilizing earth-moving equipment to restore/rebuild/build the foredunes (Nordstrom and Jackson, Chap. 2, this volume; Vestergaard, Chap. 4, this volume; Hesp and Hilton, Chap. 5, this volume; Feagin, Chap. 6, this volume; Jones et al. 2010). In addition, the negative impact of human actions also needs to be managed by means of controlling access to the beaches and avoid human trampling (Acosta et al. Chap. 12; Kutiel, Chap. 11; Hesp et al. 2010; Pye and Neal 1994), restricting beach raking that eliminates plants and propagules, and limiting driving on beaches and dunes (Kutiel, Chap. 11, this volume).

It is important to bear in mind that commonly, the restoration of the morphology and vegetation on foredunes may take 10 years or more and that, because of the dynamic nature of these systems, restoration may in fact need to be a recurrent action (Psuty and Silveira, Chap. 3, this volume; Hesp and Hilton, Chap. 5, this volume; Feagin, Chap. 6, this volume). Repeated beach nourishment is perhaps inevitable in the case of a sand-deficient environment, such as eroding shorelines (Nordstrom and Jackson, Chap. 2, this volume; Psuty and Silveira, Chap. 3, this volume; Muñoz-Pérez et al. 2001; Bezzi et al. 2009; González et al. 2009).

20.2.2 Restoring Dunefields: Vegetation vs Sand

Dunefields are very diverse and range from foredune plains, foredune–blowout complexes, parabolic dunefields to transgressive dune sheets and dunefields (Hesp 2004; Hesp and walker 2012; Hesp et al. 2011)

Restoration of dunefields can be very generally grouped into two contrasting sets of actions: the reactivation of stabilized dunes and the stabilization of mobile dunes. Usually, the goal is to restore the heterogeneity of features in the geo-temporal context (Psuty and Silveira, Chap. 3, this volume; Hesp and Hilton, Chap. 5, this volume; Arens et al. 2004), and improve/change the environment for psammophyte species including plants and animals (Pickart, Chap. 10, this volume; van der Hagen et al. 2008). Needless to say, human needs are also a common reason for restoration.

Drivers influencing dune stabilization can be direct and human-induced (e.g., removal and/or flattening of dunes; stabilization by urban, industrial and farming infrastructure; plantations of grasses, herbs, shrubs or trees; introduction or removal of grazing, off-road vehicle activity, nutrient enrichment due to pollution); indirect and external (e.g., changes in sediment supply, salt spray, and water availability and water table heights, nutrient enrichment, and climate change); and internal (such as soil development, and grazing) (Rhind et al., Chap. 8, this volume; Arens and Geelen 2006).

From perhaps the earliest historic times humans have destroyed, modified or changed dune systems, both continental and coastal, often because of farming activities and forest felling. The creation of drifting sands was enhanced in some cases by climatic events (e.g., the Little Ice Age in Europe). Humans then tended to view *all* drifting sands as bad and attempted to stabilize them, often with considerable success, but commonly by introducing exotic species and creating mono-specific stands (Muñoz-Reinoso et al., Chap. 9, this volume; Arens et al., Chap. 7, this volume; Moreno-Casasola et al., Chap. 14, this volume; Pickart, Chap. 10, this volume; Grootjans et al. 2001; van der Meulen and Salman 1996; Bossuyt et al. 2007). This trend continued into the late 1900 s in some countries, and, led by agriculturalists and soil conservationists in particular, many exotic species were introduced into coastal dunes, and then spread from those sites, becoming invasive species.

Stabilization (both natural and artificially induced) of the natural dunes results in the loss of space or habitat for the native species (Hesp and Hilton, Chap. 5, this volume; Pickart, Chap. 10, this volume); the rare and uncommon obligate or semi-obligate psammophytes (Rhind et al., Chap. 8, this volume); the dune slack hygrophytes (Grootjans et al., Chap. 15, this volume); as well as many other impacts. When dunes become stabilized, diversity decreases because the early successional stages disappear, and certain landform units may not be formed (e.g., deflation basins and plains, wet slacks, gegenwalle ridges, precipitation, and trailing ridges, etc.) resulting in habitat and diversity losses or changes (Arens and Geelen 2006; Arens et al., Chap. 7, this volume; Rhind and Jones 2010; Rhind et al., Chap. 8, this volume; Grootjans et al. 2001). It must be recognized that stabilization is a perfectly natural evolutionary trend for coastal dunes in the absence of human impact or interference. For example, foredunes can eventually evolve from incipient foredunes dominated by a few pioneer species to stabilized relict foredunes covered in tree and shrub species. Transgressive dunefields evolve from highly mobile systems with minimal plants present, through various evolutionary steps or phases. At some point the dunefields may have active interdunes, precipitation, and trailing ridges, deflation plains and slacks, nebkha, etc., with many rare and endemic species present. At a later stage these may all disappear as large-scale natural stabilization takes place (e.g., Martinho et al. 2010; Hesp 2011). We must be careful not to view this as ecologically or geomorphologically "bad" just because the natural diversity has decreased or changed. However, The negative impact of the human-driven stabilization of dunescapes has been demonstrated in many dune systems of the world, such as Israel (Kutiel, Chap. 11, this volume), Wales (Rhind et al., Chap. 8, this volume), the Netherlands (Grootjans et al., Chap. 15, this volume; Arens et al., Chap. 7, this volume); New Zealand (Hesp and Hilton, Chap. 5, this volume), the USA (Pickart, Chap. 10, this volume), Spain (Muñoz-Vallés et al. 2011), South Africa, and many other countries.

Note that stabilization may also occur as a result of a *reduction* in human activities, such as occurred in Israel. Here, when the State of Israel was established in 1944, grazing was banned and exploitation of coastal vegetation was stopped. Bedouin nomads, who intensively used the coastal dune vegetation for grazing and fuel, moved elsewhere. The result was rapid dune stabilization (in a few decades) and the loss of many endemic psammophytic species (Kutiel, Chap. 11, this volume). Coastal dunes have also become artificially stabilized as a result of atmospheric nitrogen deposition from atmospheric pollution, which enhances the growth of a few species (Kooijman and de Haan 1995).

Different plant species have been used in stabilization actions, such as grasses (*Ammophila arenaria*, or marram grass), shrubs (*Acacia karroo*, *Acacia saligna*, *Retama monosperma*), and trees (*Pinus* spp., *Eucalyptus* spp., *Casuarina* spp.).

20.2.3 Reactivating Stabilized Dunes

Because of the perceived negative impact of artificial stabilization of dunes on diversity and the formation/evolution of geomorphological units, the most frequent restoration action currently taking place on parabolic and transgressive dunefields is re-mobilization. In a way, restoration actions through the application of a "desertification process" aimed at reducing vegetation cover sound quite paradoxical. Nevertheless, it should be emphasized that in this case, a truly healthy dune system according to some (but not all) requires all successional stages, including mobile dunes. When all successional seres occur in a coastal dune system, biodiversity increases and functionality is strengthened.

A range of options to counter stabilization have been used and again, similar to restoration actions on foredunes, they range from low-impact and small-scale, to high-impact and large-scale. In terms of low-impact and small-scale actions, local disturbance can be promoted by increasing the levels of grazing pressure through encouraging rabbits, or by the use of domestic stock such as cattle, sheep or horses. Heavy grazing can, however, result in severe destabilization (Rhind et al., Chap. 8, this volume) and changes in plant communities (Zunzunegui et al., 2012), but when used at perceived "proper" intensities, may reactivate the system, as was found by Grootjans et al. (Chap. 15, this volume, and 2001) and López Rosas et al. (Chap. 16, this volume)

Elimination of exotic species is also a difficult task. Sometimes they have been sprayed with herbicides (Hesp and Hilton, Chap. 5, this volume; Hilton et al. 2009), although some invasives, such as marram grass (*Ammophila arenaria*) are difficult to eradicate, even with continued application of herbicides and mechanical removal. In contrast to the above, softer techniques have also proven to be effective. For instance, in Northwestern USA, marram grass was removed by hand. This involved painstaking labor, with many volunteer-hours involved. But in the end, biodiversity increased naturally once the dominant exotic grass was eliminated (Pickart, Chap. 10, this volume). In the Netherlands, remobilization has been achieved by sod cutting, as well as hand removal of re-sprouting roots for a number of years (Grootjans et al. 2001). Intensive volunteer work has also been necessary. Inhibitory woody vegetation has been removed by pruning and felling (Muñoz-Reinoso et al., Chap. 9, this volume), but also with bulldozers (Kutiel, Chap. 11, this volume).

Larger scale, more invasive options to reactivate coastal dunes and restore original vegetation include mechanical disturbances, such as the removal of alien soils and materials and the creation of new dunes (Hesp and Hilton, Chap. 5, this volume; van Aarde et al. 1998), and topsoil stripping and deep ploughing (Rhind et al., Chap. 8, this volume; Graham and Haynes 2004). The latter inverts the soil profile by burying any surface nutrients and unwanted seeds, while exposing low fertility subsoil, and inhibits the germination of undesired species (Jones et al. 2010). Another high impact large-scale mobilization action includes beach and shoreface nourishment in order to counteract shoreline erosion. This strategy has proven to be very successful in The Netherlands (Arens et al., Chap. 7, this volume), where the coastline retreat has been reduced and the transfer of sand into the dunes has increased considerably. A potential drawback of these activities is the quality of the sand used for nourishment. If the nourished sand differs from the original sand, e.g., in grain size, carbonate content or mineralogical compounds, there may be ecological consequences affecting species colonization and growth. The bottom of the ocean floor, where sediment is removed for beach nourishment, may also be severely affected after these dredging activities, and some would argue that this is not a medium- to long-term sustainable action, as coastline retreat is likely to eventually occur.

Once blowouts are reactivated the dune systems may become destabilized (Rhind et al., Chap. 8, this volume; Arens et al. 2004). However, maintenance of coastal dune mobility can be a long-term project, requiring periodic human intervention. In this sense, Arens and Geelen (2006) found that even extensively destabilized areas (tens of hectares) are likely to re-stabilize within a few decades and that new measures to reduce stabilization may be required every 10 or 20 years (see Arens et al, Chap. 7, this volume). Lastly, it is very important to bear in mind that any attempt to create dunes where they do not exist, or to increase the dynamism of stabilized dunes, must have a strong public information component to demonstrate their feasibility and increase their acceptability. When people are informed, they may even be willing to pay for the remobilization of dunes, as well as the containment and elimination of exotic species (Lehrer et al., Chap. 17, this volume).

20.2.4 Stabilizing Active Dunes

Having shown the negative effects of overstabilization, it is necessary to also consider that stabilization and re-vegetation can sometimes be a required action. This is the case when coastal dunes are created de novo, artificially, and they need to be planted with preferably native plants in order to enhance further sand accumulation and dune growth (Vestergaard, Chap. 4, this volume; Gallego-Fernández et al. 2011). Sometimes, cities may develop on coastal dunes or are surrounded by them, as occurs with the Port of Veracruz, on the Gulf of Mexico, Mexico. For centuries, the Port of Veracruz has suffered the consequences of its unfortunate location, with sand blowing into the city from neighboring mobile transgressive dunes when the strong winds blow and the weather is relatively dry during the winter months. In this case, the stabilization of mobile dunes was seen as a societal need (Moreno-Casasola et al. 2008). The relevant fact is, in these cases, how restoration–revegetation was performed.

In Denmark, an artificial dune was built with earth-moving machinery and was then planted with native species and natural regeneration was allowed to occur (Vestergaard, Chap. 4, this volume). In contrast, the dunes surrounding the city of Veracruz were stabilized with the exotic tree *Casuarina equisetifolia*. Usually, this is a very aggressive invasive species that does not promote natural regeneration (Mailly and Margolis 1992; Moreno-Casasola et al., Chap. 14, this volume). Nevertheless, Moreno-Casasola et al. (Chap. 14, this volume) found that, under high moisture conditions and if tropical rainforest remnants occur in the vicinity, the original tropical forest can, in fact, regenerate beneath the shade of the exotic trees. This is a very peculiar situation and should not be interpreted as a sign that *Casuarina* is a good option for dune re-vegetation outside its native distribution range (Australia). Native species should always be the preferred plants in any restoration project. Interestingly, in spite of the overwhelming evidence of the negative impacts of artificial stabilization, coastal dune forestation programs are still commonplace in many countries.

Building artificial dunes and enhancing the arrival and growth of vegetation has also been promoted after heavy mineral dune mining (Lubke, Chap. 13, this volume; Zeppelini et al. 2009; Grainger et al. 2011). For example, in Namibia, nets were used as wind traps for rehabilitation following diamond mining acting to trap sand. In addition, top soil was spread onto the site and native plants were transplanted and allowed to follow their natural successional sequence (Lubke, Chap. 13, this volume).

20.2.5 Restoring Wet Slacks

Wet slacks (also termed deflation plains and basins, ponds and wetlands) are located in the lower parts of the coastal dune system. Here, the water table is close to the surface of the sand and thus, slacks may become flooded during the rainy season, creating interdune ponds and lagoons (ephemeral or permanent).

Wet slacks can be severely affected by artificial fluctuations in water tables because of the production of drinking water for human populations (e.g., in the Netherlands), for watering crops, and also because of high atmospheric nitrogen deposition from agricultural lands and industrial areas. The result is a decline in species diversity, and the local extinction of endemic species. In this case, restoration activities involved rewetting wet slacks, and re-establishing the hydrological regime; mowing, grazing, and sod-cutting (Grootjans et al., Chap. 15, this volume; López Rosas et al., Chap. 16, this volume; Grootjans et al. 2001, 2002). Once the hydrological conditions are restored, restoration projects in slacks are generally successful, especially when the top soil is removed. Nevertheless, repeated maintenance is also necessary (López Rosas et al., Chap. 16, this volume).

20.3 Who Pays for Coastal Dune Restoration?

Coastal dunes and beaches are amongst the most intensely used natural ecosystems, because of the associated economic benefits. They offer a wide array of ecosystem services to society and are exploited by human society throughout the world. No doubt, there are many interests that acknowledge the need to restore those dune systems that are degraded as efficiently as possible and in the best possible way: beaches are nourished every year; coastal dunes are re-mobilized or stabilized. Because humans gravitate to the coast, with 40 % of the world population living at or near the coast, and because coastal tourism is significant, human intervention in and on coastal ecosystems is constantly high.

Thus, it is obvious that there is a societal need at times and in certain places to restore and rehabilitate coastal ecosystems, such as beaches and coastal dunes. Stakeholders (hotel and private property owners) as well as the general public tend to benefit from these restoration practices. Who pays for them? The restoration examples presented in this book show that restoration and rehabilitation are mostly financed by Government authorities (Table 20.1). Restoration of dunes and the re-mobilization of stabilized dunes is frequently performed in public parks/conservation regions and financed by government agencies. Private funds for restoration and rehabilitation are less frequent. For example, in South Africa the mining companies financed restoration activities. In Galveston (Texas), it was the stakeholders and local government agencies who paid for rehabilitation after every hurricane landing on this island (Feagin, Chap. 6, this volume). As human impacts increase, so will the need for more funds for restoration and rehabilitation.

20.4 Discussion

Independently of how effective the restoration actions were, what we have learned from numerous restoration projects at dune sites is that we cannot fully compensate for the ecological functioning of natural processes, such as, for example, intensive aeolian transport, local blowout development and maintenance, a regular supply of nonpolluted groundwater, or shoreline protection. As much as possible, dunes should be allowed to evolve and move freely, or at least to the extent possible within the constraints of human infrastructure (on developed coasts). The heterogeneity of these systems needs to be maintained in order to protect their integrity and conservation values.

Coastal dune restoration must take into account the spatial and temporal scale of their evolution, landforms, structures, and functions, and therefore dune restoration plans should have a regional approach, incorporating landscape-scale processes (Gallego-Fernández et al. 2011). In general, coastal dune restoration practices require the following actions. Independently of the goals for restoration (rehabilitation, remobilization, stabilization), the factors and conditions that are affecting coastal beach and dune dynamics, development, and evolution, need to be determined so that these factors can be addressed. Then, it can be decided what are the best actions that need to be taken. Natural sediment and vegetation dynamics should be allowed to occur wherever possible. In order to allow this we need to protect dune systems far better, and one priority should be to work toward the creation of many more dune conservation areas around the world. In addition, there are many actions we can take to better protect beach–dune system dynamics: beach raking and driving on beaches and dunes should be eliminated; reducing and preferably eliminating pollution should be of paramount importance; protecting endangered species, restricting or eliminating invasive species should be a primary objective; construction on foredunes should be declared illegal; and the creation of set-back zones where construction is forbidden should become a common planning requirement. Chiefly, natural aeolian processes and natural vegetation dynamics should be allowed to function wherever and whenever possible. By accepting more natural system dynamics and natural mobility, coastal dunes can be more functional, better preserved, and protect more efficiently.

Restoration of coastal dunes and recovery of characteristic dune species are difficult tasks. Long-term monitoring of remobilization actions, for example, has shown that a single restoration intervention is not sufficient to restore the large-scale landscape forming process or to eliminate invasive species (Kutiel, Chap. 11, this volume; Hesp and Hilton, Chap. 5, this volume). In this sense, Arens et al. Chap. 7, this volume argue that for a number of years, a certain form of maintenance, such as the removal of re-growing roots, is necessary to get the dune moving. In addition, recent evidence (Arens et al. 2004) shows that mobilization of coastal dunes is more effective when dunes are connected to the coastline (Psuty and Silveira, Chap. 3, this volume; Arens et al., Chap. 7, this volume). Thus, it seems that foredune erosion, with sediment being transferred to inland coastal dunes, might "restart the engine" of coastal dune mobility (Arens et al., Chap. 7, this volume).

An important question here is whether there is equilibrium between the protection of human assets and the dynamic nature of coastal dunes. How much should they be allowed to move? Is it possible to nourish our coasts, prevent shoreline erosion, allow sand movement, and protect human infrastructure and lives? Can we have it all? This is certainly an important challenge for the coming years. While working on this, it is essential to bear in mind that there is no equilibrium form or dimension in a functional dune system: each one is unique and has its own dynamic equilibrium that changes in time and space.

Restoration is not always a real possibility. When coastal dunes and beaches are very degraded, as occurs on urban and developed coasts, rehabilitation may be the only viable option. In this case, the system is not self-sustainable and requires periodic human intervention that repairs and maintains the integrity of the system. When sediment is insufficient, as occurs on eroding coasts, rebuilding a dune is not really restoration, but an artificial creation of a sand ridge that is unsustainable over time (Psuty and Silveira, Chap. 3, this volume). Adaptive management, with continued human input, is critical where space is restricted and long-term erosion continues. Sustainability of natural features in developed areas requires humans to act as intrinsic agents of landform change.

20.5 Conclusions

Coastal dune restoration efforts require the combination and integration of different criteria, including ecological, geomorphological, and social, so that we can maximize the goods and services that coastal dunes can provide (Nordstrom and Jackson, Chap. 2, this volume; Lithgow et al., Chap. 19, this volume; Pérez-Maqueo et al., Chap. 18, this volume). A theoretical and empirical catalog of the actions needed to perform adequate coastal dune restoration should consider:

1. Assessing the factors that are affecting the dune systems, and in particular, understanding the evolutionary state of each dune system.
2. Eliminating or reducing factors of tension.
3. Creating the landforms or the conditions necessary for natural landform creation where they have been eliminated or truncated.
4. Allowing these new or modified landforms to function as natural dunes by allowing them (or some portions) to be dynamic, but also to evolve and change over time.
5. Eliminating exotics or invasive species.
6. Favoring dune and vegetation evolution through time by following adaptive management strategies.
7. Last but not least important, convincing society to accept coastal dunes, with their natural dynamism, as a natural, and very important feature of the coast.

Finally, the ever-increasing human population (especially at the coast) (Martínez et al. 2007), and human-related environmental changes, such as climate change and potential sea level rise, coupled with possible increased storminess, will add more stress to the world's coastal beaches and dunes. Humans and natural ecosystems will be increasingly affected by these changes. More than ever, humans will benefit from the protection gained from the natural functioning of coastal dunes. Restoration of these ecosystems will thus become increasingly important as human impact (direct and indirect) destroys or disrupts them. Further research on the dynamics and functionality of coastal dunes and how ecosystem services such as protection are related will become increasingly relevant, and restoration and conservation will likely become more necessary on an increasingly crowded planet.

Acknowledgments We are very grateful to all the authors of the chapters in this book who kindly contributed their chapters and shared their restoration experiences. We are also very grateful to Lawrence R. Walker for his insightful comments made to earlier versions of this chapter.

References

Acosta ATR, Jucker T, Prisco I, Santoro R (2012) Passive recovery of Mediterranean coastal dunes following limitations to human trampling. In: Martínez ML, Gallego-Fernández JB, Hesp PA (eds) Restoration of coastal dunes. Springer, Berlin, Chapter 7, this volume

Arens SM, Geelen LHWT (2006) Dune landscape rejuvenation by intended destabilisation in the Amsterdam water supply dunes. J Coastal Res 23:1094–1107

Arens SM, Slings Q, de Vries CN (2004) Mobility of a remobilised parabolic dune in Kennemerland, The Netherlands. Geomorphology 59:175–188

Arens SM, Slings QL, Geelen LHWT, van der Hagen HGJM (2012) Restoration of dune mobility in the Netherlands. In: Martínez ML, Gallego-Fernández JB, Hesp PA (eds) Restoration of coastal dunes. Springer, Berlin, Chapter 7, this volume

Bezzi A, Fontolan G, Nordstrom KF, Carrer D, Jackson NL (2009) Beach nourishment and foredune restoration: practices and constraints along the Venetian shoreline Italy. J Coastal Res 56:287–291

Bossuyt B, Cosyns E, Hoffmann M (2007) The role of soil seed banks in the restoration of dry acidic dune grassland after burning of Ulex europaeus scrub. Appl Veg Sci 10:131–138
Everard M, Jones L, Watts B (2010) Have we neglected the societal importance of sand dunes? An ecosystem services perspective. Aquatic Conserv 20(4):476–487
Feagin R (2012) Foredune restoration before and after Hurricanes: Inevitable destruction, certain reconstruction. In: Martínez ML, Gallego-Fernández JB, Hesp PA (eds) Restoration of coastal Dunes. Springer, Berlin, Chapter 6, this volume
Gallego-Fernández JB, Sánchez IA, Ley C (2011) Restoration of isolated and small coastal sand dunes on the rocky coast of Northern Spain. Ecol Eng 37:1822–1832
González M, Medina R, Losada M (2009) On the design of beach nourishment projects using static equilibrium concepts: application to the Spanish coast Spain. Coastal Eng 57:227–240
Graham MH, Haynes RJ (2004) Organic matter status and the size, activity and metabolic diversity of the soil microflora as indicators of the success of rehabilitation of mined sand dunes. Biol Fertil Soils 39(6):429–437
Grainger MJ, van Aarde RJ, Wassenaar TD (2011) Landscape composition influences the restoration of subtropical coastal dune forest. Restor Ecol 101(19):111–120
Grootjans AP, Everts H, Bruin K, Fresco L (2001) Restoration of wet dune slacks on the Dutch Wadden Sea Islands: recolonization after large-scale sod cutting. Restor Ecol 9(2):137–146
Grootjans AP, Geelen HWT, Jansen AJM, Lammerts EJ (2002) Restoration of coastal dune slacks in the Netherlands. Hydrobiologia 478:181–203
Grootjans A, Dullo BW, Kooijman A, Bekker R, Aggenbach C (2012) Restoration of dune vegetation in the Netherlands. In: Martínez ML, Gallego-Fernández JB, Hesp PA (eds) Restoration of coastal Dunes. Springer, Berlin, Chapter 15, this volume
Hanson H, Brampton A, Capobianco M, Dette HH, Hamm L, Laustrup C, Lechuga A, Spanhoff R (2002) Beach nourishment projects, practices, and objectives—a European overview. Coast Eng 47(2):81–111
Hesp PA (1991) Ecological processes and plant adaptations on coastal dunes. J Arid Environ 21:165–191
Hesp PA (2002) Foredunes and blowouts: initiation, geomorphology and dynamics. Geomorphology 48:245–268
Hesp PA (2004) Coastal dunes in the tropics and temperate regions: location, formation, morphology and vegetation processes. In: M. Martínez, N. Psuty (eds) Coastal Dunes, ecology and conservation. Springer, Berlin, pp 29–49 (Ecological studies v. 171)
Hesp PA (2011) Dune coasts. In: Wolanski E, McLusky DS (Eds) Treatise on Estuarine and coastal science, vol 3. Academic Press, Waltham, pp 193–221
Hesp PA, Hilton MJ (2012) Restoration of foredunes and trangressive dunefields: case studies from New Zealand. In: Martínez ML, Gallego-Fernández JB, Hesp PA (eds) Restoration of coastal Dunes. Springer, Berlin (in press)
Hesp PA, Thom BG (1990) Geomorphology and evolution of transgressive dunefields. In: Nordstrom K, Psuty N, Carter W (eds) Coastal dunes: processes and morphology. Wiley, New York, pp 253–288
Hesp PA, Walker IJ (2012) Treatise on geomorphology, Chapter 11.17 Aeolian environments: coastal dunes. In: Shroder J, Baas ACW (eds) Treatise on geomorphology. Elsevier, Amsterdam (in press)
Hesp PA, Schmutz P, Martinez ML, Driskell L, Orgera R, Renken K, Rodriguez-Revelo N, Jimenez-Orocio OA (2010) The effect on coastal vegetation of trampling on a parabolic dune. Aeolian Res 2:101–111
Hesp PA, Martinez ML, Miot da Silva G, Rodríguez-Revelo N, Gutierrez E, Humanes A, Laínez D, Montaño I, Palacios V, Quesada A, Storero L, González Trilla G, Trochine C (2011) Transgressive dunefield landforms and vegetation associations, Doña Juana, Veracruz, Mexico. Earth Surf Process Landf 36(3), 15: 285–295
Hilton M, Woodley D, Sweeney C, Konlechner T (2009) The development of a prograded foredune barrier following ammophila arenaria eradication, Doughboy Bay, Stewart Island. J Coastal Res 56:317–321

Jones MLM, Norman K, Rhind PM (2010) Topsoil inversion as a restoration measure in sand dunes, early results from a UK field-trial. J Coastal Conserv 14(2):139–151

Kooijman AM, de Haan MWA (1995) Grazing as a measure against grass encroachment in Dutch dry dune grassland: effects on vegetation and soil. J Coastal Conserv 1:127–134

Kutiel PB (2012) Restoration of coastal sand dunes for conservation of biodiversity: the Israeli experience. In: Martínez ML, Gallego-Fernández JB, Hesp PA (eds) Restoration of coastal dunes. Springer, Berlin, Chapter 11, this volume

Lehrer D, Becker N, Kutiel PB (2012) The value of coastal sand dunes as a measure to plan an optimal policy for invasive plant species: the case of the *Acacia saligna* at the Nizzanim LTER coastal sand dune nature reserve, Israel. In: Martínez ML, Gallego-Fernández JB, Hesp PA (eds) Restoration of coastal dunes. Springer, Berlin, Chapter 17, this volume

Lithgow D, Martínez ML, Gallego-Fernández JB (2012) Multicriteria analysis to implement actions leading to coastal dune restoration. In: Martínez ML, Gallego-Fernández JB, Hesp PA (eds) Restoration of coastal dunes. Springer, Berlin, Chapter 19, this volume

López RA, Marcomini SC (2006) Monitoring the foredune restoration by fences at Buenos Aires coast. J Coastal Res 39:955–958

López Rosas H, Moreno-Casasola P, López Barrera F, Sánchez-Higueredo LE, Espejel-González VE, Vázquez J (2012) Interdune wetland restoration in central Veracruz, Mexico: plant diversity recovery mediated by hydroperiod. In: Martínez ML, Gallego-Fernández JB, Hesp PA (eds) Restoration of coastal dunes. Springer, Berlin, Chapter 16, this volume

Lubke RA (2012) Restoration of dune ecosystems following mining in Madagascar and Namibia: contrasting restoration approaches adopted in regions of high and low human population density. In: Martínez ML, Gallego-Fernández JB, Hesp PA (eds) Restoration of coastal dunes. Springer, Berlin, Chapter 13, this volume

Mailly D, Margolis HA (1992) Forest floor and mineral soil development in Casuarina equisetifolia plantations on the coastal sand dunes of Senegal. For Ecol Manag 55(1):259–278

Martínez ML, Intralawan A, Vázquez G, Pérez-Maqueo O, Sutton P, Landgrave R (2007) The coasts of our world: ecological, economic and social importance. Ecol Econ 63:254–272

Martínez ML, Hesp PA, Gallego-Fernández JB (2012) Coastal dunes: human impact and need for restoration. In: Martínez ML, Gallego-Fernández JB, Hesp PA (eds) Restoration of coastal dunes. Springer, Berlin, Chapter 1, this volume

Martinho CT, Hesp PA, Dillenburg SR (2010) Morphological and temporal variations of transgressive dunefields of the Northern and Mid-littoral Rio Grande do Sul coast, Southern Brazil. Geomorphology 117:14–32

Miller DL, Yager L, Thetford M, Schneider M (2003) Potential use of Uniola paniculata rhizome fragments for dune restoration. Restor Ecol 11(3):359–369

Moreno-Casasola P, Martínez ML, Castillo-Campos G (2008) Designing ecosystems in degraded tropical coastal dunes. Ecoscience 15(1):44–52

Moreno-Casasola P, Martínez ML, Castillo-Campos G, Campos A (2012) The impacts on natural vegetation following the establishment of exotic *Casuarina* plantations. In: Martínez ML, Gallego-Fernández JB, Hesp PA (eds) Restoration of coastal dunes. Springer, Berlin, Chapter 14, this volume

Muñoz-Pérez JJ, López de San Román-Blanco B, Gutiérrez-Mas JM, Moreno L, Cuena GJ (2001) Cost of beach maintenance in the Gulf of Cadiz (SW Spain). Coastal Eng 42(2):143–153

Muñoz-Reinoso JC, Saavedra-Azqueta C, Redondo-Morales I (2012) Restoration of Andalusian coastal juniper woodlands. In: Martínez ML, Gallego-Fernández JB, Hesp PA (eds) Restoration of coastal dunes. Springer, Berlin, Chapter 9, this volume

Muñoz Vallés S, Gallego Fernández JB, Dellafiore CM (2011) Dune vulnerability in relation with tourism pressure in central Gulf of Cádiz (SW Spain), a case of study. J Coastal Res 27:243–251

Nordstrom KF, Jackson NL (2012) Foredune restoration in urban settings. In: Martínez ML, Gallego-Fernández JB, Hesp PA (eds) Restoration of coastal dunes. Springer, Berlin, Chapter 2, this volume

Nordstrom KF, Lampe R, Vandemark LM (2000) Reestablishing naturally functioning dunes on developed coasts. Environ Manag 25(1):37–51

Nordstrom KF, Jackson NL, Bruno MS, de Butts HA (2002) Municipal initiatives for managing dunes in coastal residential areas: a case study of Avalon New Jersey, USA. Geomorphology 47:137–152

Pérez-Maqueo OM, Martínez ML, Lithgow D, Mendoza-González G, Feagin RA, Gallego-Fernández JB (2012) The coasts and their costs. In: Martínez, ML, Gallego-Fernández JB, Hesp PA (eds) Restoration of coastal dunes. Springer, Berlin, Chapter 18, this volume

Peterson CH, Hickerson DHM, Johnson GG (2000) Short-term consequences of nourishment and bulldozing on the dominant large invertebrates of a sandy beach. J Coastal Res 16(2):368–378

Pickart AJ (2012) Dune restoration over two decades at the Lanphere and Ma-le'l Dunes in Northern California. In: Martínez ML, Gallego-Fernández JB, Hesp PA (eds) Restoration of coastal dunes. Springer, Berlin, Chapter 10, this volume

Psuty NP, Silveira TM (2012) Restoration of coastal foredunes, A geomorphological perspective: examples from New York and from New Jersey, USA. In: Martínez, ML, Gallego-Fernández JB, Hesp PA (eds) Restoration of coastal dunes. Springer, Berlin, Chapter 3, this volume

Pye K, Neal A (1994) Coastal dune erosion at Formby point, North Merseyside, England: causes and mechanisms. Mar Geol 119:39–56

Rhind P, Jones R (2010) A framework for the management of sand dune systems in wales. J Coastal Conserv 13:15–23

Rhind P, Jones R, Jones L (2012) The impact of dune stabilization on the conservation status of sand dune systems in Wales. In: Martínez ML, Gallego-Fernández JB, Hesp PA (eds) Restoration of coastal dunes. Springer, Berlin, Chapter 8, this volume

Van Aarde RJ, Smit AM, Claassens AS (1998) Soil characteristics of rehabilitating and unmined coastal dunes at Richard Bay, KwaZulu-Natal South Africa. Restor Ecol 6(1):102–110

Van der Hagen HJML, Geelen LHWT, de Vries CN (2008) Dune slack restoration in Dutch mainland coastal dunes. J Nat Conserv 16:1–11

Van der Meulen F, Salman AHPM (1996) Management of Mediterranean coastal dunes. Ocean Coast Manag 30:177–195

Vestergaard P (2012) Natural plant diversity development on a man-made dune system. In: Martínez ML, Gallego-Fernández JB, Hesp PA (eds) Restoration of Coastal dunes. Springer, Berlin, Chapter 4, this volume

Webb CE, Oliver I, Pik AJ (2000) Does coastal foredune stabilization with ammophila arenaria restore plant and arthropod communities in Southeastern Australia? Restor Ecol 8(3):283–288

Wilcock FA, Carter RWG (1977) An environmental approach to the restoration of badly eroded sand dunes. Biol Conserv 77:279–291

Zeppelini D, Bellini BC, Creao-Duarte AJ, Medina-Hernández MI (2009) Collembola as bioindicators of restoration in mined sand dunes of Northeastern Brazil. Biodivers Conserv 18(5):1161–1170

Zunzunegui M, Esquivias MP, Oppo F, Gallego-Fernández JB (2012) Interspecific competition and livestock disturbance control the spatial pattern of two coastal dune shrubs. Plant Soil 354(1–2):299–309

[illegible]

Glossary

Blowout A saucer-, cup-, bowl-, or trough-shaped depression or hollow formed by wind erosion in a pre-existing sand deposit. The adjoining accumulation of sand, the depositional lobe, derived from the depression, and possibly other sources, is normally considered part of the blowout landform.

Established Foredune Evolves from an incipient foredune, and either occupies the seaward-most position at the rear of the beach or is situated behind an incipient foredune. It is commonly distinguished by the growth of intermediate, sometimes woody, plant species, and by its greater morphological complexity, height, width, age, and geographical position.

Foredune (or fore-dune) A shore-parallel dune ridge formed on the backshore by aeolian sand deposition within vegetation.

Foredune plains Comprise two or more (typically more) former foredunes. They are formed where progradation has resulted in the multiple formations of foredunes and swales over time.

Incipient foredune A new or developing foredune that forms by aeolian sand deposition within pioneer plant communities.

Parabolic dune Also termed U-dunes, upsiloidal dunes, and hairpin dunes. Typically U- or V-shaped dune, characterized by short to elongate trailing ridges that terminate downwind in U- or V-shaped depositional lobes.

Relict foredune A stable, non-accretionary, former foredune that has become isolated from accretionary or erosional processes by the seaward formation of a new foredune. In some cases, a relict foredune has been termed a "beach ridge."

Transgressive dunefield Aeolian sand deposit formed by the downwind or alongshore movement of sand over vegetated to semi-vegetated terrain. Such dunefields may range from quite small (hundreds of meters in their alongshore

M. L. Martínez et al. (eds.), *Restoration of Coastal Dunes*, Springer Series on Environmental Management, DOI: 10.1007/978-3-642-33445-0, and landward extent) to very large fields that can be similar in size to some small to moderate-sized desert dunefields. They may be low, rolling, blowout-dominated, machair-type dunefields, low plains dominated by sand sheets, erosional knobs, and nebkha fields, or very large dunefields dominated by extensive deflation plains and transverse, barchan, and barchanoid dune systems.

Index

M. L. Martínez et al. (eds.), *Restoration of Coastal Dunes*, Springer Series on Environmental Management, DOI: 10.1007/978-3-642-33445-0,

The manufacturer's authorised representative in the EU is Springer Nature Customer Service Centre GmbH, Europaplatz 3, 69115 Heidelberg, Germany. If you have any concerns regarding our products, please contact ProductSafety@springernature.com

Printed and bound by CPI Group (UK) Ltd, Croydon, CR0 4YY
30/09/2026
02225643-0011